Bridging Disciplines: Interdisciplinary Approaches for Transforming Education

Edited by

Tahmeena Khan
Department of Chemistry, Integral University
Lucknow, Uttar Pradesh 226026, India

Manisha Singh
Department of Education, Integral University
Lucknow, Uttar Pradesh 226026, India

&

Saman Raza
Department of Chemistry, Isabella Thoburn College,
Lucknow, Uttar Pradesh 226007, India

Bridging Disciplines: Interdisciplinary Approaches for Transforming Education

Editors: Tahmeena Khan, Manisha Singh & Saman Raza

ISBN (Online): 979-8-89881-648-3

ISBN (Print): 979-8-89881-649-0

ISBN (Paperback): 979-8-89881-650-6

First published in 2026.

Bentham Science Publishers Pte. Ltd.
No. 9 Raffles Place
Office No. 26-01
Singapore 048619
Singapore
Email: subscriptions@benthamscience.net

CONTENTS

FOREWORD

I take this opportunity to write a foreword to the book titled "Bridging Disciplines: Interdisciplinary Approaches for Transforming Education", edited by Dr. Tahmeena Khan, Dr. Manisha Singh, and Dr. Saman Raza, with whom I have been academically associated for the last 15 years. The book is a wonderful amalgamation of research and academics in the field of Science, Technology, Engineering, and Mathematics. This work is a masterpiece in emphasising the relevance of STEM in tackling global challenges across various areas of development and in sustaining human civilisation. It addresses the challenges of various life arenas, including science, Commerce, Technology, and achieving Developmental goals, as well as managing eco-balance and work-life balance, thereby covering the intersection of science and technology with sociology, economics, culture, and politics of a particular nation. The values imparted by research across different aspects of STEM shall provide a holistic learning approach for future generations. This has the potential to bridge the gap between education, rationalism, social norms, economics and foreign relations.

The study of STEM has imbued the application of Artificial Intelligence and its role in teaching and learning. The book contains contributions on diverse topics ranging from interdisciplinary aspects, integration of diverse fields, analytical creativity, environmental, cultural, social, and economic impacts, and an education system imparting knowledge in STEM. This book highlights an interdisciplinary approach to the field of education and emphasises the existence of a multi-faceted education and social system of our nation.

I appreciate the efforts of all the contributing authors, especially young minds who have conceptualised and written the chapters, throwing light on newer aspects of STEM. The growing need for interdisciplinary research has created a requirement for such an education system. The editors should be applauded for bringing out a collection that breaks all the traditional boundaries and sets up a new and deeper meaning towards understanding life. I wish this brainchild of the editors would ignite more minds towards the role of STEM in overall societal development.

Dr. Nidhi Mishra
Department of Applied Sciences
Indian Institute of Information Technology, Allahabad
Uttar Pradesh, India

PREFACE

In a world that is ever-changing and increasingly complex, a field such as education, which encompasses various disciplines, must work to prepare students to understand the connections between this complex world and to meet the uncertain demands of the future, as well as the pressing issues of the present. There is no doubt that traditional discipline-specific teaching methods are the foundation of the education system, but they are unable to meet the changing demands of today's education. There is a need for an approach that fosters connections between various disciplines and equips students with adaptable abilities, integrative thinking, and a collaborative mindset, essential for addressing today's world and its complexities. One such approach that may fulfil this need is addressed in this book titled " Bridging Disciplines: Interdisciplinary Approaches for Transforming Education." The book is a compilation of 14 chapters from contributions by different experts in their respective areas and therefore offers a rich account of insights, new frontiers, and interdisciplinarity across disciplines. Each chapter is constructed to be self-contained, allowing readers to dive in and out as their understanding permits.

The book begins with an introduction to interdisciplinarity in education, where the chapter presents various related concepts, including interdisciplinary, multidisciplinary, and bridging disciplines, to enable readers to achieve a clear conceptual understanding upon completing this chapter. The second chapter, Bridging the Barrier Between Science and Arts, aims to reduce the gap between the two major categories of disciplines, *viz.*, science and arts. The third chapter discusses various pedagogical strategies for Interdisciplinary Learning in a general domain. The subsequent chapters cover a broad spectrum of topics, written in a way that each chapter draws upon insights from various fields, including the STEAM paradigm, Commerce, Social Studies, Languages, Biological Science, Chemical Science, Mathematics, Environmental Science, and Architectural Studies. The last two chapters provide insight into creativity and AI tools in interdisciplinary research.

The revolutionary potential of interdisciplinary approaches in education is explored in this book, which advocates for instructional models that break down silos, foster holistic understanding, and stimulate creativity by integrating multiple fields of knowledge. All the chapters are presented herein with the objective of rethinking curriculum, pedagogy, and learning methodologies.

We aim to provide a comprehensive account of insights to the major stakeholders in the field of education, including educators, researchers, policymakers, and students, while promoting interdisciplinarity. Our vision for this book is to empower stakeholders with an education that is both meaningful and skilful, capable of standing up to global competition. This book would be helpful for students, teachers, researchers, and academicians who seek to integrate across different disciplines.

We anticipate that educators, researchers, policymakers, and students will find inspiration and practical guidance within these pages. As we go towards more inclusive, adaptable, and future-oriented educational settings, may this book function as both a resource and a stimulus for continuous experimentation and discourse.

As the editors of this book, we would like to thank the contributing authors for their time and expertise. We also want to thank the readers whose curiosity and commitment to advancing education through technology drive our ongoing investigation of this fascinating intersection.

Tahmeena Khan
Department of Chemistry, Integral University
Lucknow, Uttar Pradesh 226026, India

Manisha Singh
Department of Education, Integral University
Lucknow, Uttar Pradesh 226026, India

&

Saman Raza
Department of Chemistry, Isabella Thoburn College
Lucknow, Uttar Pradesh 226007, India

List of Contributors

Abdul Rahman Khan	Department of Chemistry, Integral University, Lucknow, Uttar Pradesh 226006, India
Aleena Alam Khan	Department of Humanities and Social Sciences, Integral University, Lucknow, Uttar Pradesh 226026, India
Anukriti Pandey	Department of English and Foreign Languages, Dr. Shakuntala Misra National Rehabilitation University, Lucknow, Uttar Pradesh 226017, India
Anushka Pandey	Department of Chemistry, Integral University, Lucknow, Uttar Pradesh 226006, India
Arbind K Jha	School of Education, Indira Gandhi National Open University (IGNOU), New Delhi, Delhi 110068, India
Arpita Tripathi	Colvin Talukedar College, Lucknow, Uttar Pradesh 226007, India
Bushra Saiyeda	Department of Humanities and Social Sciences, Integral University, Lucknow, Uttar Pradesh 226026, India
Ekta Pal	Department of Commerce, St.John's College, Agra, Uttar Pradesh 282002, India
Kashifa Khatoon	Department of Biosciences, Integral University, Lucknow, Uttar Pradesh 226026, India
Khurram Ashraf	Department of Architecture, Integral University, Lucknow, Uttar Pradesh 226006, India
Leena Rajak	Department of Education, Babasaheb Bhimrao Ambedkar University, Lucknow, Uttar Pradesh 226025, India
Manisha Singh	Department of Education, Integral University, Lucknow, Uttar Pradesh 226026, India
Martha Sharma	Department of Management, St.John's College, Agra, Uttar Pradesh 282002, India
Michael Tarance Suraj	Department of Education, Babasaheb Bhimrao Ambedkar University, Lucknow, Uttar Pradesh 226025, India
Mohammad Imran Ahmad	Department of Chemistry, Integral University, Lucknow, Uttar Pradesh 226006, India
Pallavi Awasthi	Department of Biosciences, Integral University, Lucknow, Uttar Pradesh 226026, India
Prashant Kumar Astalin	School of Education, Uttar Pradesh Rajarshi Tandon Open University, Prayagraj, Uttar Pradesh 211021, India
Priya Mishra	Department of Chemistry, Isabella Thoburn College, Lucknow, Uttar Pradesh 226007, India
Reena Vishvakarma	Department of Biosciences, Integral University, Lucknow, Uttar Pradesh 226026, India
Sakshi Gupta	Department of Chemistry, Isabella Thoburn College, Lucknow, Uttar Pradesh 226007, India

Saman Raza	Department of Chemistry, Isabella Thoburn College, Lucknow, Uttar Pradesh 226007, India
Sangeeta Chauhan	School of Education, Uttar Pradesh Rajarshi Tandon Open University, Prayagraj, Uttar Pradesh 211021, India
Satya	Department of Chemistry, Isabella Thoburn College, Lucknow, Uttar Pradesh 226007, India
Seema Joshi	Department of Chemistry, Isabella Thoburn College, Lucknow, Uttar Pradesh 226007, India
Shahla Tanveer	Department of Chemistry, Integral University, Lucknow, Uttar Pradesh 226026, India
Shifa Suhail	Department of Biosciences, Integral University, Lucknow, Uttar Pradesh 226026, India
Shweta Pandey	Chandigarh College of Pharmacy, Mohali, Punjab 140307, India
Sonu Bara	Department of Education, Babasaheb Bhimrao Ambedkar University, Lucknow, Uttar Pradesh 226025, India
Swati Sharma	Department of Biosciences, Integral University, Lucknow, Uttar Pradesh 226026, India
Tahir Hussain Ansari	Department of Humanities and Social Sciences, Integral University, Lucknow, Uttar Pradesh 226026, India
Tahmeena Khan	Department of Chemistry, Integral University, Lucknow, Uttar Pradesh 226006, India
Venika Dhoonter	Chandigarh College of Pharmacy, Mohali, Punjab 140307, India
Waseem Zahra	Department of B.Ed., Mahila Vidyalaya Degree College, Lucknow, Uttar Pradesh 226018, India
Zeba Nisar	Department of Architecture, Integral University, Lucknow, Uttar Pradesh 226006, India

CHAPTER 1

Interdisciplinarity in Education: Beyond Traditional Boundaries

Manisha Singh[1,*], Waseem Zahra[2], Arbind K Jha[3], Tahmeena Khan[4] and **Saman Raza[5]**

[1] *Department of Education, Integral University, Lucknow, Uttar Pradesh 226026, India*

[2] *Department of B.Ed., Mahila Vidyalaya Degree College, Lucknow, Uttar Pradesh 226018, India*

[3] *School of Education, Indira Gandhi National Open University (IGNOU), New Delhi, Delhi 110068, India*

[4] *Department of Chemistry, Integral University, Lucknow, Uttar Pradesh 226026, India*

[5] *Department of Chemistry, Isabella Thoburn College, Lucknow, Uttar Pradesh 226007, India*

Abstract: Progress in education is best achieved by using the best of every discipline. A field such as education can never progress in isolation. This chapter emphasizes several key points, including the concept of interdisciplinary, the differences between related terms such as interdisciplinary, multidisciplinary, and bridging disciplines. The chapter highlights the impact of an interdisciplinary approach in school and higher education. Further, it highlights the benefits of an interdisciplinary approach among two major stakeholders: students and teachers. Additionally, the chapter explores the characteristics of a multidisciplinary approach and the strategies, such as contextualising, conceptualising, and problem-centred teaching, employed in interdisciplinary teaching. The primary objective of this chapter is to provide insight into the concept of interdisciplinary education and its various dimensions.

Keywords: Education, Interdisciplinary thinking, Interdisciplinary, Multidisciplinary, Strategies, Teaching-learning.

INTRODUCTION

No discipline from its very origin has progressed in isolation. The disciplines have always been interconnected with each other. Historically, Isaac Newton's Principia Mathematica, a fundamental work of natural philosophy, significantly influenced seventeenth-century intellectual communities through its innovative methodology for knowledge creation and dissemination. Researchers in both the scientific and social sciences acknowledged that for their work to be deemed

* **Corresponding Author Manisha Singh:** Department of Education, Integral University, Uttar Pradesh 226026, India; E-mail: singh.manisha123@gmail.com

comprehensive, it must conform to the principles outlined in the Principia. In the eighteenth century, Newton's Principia initiated this transformation, which facilitated the Age of Enlightenment. This period's results impacted the intellectual shift and influenced the nineteenth century, known as the Age of Reason. It was the distinguished Anglican historian of science, William Whewell, who metaphorized natural philosophy as “science,” thereby marking the nineteenth century as the Age of Science. With the ongoing revolutions, the social sciences emerged as disciplines that involve scientific understanding of social processes. These preliminary investigations established the foundation for the Social Sciences of the nineteenth century. Eventually, August Comte sought to provide a framework for social physics and established the methodology termed Positivism. As a result, academic departments focused on the Physical, Natural, and Social Sciences, in addition to those for the Arts and Humanities, began to be established at universities and colleges in Europe and the Americas. Consequently, the majority of fields sprang from philosophy. Philosophy has historically been categorized into two principal branches: Natural Philosophy and Moral Philosophy. These branches developed into separate disciplines, one pertaining to science and the other to non-science. Interdisciplinary research and education inherently facilitate the augmentation of knowledge, propelling societal transformation. The generation of knowledge across disciplines yields substantial, influential, and non-linear results, unlike the more constrained impacts observed inside conventional academic confines. Knowledge produced at the intersections of diverse disciplines is often exceptionally inventive, rejuvenating, and integrative (Phillips, 1976). Interdisciplinary learning is a pedagogical approach that intentionally integrates methodologies and terminologies from multiple academic disciplines. This synthesis serves to explore a central theme, complex issue, or multifaceted problem (Martín-Páez *et al.* 2016). Aligning with this perspective, the Collins English Dictionary defines “interdisciplinary” as encompassing two or more distinct fields of study. The term “interdisciplinary” is used throughout the book generally to denote the full interdisciplinary continuum. The related terms, such as multidisciplinary, interdisciplinary, and transdisciplinary, are utilized as defined by Klein (Kolmos *et al.*, 2024). In the work of Choi, multidisciplinary is defined as “a process for providing a juxtaposition of disciplines that is additive, not integrative,” interdisciplinary as “a synthesis of two or more disciplines [that] establishes a new level of discourse and integration of knowledge,” and transdisciplinary as an approach that “provides holistic schemes that subordinate discipline, looking at the dynamics of whole systems” (Raggi *et al.*, 2024). A key feature of effective interdisciplinary practice is the synergistic combination of disciplines, which fosters a deeper understanding and sustained knowledge retention. Educators collaborate to weave together topics, content, and activities, ensuring a cohesive and enriching learning experience. This collaborative approach extends to the design of both instruction and assessment, promoting a holistic view of student learning. No man is an Island, entire of itself fits well for disciplines, which means no discipline can be compartmentalized in itself without any interconnectedness with others. The disciplines progress in connection with each other. There may be cases where sub-disciplines within a primary discipline act as separate ones, as they belong to two entirely different disciplines. It appears that a particular discipline may stand in opposition to other disciplines, but there are well-known disciplines recognized for their bridging nature. The bridging nature of a discipline with other disciplines leads to many related yet distinct concepts, such as interdisciplinary, multidisciplinary, and Bridging disciplines. There have been conceptual categorizations that distinguish between "interdisciplinary," "transdisciplinary," "conceptual interdisciplinary," and other such categories (Kolmos *et al.*, 2024); (Raggi *et al.*, 2024); (Lattuca & Voigt, 2012); (Kockelmans, 1979); (Newell, 1992).

a. **Interdisciplinary/Transdisciplinary:** The true potential of interdisciplinary research lies in transcending traditional disciplinary boundaries. By developing new theoretical frameworks and methodologies, researchers can create a synergistic approach that surpasses the limitations of individual fields. This transformative process enables the development of holistic solutions to complex challenges (Newell, 1992; Repko, 2008).
b. **Multidisciplinary or cross-disciplinary** is a collaboration between academic disciplines that offers unique opportunities for innovation. When experts from diverse fields find common ground, they can leverage their distinct perspectives and methodologies to address multifaceted problems. The Santa Fe Institute exemplifies this approach, fostering collaboration across disciplines to generate novel insights (Scharte, 2025).
c. **Cross-disciplinarity** is the study of a subject by applying techniques from fields unrelated to the subject. In cross-disciplinary communication, the boundaries of disciplines are surpassed, although neither methodologies nor objectives are altered, whereas interdisciplinarity amalgamates the practices and foundations of all participating disciplines.
d. **Bridging disciplines:** Specific disciplines inherently possess a broader scope, facilitating connections between seemingly disparate fields. For example, disciplines such as anthropology and geography act as a bridge between the natural sciences, social sciences, and humanities, providing a comprehensive understanding of the world and human society (Youngblood, 2007).

LEVELS OF INTEGRATION AND COLLABORATION

To understand the collaboration towards interdisciplinary education, it is necessary to explore different levels of integration that lead to other approaches, such as multidisciplinary and interdisciplinary.

Multidisciplinary Approach

A comprehensive understanding of a concept is best achieved through interdisciplinary examination, drawing upon diverse perspectives from various fields. This involves studying a theme through the lens of multiple disciplines. For instance, agriculture can be analyzed from the perspectives of chemistry, biology, history, geography, and economics. Similarly, understanding climate change requires consideration of historical, political, economic, ethical, and scientific viewpoints. This approach to curriculum integration emphasizes the unique perspectives that different disciplines offer in illuminating a topic, theme, or issue. A multidisciplinary curriculum involves studying the same subject from the perspective of multiple disciplines. While the terms "multidisciplinary" and "cross-disciplinary" are often used synonymously to describe efforts to bridge boundaries between disciplines, a multidisciplinary approach specifically involves drawing on multiple fields to explore problems that extend beyond conventional disciplinary boundaries. This leads to solutions grounded in a novel understanding of complex situations. In essence, the multidisciplinary approach is a method of curriculum integration that accentuates the diverse perspectives that different disciplines can bring to bear on a theme, subject, or issue. Implementing a multidisciplinary curriculum often necessitates collaboration among educators from various fields to ensure proper execution and a holistic understanding of the subject matter.

Interdisciplinary

For a deeper understanding of complex concepts, an interdisciplinary approach appears to be crucial. Isolating disciplines hinders the development of learning that is developmentally appropriate, resulting in an incomplete grasp of the subject matter. Synthesizing multiple perspectives is vital; for instance, Environmental Studies (EVS) should be grounded in a student's immediate surroundings to facilitate understanding for those in grades 3-5. Similarly, a proper conceptual understanding of mathematics must be rooted in real-world examples drawn from science, social science, geography, and history. Without these connections, students develop only a procedural understanding, which is insufficient for genuine learning and understanding. The activities, such as projects and model making, are, by their very nature, interdisciplinary. They foster an understanding of overarching themes and ideas that transcend individual disciplines. Such approaches focus on both the process and meaning, rather than mere product and content, integrating four main components of knowledge: content, theories, methodologies, and perspectives from various fields. By dissolving the boundaries between conventional disciplines, teaching and learning can be organized around constructing meaning within the context of real-world problems or themes. Interdisciplinary learning involves the integration of two or more branches of knowledge, expanding students' ability to address problems that do not fit neatly into a single subject area. It also transforms the learning process by encouraging students to synthesize multiple perspectives, moving beyond passive acceptance of information. This approach avoids compartmentalizing learning and can be effectively implemented by a single teacher. The interdisciplinary approach believes in working together of different disciplines with a positive impact on all the integrating disciplines. It stands on the assumption that many problems and doubts cannot be wholly understood or solved by a single discipline, thereby leading to collaboration and communication among experts from various fields.

Features of Interdisciplinary Approach

Different features of the Interdisciplinary approach, particularly in education, are outlined in Fig. (**1**), and each feature is separately discussed below.

Integration of Knowledge and Methods: Interdisciplinary approaches combine theories, concepts, and methodologies from various disciplines to create a holistic picture for a better understanding of a problem. This integration allows researchers and educators to approach topics from multiple perspectives. This integration enables a more comprehensive understanding of a topic or problem than what is typically achieved within a single discipline (Kolmos *et al.*, 2024).

Collaboration: It emphasizes collaboration among experts from different fields. This teamwork brings diverse perspectives and expertise, enriching the research process and leading to more thorough analyses and innovative outcomes (National Academy of Sciences, 2005).

Flexible Methodology: Interdisciplinary approaches often employ mixed methodology, which is a combination of qualitative and quantitative research method. This flexibility enables researchers to select the most suitable techniques from various disciplines to address the complexities of their research questions (Repko, 2008).

Problem-Focused Orientation: The interdisciplinary approach is typically geared toward solving real-world problems. It seeks to address complex issues that cannot be understood or resolved through the lens of a single discipline, such as climate change, public health, or social justice (Maxey, 2009).

Innovative Thinking: By crossing traditional boundaries, interdisciplinary research promotes new ways of thinking and encourages creativity. It can lead to the development of novel theories and solutions that are not constrained by existing disciplinary frameworks (Savin-Baden & Major, 2023).

Cultural and Contextual Awareness: Interdisciplinary research is a discourse in itself that often considers the cultural, social, and contextual factors. This awareness leads to enhancing the meaning-making and the applicability of research findings (Hladchenko & Kriuchkova, 2020).

Communication Across Disciplines: Effective interdisciplinary work requires strong communication skills. Researchers must be able to translate concepts and findings across different fields, facilitating dialogue and understanding among team members from diverse backgrounds (Trench, 2023).

Focus on Systemic Change: The interdisciplinary approach often aims for systemic transformation, recognizing that solutions may require changes at multiple levels—individual, organizational, and societal (Meadows, 2008).

Continuous Learning: Engaging in interdisciplinary research fosters a culture of continuous learning. Researchers are encouraged to expand their knowledge and expertise beyond their primary discipline, enhancing their skills and adapting to new challenges (Lawrence, 2020).

Emergent Understanding: The findings of interdisciplinary research can lead to emergent understandings that reveal new relationships and dynamics, offering insights that are often missed in single-disciplinary studies (Shanta & Wells, 2022).

Application-Oriented: Many interdisciplinary efforts focus on practical applications and real-world issues, leading to actionable solutions. This focus on application helps bridge the gap between theory and practice.

Communication Skills: Successful interdisciplinary work requires strong communication skills, as individuals must convey their ideas and findings to colleagues from other fields. The ability to discuss complex concepts in an accessible way is essential.

Shared Goals and Objectives: Interdisciplinary projects often centre on shared goals. This common mission can help unify diverse perspectives and foster teamwork among participants.

Respect for Diversity: Embracing different disciplines involves respecting diverse methodologies, languages, and epistemologies. This respect is crucial for productive dialogue and collaboration.

Reflexivity and Meta-cognition: Participants reflect on their disciplinary assumptions and biases, making them explicit for negotiation and discussion. Enhances critical thinking and deepens understanding of epistemological diversity (Lattuca, 2003).

Methodological Pluralism: It draws on quantitative, qualitative, and mixed methods as appropriate, with methodological choices driven by the needs of the problem rather than disciplinary tradition, yielding richer datasets and triangulated evidence (Tress & Fry, 2006).

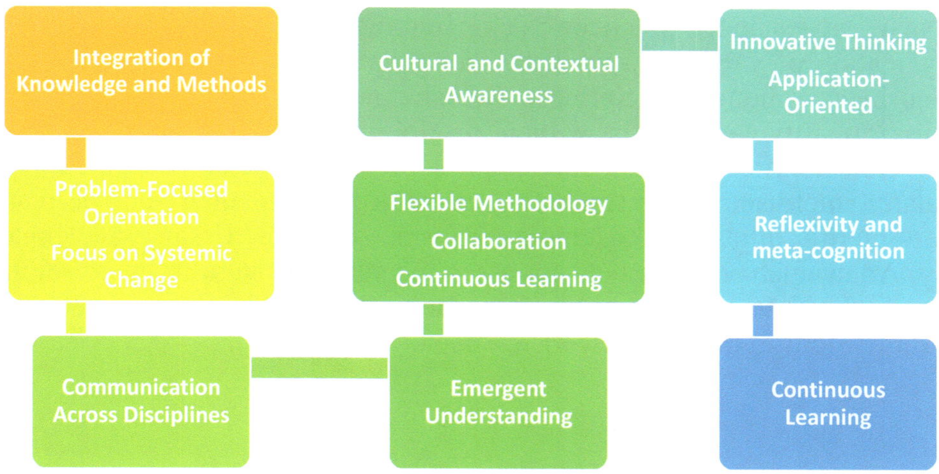

Fig. (1). Characteristics of the interdisciplinary approach.

The idea of integrating or connecting school topics with various areas with a motive to provide a more cohesive and successful teaching and learning strategy has drawn much attention in recent years (Billion-Kramer *et al.*, 2024). The integrated and multidisciplinary curricular approach in technology education aims to help students understand and value the interconnectedness of academic topics and the cumulative nature of their learning (Pang & Good, 2000). Teachers from various disciplines can work together and support one another while stepping up their instruction by utilising interdisciplinary approaches.

Interdisciplinarity in School Education

Interdisciplinarity in schools focuses on the utilization of knowledge from a broader educational perspective. The aim is to create optimal conditions that facilitate the integrative process in students' minds. They learn not only the particular concept but also develop ways to integrate different knowledge for maximum progress. This requires an adjustment of educational understanding at the curricular, teachers, didactic, and pedagogical levels (Lenoir & Sauvé, 1998). Teachers can encourage and support the use of interdisciplinary techniques in the classroom by following certain guidelines. The six essential criteria that characterize quality school interdisciplinarity are delineated by Lenoir and Hasni (2016). The authors identified three distinct interdisciplinary perspectives (logics), stating that “these three logics of meaning, functionality, and affectivity are complementary and should be interwoven when employing an interdisciplinary approach in teaching practices”; (1) interdisciplinarity is not simply an amalgamation of various disciplinary viewpoints; (2) there are no hierarchical relationships among disciplines, which are regarded as equal; (3) interdisciplinarity cannot exist without disciplines; and (4) mutual collaboration among educators is important for integration. They developed a paradigm that articulates interdisciplinary teaching methodologies and their four derivatives: pseudo-interdisciplinarity, hegemony, holism, and eclecticism, with enhanced clarity. An evaluative framework for assessing students' work in interdisciplinary settings by Boix-Mansilla and Duraising (Boix-Mansilla & Duraising, 2007). The three dimensions in their framework are depicted in Fig. (**2**).

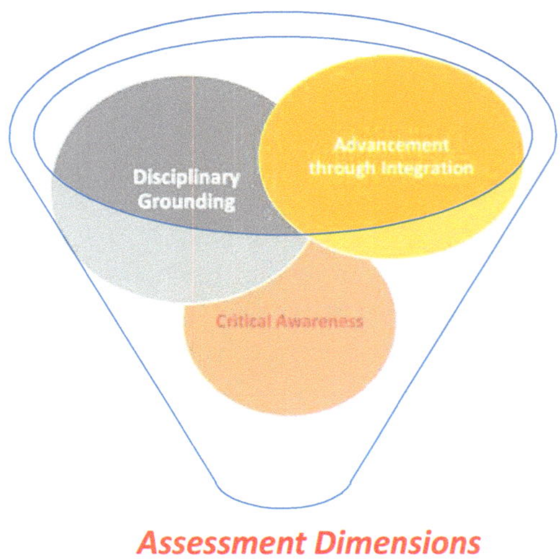

Fig. (2). Assessment dimensions.

- Disciplinary Grounding: It is a must to explore the roots of the discipline. This provides knowledge of different methodologies and helps to connect these disciplines for a better understanding of a concept. Furthermore, this comprehension safeguards against subpar transdisciplinary initiatives.
- Advancement through integration: All multidisciplinary efforts should strive to enhance students' comprehension.
- Critical awareness: It is essential to evaluate students' progress, but they should also be judged on their capacity to contemplate the objectives, methods, and constraints of their work.

Interdisciplinarity in Higher Education

Preparing students to tackle complex problems in professional and scientific settings is becoming increasingly important in higher education (Wang *et al.*, 2025); (Roehler, Fear, & Herrmann, 1998). A cross-disciplinary approach is believed to facilitate a comprehensive understanding, and interdisciplinarity can assist in tackling contemporary complex issues (Newell, 2009). This conviction has led to the increased popularity of multidisciplinary higher education throughout time (Newell, 2009). The conventional higher education approach focuses on knowledge of a particular discipline, which is primarily theoretical and general, whereas the interdisciplinary approach works towards developing higher-level skills and making an individual interdisciplinary competent. Boundary-crossing talents encompass the capacity to navigate complexity, synthesise knowledge from other domains, and alter perspectives. Interdisciplinarity is integrative, as opposed to additive, like multidisciplinary; through integration, knowledge from multiple disciplines is compared and transformed (Kolmos *et al.*, 2024). The amalgamation of knowledge is regarded as the quintessential attribute of interdisciplinarity. The capacity to synthesise or integrate is deemed a crucial educational outcome of interdisciplinarity in higher education. Consequently, this learning outcome is referred to as interdisciplinary understanding or interdisciplinary thinking. Martín-Páez and his associates characterise interdisciplinary understanding as “the ability to synthesise knowledge and methodologies from two or more disciplines or recognised fields of expertise to achieve cognitive advancement—such as clarifying a phenomenon, addressing a problem, or creating a product—in ways that would have been unlikely or impossible through a singular disciplinary perspective.” The performance-oriented view of comprehension is the foundation of this approach, suggesting that individuals understand a subject when they can accurately and adaptively apply it in novel contexts (Martín-Páez *et al.*, 2019). Students may benefit to the fullest when they think in terms of interdisciplinarity, and for that, inculcating a thinking style known as “interdisciplinarity thinking” is essential.

What is Interdisciplinary Thinking?

The ability to transit between disciplinary views and create significant connections across disciplines is one of the many subskills that comprise interdisciplinary thinking, a complex cognitive skill (Van Merriënboer, 1997). Interdisciplinary thinking does not arise spontaneously; students may need a good amount of time to attain a sufficient level of proficiency in its application. This approach synthesises concepts, theories, and methods from other disciplines to address complex issues, foster creativity, and deepen knowledge. This cognitive approach transcends conventional limits, fostering cooperation and diverse viewpoints.

Dimensions of Interdisciplinary Thinking: There are various dimensions associated with interdisciplinary thinking, which are illustrated in Fig. (**3**) and briefly discussed below.

Analysis and Synthesis of Various Knowledge: It allows individuals to evaluate and integrate theoretical perspectives from fields including science, humanities, engineering, and social sciences. This facilitates a more comprehensive knowledge of intricate topics.

Problem-solving Approach: (PSA) PSA frequently addresses complex real-world issues necessitating a diversified strategy. This characteristic highlights the importance of employing diverse methodologies and concepts to identify effective solutions.

Critical Thinking: Interdisciplinary thinking fosters the critical examination of concepts and assumptions across multiple fields. It encourages individuals to scrutinise the legitimacy of diverse viewpoints and cultivates open-mindedness (Thornhill-Miller *et al.*, 2023).

Flexibility and Adaptability: Individuals who practice interdisciplinary thinking must exhibit adaptability in their methods as they navigate multiple paradigms and procedures. This adaptability is essential in a world where issues are frequently interrelated and dynamic (Nicolescu, 2014).

Innovative Thinking: Interdisciplinary thinking fosters innovation by integrating concepts from multiple fields. The intersection of ideas can yield innovations and creative solutions that may not arise within a single discipline (Lattuca, Voigt, & Fath, 2004).

Systemic Perspective: Interdisciplinary thinking fosters a holistic perspective on challenges, prompting individuals to recognise the broader context. This viewpoint is essential for comprehending the interconnections among diverse elements within any specific setting (Meadows, 2008).

Contextual and Ethical Awareness: It situates information within broader social, cultural, and ethical contexts. Examines the ramifications and obligations of practical knowledge. It is concerned with contextual awareness and ethical utility.

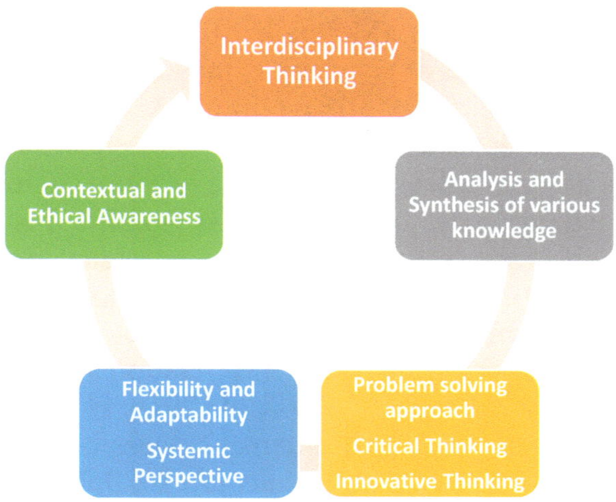

Fig. (3). Dimensions of interdisciplinarity thinking.

Benefits for Students

Motivation appears to be positively influenced by the use of multidisciplinary sequences; (Cuervo, 2018); (Tonnetti & Lentillon-Kaestner, 2023); (Hawkey, James, & Tidmarsh, 2019); (Ríordáin, Johnston, & Walshe, 2016); (Vieira, Dias, & Chediak, 2018). The interdisciplinary sequences, when implemented, align with the students' needs and interests. It enables them to select their learning trajectories and learning sources, and to participate in the construction of knowledge equally; additionally, interest in a discipline can occasionally lead to interest in another discipline (Holmbukt & Larsen, 2016); (McPhail, 2018); (Michelsen & Sriraman, 2009); (Moss, Godinho, & Chao, 2019); (Pountney & McPhail, 2019); (Tammaro, D'Alessio, & Petolicchio, 2017). Numerous studies have indicated that students appear to derive greater enjoyment from learning in interdisciplinary sequences compared to conventional lessons, and this enjoyment is both communal and varied (Holmbukt & Larsen, 2016); (Moss, Godinho, & Chao, 2019); (Tammaro, D'Alessio, & Petolicchio, 2017); (Kneen, Breeze, Davies-Barnes, John, & Thayer, 2020); (Vega & Schnackenberg, 2004). By promoting a deeper understanding of academic subjects, interdisciplinary sequences seem to improve academic performance (Bollati *et al.*, 2017). Students and educators argue that interdisciplinary methods are inherently holistic in their nature and approach (Cuervo, 2018); (Fenwick, Minty, & Priestley, 2013); (Harris & de Bruin, 2018). These approaches create interconnections among disciplines, improve the understanding of disciplinary logics, and promote the transfer of information and skills across areas (Ríordáin, Johnston, & Walshe, 2016); (Bollati *et al.*, 2017). Interdisciplinary sequences can augment deep learning and provoke student cognition (Cuervo, 2018); (McPhail, 2018); (Pountney & McPhail, 2019). Ultimately, they appear to have a positive impact on student performance in summative examinations (Cuervo, 2018); (Holmbukt & Larsen, 2016); (Bollati *et al.*, 2018). It appears that interdisciplinary sequences enhance individual skills by engaging students as active learners. Interviews with educators and students reveal that these methodologies offer numerous benefits, fostering autonomous, empathetic, responsible, respectful, and creative individuals while simultaneously enhancing students' self-confidence and critical thinking skills. In addition to strengthening social ties between students and between professors and students, interdisciplinary approaches improve students' relational abilities by encouraging collaboration and group projects (Hawkey *et al.*, 2019; Holmbukt & Larsen, 2016; McPhail, 2018; Moss *et al.*, 2019; Tammaro, D'Alessio *et al.*, 2017; Kneen etal al., 2020; Harris & de Bruin, 2018; Braskén *et al.*, 2020; Manalo & Fukuda, 2024; Queiruga *et al.*,, 2021; Auinger, 2025; Shvets *et al.*, 2024).

Benefits for Teachers

According to (Hawkey, James, & Tidmarsh, 2019), one of the primary sources of ***motivation for educators*** is interdisciplinary. This is a demanding, conventional-disrupting strategy that invigorates stakeholders (Ríordáin, Johnston, & Walshe, 2016); (Michelsen & Sriraman, 2009); (Harris & de Bruin, 2018); (Michelsen & Sriraman, 2007); (Moser, Ivy, & Hopper, 2019). This method enables educators to recognise the ***advantages for students and fosters student participation***. The capacity of educators to monitor their pupils' progress enhances motivation (Hardré *et al.*, 2013). This method enables educators to cultivate improved relationships with their peers by fostering an environment of trust (McPhail, 2018); (Fenwick, Minty, & Priestley, 2013); (Auinger, 2025). Interdisciplinary teaching appears to enhance instructors' professional development through collaboration with peers across several disciplines. The expertise and professional experience of each teacher are valuable assets that persist beyond the interdisciplinary experience (Tonnetti & Lentillon-Kaestner, 2023); (Ríordáin, *et al.*, 2016); (Holmbukt & Larsen, 2016); (McPhail, 2018); (Michelsen & Sriraman, 2009); (Tinnell, Tretter, Thornburg, & Ralston, 2019). The educators remain well-informed and seek many opportunities through interdisciplinary techniques to enhance their expertise and knowledge, particularly in areas beyond their teaching scope. Interdisciplinary instruction enables educators to establish and comprehend relationships within several disciplines (Hawkey, James, & Tidmarsh, 2019); (Hardré *et al.*, 2013); (Braskén, Hemmi, & Kurtén, 2020); (Manalo & Fukuda, 2024); (Auinger, 2025); (Moser, *et al.*, 2019); (Tinnell, Tretter, Thornburg, & Ralston, 2019). Ultimately, it provides diverse instructional methods grounded in an approach that fosters real-world linkages, allowing educators to dedicate time to children who face the most significant challenges (Tinnell, *et al.*, 2019).

Strategies Towards Inter-Disciplinary Teaching

Contextualising, conceptualising, and problem-centring are three multidisciplinary teaching strategies (Fig. **4**) (Li *et al.*, 2025). Assume that the foundational epistemology in multidisciplinary endeavours is derived from the humanities. Contextualizing—that is, placing facts and concepts within a cultural, historical, or ideological framework—is probably the method used to integrate disciplinary material in that situation. The scientific method creates the standards for integration, while the conceptualisation of work takes place. Ultimately, if the principles and methods of investigation correspond with applied sciences or creative product development, the integrative process will emerge as a problem-oriented investigation of urgent or tangible issues.

The initial tactic, contextualising, involves integrating disciplinary content into the framework of temporal, cultural, and personal experiences. Consequently, contextualisation can manifest in various forms depending on the specific setting. Fundamental metaphysical convictions and individual or societal philosophies may serve as an alternative contextual framework, providing a philosophical metaphysical context. Another context includes systems of knowledge, various sources of knowledge, and methodologies of thinking, with a focus on epistemology. The second strategy, conceptualisation, involves identifying essential concepts that are vital across several disciplines (*e.g.*, change, linearity) and establishing a rigorous quantitative relationship among them. The ability to abstract physical stuff to its mathematical or empirical essence is a skill that students in conceptualising courses possess. They discover that different systems of notation and symbolic representation in science unveil basic patterns and processes. The third strategy, problem-solving, involves employing knowledge and methodologies from diverse fields, including biology, chemistry, political science, and economics, to address complex real-world challenges such as water pollution and genetic engineering, which necessitate interdisciplinary solutions. This method is defined by action-oriented applied social sciences or allied disciplines that seek to deliver concrete outputs (products, technologies, policies, processes) to improve the human situation. The aim of these disciplines is not primarily to enhance self-awareness or understanding of the natural world, as observed in the humanities and basic sciences, but rather to use this knowledge towards action, creation, and social change.

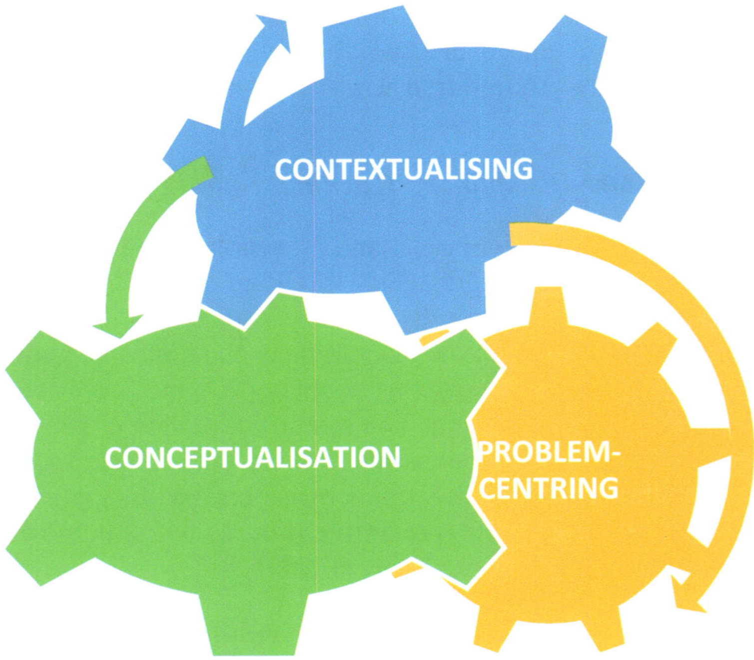

Fig. (4). Strategies of interdisciplinary teaching.

CONCLUDING REMARKS

A single discipline may not adequately address complicated situations, yet interdisciplinary approaches offer enormous value and potential. From the philosophical perspective, disciplines such as the natural sciences and the moral sciences have their origin in one discipline, viz., philosophy; hence, an organic bond exists between these disciplines (Singh *et al.* 2024). By synthesising viewpoints, approaches, and knowledge from diverse disciplines, we cultivate a more comprehensive and nuanced understanding of issues. This collaboration not only promotes invention but also encourages a more comprehensive approach to problem-solving. In confronting the escalating interrelated global challenges—such as climate change, public health emergencies, and social inequality—interdisciplinary collaboration is imperative. It promotes diverse perspectives and fosters innovative ideas that may be overlooked within the confines of conventional disciplines. Furthermore, interdisciplinary collaboration fosters critical thinking and adaptation, preparing individuals to traverse a continually changing knowledge landscape. The pedagogies such as STEAM education, which is an amalgamation of different disciplines, have a positive impact on students' engagement, clarity of concepts, and understanding (Singh, Azad, Qayyoom, & Khan, 2024); (Raza, Khan, & Singh, 2024). In conclusion, adopting interdisciplinary approaches is essential for advancing knowledge and facilitating constructive change. By promoting collaboration across boundaries, we open new potential that can greatly improve our capacity to address the urgent challenges of our era, ultimately resulting in a more innovative and sustainable future.

REFERENCES

Auinger, A. (2025). In Search Of An Elusive Object: Interdisciplinary Teaching In Secondary School From The Teaching Perspective. *Technology Journal of Management, Accounting and Economics, 13*(1), 1-32.

Billion-Kramer, T., Rehm, M., Tittel, C. (2024). Guter und effektiver Chemieunterricht. *Fachdidaktik Chemie in Theorie und Praxis,* Berlin, Heidelberg: Springer.33-69.

Bollati, I.M., Gatti, C., Pelfini, M.P., Speciale, L., Maffeo, L., Pelfini, M. (2018). Climbing walls in Earth Sciences education: an interdisciplinary approach for the secondary school (1st level). *Rendiconti Online della Società Geologica Italiana, 44*, 134-142.
[http://dx.doi.org/10.3301/ROL.2018.19]

Braskén, M., Hemmi, K., Kurtén, B. (2020). Implementing a multidisciplinary curriculum in a Finnish lower secondary school – The perspective of science and mathematics. *Scand. J. Educ. Res., 64*(6), 852-868.
[http://dx.doi.org/10.1080/00313831.2019.1623311]

Cuervo, L. (2018). Study of an interdisciplinary didactic model in a secondary education music class. *Music Educ. Res., 20*(4), 463-479.
[http://dx.doi.org/10.1080/14613808.2018.1433148]

Fenwick, A.J.J., Minty, S., Priestley, M. (2013). Swimming against the tide: A case study of an integrated social studies department. *Curric. J., 24*(3), 454-474.
[http://dx.doi.org/10.1080/09585176.2013.805658]

Hardré, P.L., Ling, C., Shehab, R.L., Nanny, M.A., Nollert, M.U., Refai, H., Ramseyer, C., Herron, J., Wollega, E.D. (2013). Teachers in an interdisciplinary learning community: Engaging, integrating, and strengthening K-12 education. *J. Teach. Educ., 64*(5), 409-425.
[http://dx.doi.org/10.1177/0022487113496640]

Harris, A., de Bruin, L. (2018). An international study of creative pedagogies in practice in secondary schools: Toward a creative ecology. *J. Curric. Pedagog., 15*(2), 215-235.
[http://dx.doi.org/10.1080/15505170.2018.1457999]

Hladchenko, M., Kriuchkova, T. (2020). Interdisciplinarity in the context of cultural competence: The case of education. *International Journal of Education and Management Engineering, 10*(1), 13-22.

Holmbukt, T., Larsen, A.B. (2016). Interdisciplinary teaching as motivation: An initiative for change in post-16 vocational education. *Nordic Journal of Modern Language Methodology, 4*, (1).
[http://dx.doi.org/10.46364/njmlm.v4i1.325]

Kate, H., James, J., Tidmarsh, C. (2019). Using Wicked problems to foster interdisciplinary practice among UK trainee teachers. *J. Educ. Teach., 45*(4), 446-460.
[http://dx.doi.org/10.1080/02607476.2019.1639263]

Kneen, J., Breeze, T., Davies-Barnes, S., John, V., Thayer, E. (2020). Curriculum integration: the challenges for primary and secondary schools in developing a new curriculum in the expressive arts. *Curric. J., 31*(2), 258-275.

[http://dx.doi.org/10.1002/curj.34]

Kockelmans, J.J. (1979). *Interdisciplinarity and higher education.* Penn State University Press.

Kolmos, A., Holgaard, J. E., Routhe, H. W., Winther, M., Bertel, L. (2024). Interdisciplinary project types in engineering education. *European Journal of Engineering Education, 49*(2), 257-282.

Lattuca, L. (2003). Creating interdisciplinarity: Grounded definitions from the college and university faculty. *Hist. Intellect. Cult., 3*(1), 1-20.

Lattuca, L.R., Voigt, L.J. (2012). Creating interdisciplinarity: Networks and novelties.. *Vanderbilt University Press.*

Lattuca, L.R., Voigt, L.J., Fath, K.Q. (2004). Does interdisciplinarity promote learning? Theoretical support and researchable questions. *Rev. Higher Educ., 28*(1), 23-48.
[http://dx.doi.org/10.1353/rhe.2004.0028]

Lawrence, R. (2020). *Creating built environments: Bridging knowledge and practice divides.* Routledge.

Lenoir, Y., Hasni, A. (2016). Interdisciplinarity in primary and secondary school: Issues and perspectives. *Creat. Educ., 7*(16), 2433-2458.
[http://dx.doi.org/10.4236/ce.2016.716233]

Lenoir, Y., Sauvé, L. (1998). Introduction. L'interdisciplinarité et la formation à l'enseignement primaire et secondaire : quelle interdisciplinarité pour quelle formation?. *Rev. Sci. Edu., 24*(1), 3-29.
[http://dx.doi.org/10.7202/031959ar]

Li, Y., Tolosa, L., Rivas-Echeverria, F., Marquez, R. (2025). Integrating AI in chemical education: Navigating UNESCO global guidelines, emerging trends, and its intersection with sustainable development goals. *ChemRxiv.*
[http://dx.doi.org/10.26434/chemrxiv-2025-wz4n9-v2]

Manalo, E., Fukuda, M. (2024). Integration of learning through the use of self-constructed diagrams: opportunities and challenges. *International Conference on Theory and Application of Diagrams,* Cham: Springer Nature Switzerland.358-365.

Martín-Páez, T., Aguilera, D., Perales-Palacios, F.J., Vílchez-González, J.M. (2019). What are we talking about when we talk about STEM education? A review of literature. *Sci. Educ., 103*(4), 799-822.
[http://dx.doi.org/10.1002/sce.21522]

Maxey, L. (2009). Choosing qualitative research: A primer for faculty and students. *Qual. Rep., 14*(4), 678-689.

Martín-Páez, T., Aguilera, D., Perales-Palacios, F. J., Vílchez-González, J. M. (2019). What are we talking about when we talk about STEM education? A review of literature. *Science Education, 103*(4), 799-822.

McPhail, G. (2018). Curriculum integration in the senior secondary school: a case study in a national assessment context. *J. Curric. Stud., 50*(1), 56-76.
[http://dx.doi.org/10.1080/00220272.2017.1386234]

Meadows, D. H. (2008). *Thinking in systems: A primer.* (D. Wright, Ed.). Chelsea Green Publishing.

Michelsen, C., Sriraman, B. (2007). Design science perspectives in mathematics and science education classroom research. *Proceedings of the 9th International History, Philosophy and Science Teaching Conference (IHPST).* Calgary, Canada.

Michelsen, C., Sriraman, B. (2009). Does interdisciplinary instruction raise students' interest in mathematics and the subjects of the natural sciences?. *ZDM Math. Educ., 41*(1-2), 231-244.
[http://dx.doi.org/10.1007/s11858-008-0161-5]

Moser, K.M., Ivy, J., Hopper, P.F. (2019). Rethinking content teaching at the middle level: An interdisciplinary approach. *Middle Sch. J., 50*(2), 17-27.
[http://dx.doi.org/10.1080/00940771.2019.1576579]

Moss, J., Godinho, S., Chao, E. (2019). Enacting the Australian Curriculum: Primary and secondary teachers' approaches to integrating the curriculum. *Aust. J. Teach. Educ., 44*(3), 24-41. [http://dx.doi.org/10.14221/ajte.2018v44n3.2]

National Academy of Sciences. (2005). Facilitating interdisciplinary research. *The National Academies Press.*

Newell, W.H. (2007). Academic disciplines and undergraduate interdisciplinary education: Lessons from the School of Interdisciplinary Studies at Miami University, Ohio. *Eur. J. Educ., 27*(3), 211-221. [http://dx.doi.org/10.2307/1503450]

Newell, W.H. (2009). Interdisciplinarity in undergraduate general education. *The Oxford Handbook on Interdisciplinarity*Oxford University Press.

Nicolescu, B. (2014). From modernity to cosmodernity: Science, culture, and spirituality.. *State University of New York Press.*

Pang, J., Good, R. (2000). A review of the integration of science and mathematics: Implications for further research. *Sch. Sci. Math., 100*(2), 73-82. [http://dx.doi.org/10.1111/j.1949-8594.2000.tb17239.x]

Phillips, D.C. (1976). Holistic thought in the social sciences. *Stanford University Press.*

Pountney, R., McPhail, G. (2019). Crossing boundaries: Exploring the theory, practice and possibility of a 'Future 3' curriculum. *Br. Educ. Res. J., 45*(3), 483-501. [http://dx.doi.org/10.1002/berj.3508]

Queiruga-Dios, M.Á., López-Iñesta, E., Diez-Ojeda, M., Sáiz-Manzanares, M.C., Vázquez-Dorrío, J.B. (2021). Implementation of a STEAM project in compulsory secondary education that creates connections with the environment (Implementación de un proyecto STEAM en Educación Secundaria generando conexiones con el entorno). *Infanc. Aprendiz., 44*(4), 871-908. [http://dx.doi.org/10.1080/02103702.2021.1925475]

Raggi, A., Leonardi, M., Arruda, M., Caponnetto, V., Castaldo, M., Coppola, G., Martelletti, P. (2024). Hallmarks of primary headache: part 1–migraine. *The Journal of Headache and Pain, 25*(1), 189.

Raza, S., Khan, T., Singh, M. (2024). AI tools for teaching-learning chemistry. *In: Artificial Intelligence: A Multidisciplinary Approach Towards Teaching and Learning.* Bentham Science Publishers. [http://dx.doi.org/10.2174/9789815305180124010011]

Repko, A.F. (2008). Interdisciplinary research: Process and theory. *Sage Publications.*

Ríordáin, M.N., Johnston, J., Walshe, G. (2016). Making mathematics and science integration happen: key aspects of practice. *Int. J. Math. Educ. Sci. Technol., 47*(2), 233-255. [http://dx.doi.org/10.1080/0020739X.2015.1078001]

Roehler, L., Fear, K., Herrmann, B.A. (1998). Connecting and creating for learning: Integrating subject matter across the curriculum and the school. *Educ. Psychol. Rev., 10*(2), 201-225. [http://dx.doi.org/10.1023/A:1022141720743]

Savin-Baden, M., Major, C. H. (2023). *Qualitative research: The essential guide to theory and practice.* Routledge.

Scharte, B. (2025). The need for general adaptive capacity—Discussing resilience with complex adaptive systems theory. *Risk Analysis, 45*(6), 1443-1452.

Shanta, S., Wells, J. G. (2022). T/E design based learning: assessing student critical thinking and problem solving abilities. *International Journal of Technology and Design Education, 32*(1), 267-285.

Shvets, T., Shestakova, S., Kryvoshlykov, S., Lohvynenko, V., Butrynovska, U. (2024). Enhancing students' social abilities via cooperative learning and project-based teaching methods: Pedagogical approaches and beneficial outcomes. *Multidisciplinary Reviews, 7*, e2024spe022.

Singh, M., Azad, I., Qayyoom, M.A., Khan, T. (2024). A study on perceptions and practices of STEAM-based education with university students. *Soc. Sci. Humanit. Open, 10*, 101162. [http://dx.doi.org/10.1016/j.ssaho.2024.101162]

Singh, M., Jha, A.K., Khan, T., Raza, S. (2024). The evolution of artificial intelligence from philosophy to new frontier. *Artificial Intelligence: A Multidisciplinary Approach Towards Teaching and Learning.* Bentham Science Publishers.
[http://dx.doi.org/10.2174/97898153051801240100005]

Stubbs, E.A., Myers, B.E. (2016). Part of what we do: Teacher perceptions of STEM integration. *J. Agric. Educ., 57*(3), 87-100.
[http://dx.doi.org/10.5032/jae.2016.0308]

Tammaro, R., D'Alessio, A., Petolicchio, A. (2017). Orienteering: Motivation, multidisciplinary and skills. A project in a secondary school in the province of Salerno. *Int. J. Humanit. Soc. Sci., 9*(5), 34-40.

Thornhill-Miller, B., Camarda, A., Mercier, M., Burkhardt, J. M., Morisseau, T., Bourgeois-Bougrine, S., Lubart, T. (2023). Creativity, critical thinking, communication, and collaboration: Assessment, certification, and promotion of 21st century skills for the future of work and education. *Journal of Intelligence, 11*(3), 54.

Tinnell, T.L., Tretter, T.R., Thornburg, W., Ralston, P.S. (2019). Successful interdisciplinary collaboration: Supporting science teachers with a systematic, ongoing, intentional collaboration between university engineering and science teacher education faculty. *J. Sci. Teach. Educ., 30*(6), 621-638.
[http://dx.doi.org/10.1080/1046560X.2019.1593086]

Tonnetti, B., Lentillon-Kaestner, V. (2023). Teaching interdisciplinarity in secondary school: A systematic review. *Cogent Education, 10*(1), 2216038.

Tonnetti, B., Lentillon-Kaestner, V. (2023). Teaching interdisciplinarity in secondary school: A systematic review. *Cogent Education, 10*(1), 2216038.
[http://dx.doi.org/10.1080/2331186X.2023.2216038]

Trench, B. (2023). Strengthening interdisciplinarity in science communication education: promise, pleasures and problems. *Jcom, 22*(06), Y02.
[http://dx.doi.org/10.22323/2.22060402]

Tress, B., Tress, G., Fry, G. (2006). Defining concepts and the process of knowledge production in integrative research. *Landsc. Urban Plan., 75*(3–4), 148-157.

Van Merriënboer, J.J.G. (1997). *Training complex cognitive skills: A four-component instructional design model for technical training.* Educational Technology Publications.

Vega, E.S., Schnackenberg, H.L. (2004). Integrating technology, art, and writing: Creating comic books as an interdisciplinary learning experience. *Association for Educational Communications and Technology, 2*(27), 818-823.

Vieira, A. A. S., Dias, L., Chediak, S. (2018). Challenges and contributions regarding integration of disciplines in the vocational education integrated to upper secondary education at ifms, nova andradina campus. *Revista Ibero-Americana de Estudos em Educação, 13*(esp1), 359-378.

Wang, M., Dede, C., Grotzer, T. A., Chen, J. (2025). Understanding and managing the complexities in situated learning in immersive virtual environments. *Educational Technology Research and Development, 73*(5), 3177-3200.

Youngblood, D. (2007). Interdisciplinary studies and the bridging disciplines: A matter of process. *J. Res. Pract., 3*(2), M18.

CHAPTER 2

Bridging the Barrier between Science and the Arts: A Novel Approach

Shweta Pandey[1], **Venika Dhoonter**[1], **Shahla Tanveer**[2] and **Mohammad Imran Ahmad**[2,*]

[1] *Chandigarh College of Pharmacy, Mohali, Punjab 140307, India*

[2] *Department of Chemistry, Integral University, Lucknow, Uttar Pradesh 226026, India*

Abstract: Recent years have witnessed a significant surge in exploring the connections between the arts and the sciences. This chapter addresses the significant rise in exploring the connections between the arts and the sciences. It analyzes the evolving dynamics of this connection and underscores the idea that a holistic approach to learning in both fields requires mutual recognition and integration. It explores the potential for the arts to enrich science teaching and learning at multiple levels: through the structural organization of subjects, the contextualization of science within STS frameworks, and the adoption of arts-inspired pedagogical practices. The rise of STEAM further underscores the evolving nature of twenty-first-century science, necessitating innovative pedagogies from the arts to bridge the widening gap between contemporary science and school science. Furthermore, this chapter conceptualizes the collaborative potential of art, science, and technology (AST) through a social-ecological lens. This framework characterizes AST collaborations through antecedent conditions (knowledge, aesthetics, interdependence, and experimentalism), process elements, and potential outcomes, offering a valuable tool for evaluation and reflection for practitioners, researchers, educators, and policymakers seeking to foster meaningful interdisciplinary engagement.

Keywords: Art and Science Integration, Interdisciplinary education, Scientific visualization, STEAM education, Synergistic interaction, Technological advancement.

INTRODUCTION

The conventional view positions science and art as contrasting forces: one objective and structured, the other subjective and intuitive. Yet, at their core, both disciplines strive to understand, interpret, and shape the world we inhabit. This artificial separation has impeded progress in both fields. This chapter presents

* **Corresponding Author Mohammad Imran Ahmad:** Department of Chemistry, Integral University, Lucknow, Uttar Pradesh 226026, India; E-mail: miahmad@iul.ac.in

Tahmeena Khan, Manisha Singh, & Saman Raza (Eds.)

"**Bridging the Barrier between Science and Arts**, a novel approach designed to integrate science and art.

As Albert Einstein wisely noted, "*The most beautiful thing we can experience is the mysterious. It is the source of all true art and science.*"

THE INTERSECTION OF ART AND SCIENCE

The intersection of art and science represents more than just a point of contact; it is a fertile ground for synergistic interaction, powerfully stimulating imagination and fostering innovation. Leonard Shlain's seminal work, *Art & Physics: Parallel Visions in Space, Time, and Light* (1991), provides a compelling illustration of this, arguing that scientific discoveries often originate in the realm of imagination, with artistic visualization preceding their scientific articulation (Shlain, 1991). This chapter reinforces this perspective, drawing on the work of Barrett (2012) and Kelley and Chang (2018), who emphasize the potential of multidisciplinary approaches, particularly digital art, to unlock novel ideas and address intricate problems (Barrett, 2012). We argue that dismantling the traditional separation between art and science is crucial for the advancement of both fields, as it nurtures the complementary skills of imaginative creation and rigorous analysis, both indispensable for revolutionary breakthroughs (Kelley & Chang, 2018).

Historical Divide between Art and Science

The historical development of academic disciplines has largely maintained a separation between art and science. This division was often justified by contrasting art's inherent subjectivity and engagement with emotion against science's commitment to objective analysis and empirical evidence. The intellectual climate of the Scientific Revolution further reinforced this compartmentalization. Nevertheless, contemporary interdisciplinary initiatives are increasingly demonstrating the powerful synergies and deeper levels of understanding that result from intentionally bridging these historically divided domains (Danto 1981). The historical journey of art and science reveals a fascinating trajectory, commencing with intertwined beginnings, as art increasingly focused on aesthetic expression and subjective interpretation, while early science started to prioritize systematic observation and categorization (Fig. **1**).

The Growing Interdisciplinary Approach in Modern Science

Modern science increasingly embraces interdisciplinary approaches, recognizing the powerful intersection of scientific inquiry and artistic creativity. This integration unlocks new avenues for innovation, enhances problem-solving capabilities, and drives progress in both technology and culture. The synergistic combination of art and science has already yielded remarkable ideas and solutions in diverse sectors such as design and education, offering a promising pathway to tackle complex global challenges (Yilmaz 2019).

Purpose and Significance of Bridging the Gap

Bridging the historical divide between art and science serves a crucial purpose in fostering innovation and enhancing our ability to solve complex problems. The benefits of cross-disciplinary collaboration are evident in Africa, where it has significantly contributed to overcoming developmental obstacles, achieving technological advancements, and deepening cultural understanding. By merging scientific rigor with artistic creativity, new perspectives emerge, ultimately boosting the creative power needed to address multifaceted international challenges (Okeke, 2015).

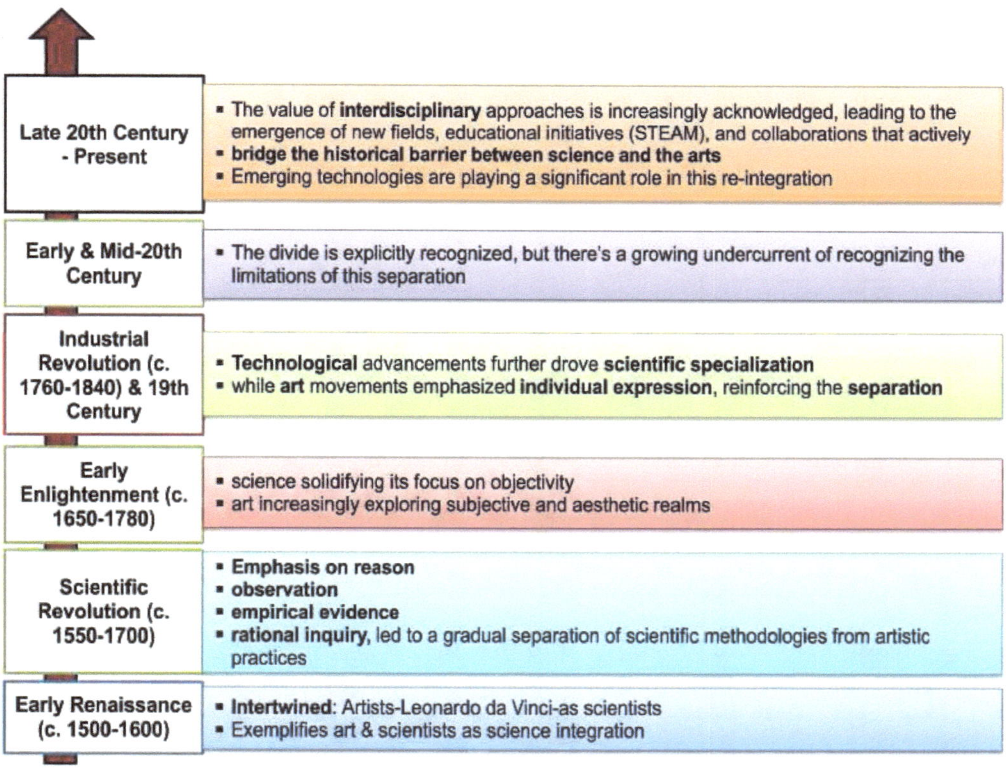

Fig. (1). Historical journey from intertwined beginnings to increasing separation.

THEORETICAL FOUNDATIONS

The quest to bridge the fields of science and art is theoretically grounded in the recognition of their shared fundamental processes: creation, observation, and expression. Challenging traditional distinctions between practitioners, cognitive science, and creativity studies emphasize the common core of researching, learning, and problem-solving inherent in both domains. This theoretical framework suggests that the true potential of both fields is realized through their integration, leading to more comprehensive and insightful approaches (Green, Beaty, Kenett, & Kaufman, 2024).

Philosophical Perspective on Art and Science

Art and science have been discussed together because of how both disciplines seek knowledge and understanding. Regarding their approaches, there seems to be a dissonance between the two fields. Art is thought of as a realm of personal reflection and intuitive creation, which is posed against the objective evaluation and empirical methods offered by science. According to Arthur Danto, both disciplines engage in the acts of representation and using imagination to challenge reality. Such an approach explains the interplay of art and science and how such a combination gives a deep and unique understanding of humanity (as cited in Goodman, 1976).

Key thinkers who advocated for integration (e.g., Leonardo da Vinci, C.P. Snow's "Two Cultures")

Integration advocates: da Vinci and C. P. Snow's "Two Cultures": For a long time, notable thinkers like Leonardo da Vinci and Edmund Husserl have advocated for a mixture of the two sides of the discipline: science and art. Husserl emphasized the need for a unifying framework of the artistic and scientific paradigms aimed at human comprehension, whereas Da Vinci used artistic lenses to approach scientific concepts. This shows that these scholars believe that integrating the two nurtures profound knowledge (Husserl, 1999).

THE ROLE OF CREATIVITY AND INNOVATION IN BOTH FIELDS

Innovation and creativity are parts of both Science and the Arts. Artistic creativity goes beyond the conventional limits, changing perspectives and prompting the use of new media. Innovation in science results in the development of new hypotheses, research, and solutions to even the most intricate problems. Creativity agnos across all fields is attributed to interdisciplinary thinking, where scientific knowledge influences creative endeavours and artistic intuition taps into science, resulting in breakthroughs that go beyond the lines of convention (Csikszentmihalyi, 1996).

CONTEMPORARY APPROACHES TO INTEGRATION

The modern approaches to art-science integration focus on promoting synergistic collaboration between the two domains. Collaboration through interdisciplinary initiatives enables scientists and artists to co-create solutions for complex problems like sustainability and technological advancement. Some of the most notable integrations include the artistic representation of data, interactive exhibitions, scientific visualizations, or transforming complex concepts into accessible and clear ideas through art. These concepts not only aid development in both domains but also offer new ideas and perspectives that go beyond traditional disciplinary boundaries (Green, Beaty, *et al.*,).

STEAM education (Science, Technology, Engineering, Arts, and Mathematics)

The addition of A for (The Arts) to STEM, forming STEAM, integrates arts into science, arguing that modern education should emphasize creativity and innovation alongside hard skills and knowledge. This teaching approach claims that participating in the arts is helpful in creating ways to solve technological and scientific problems. The integration of graphic art, logic, and imagination in STEAM creates a fuller, richer experience through education and prepares learners for many careers in the ever-changing world. The inclusion of the arts improves technical skills and also supports creative thinking and the solving modern problems (Sanz-Camarero, Ortiz-Revilla, & Greca, 2023). STEAM education, or Science, Technology, Engineering, Arts, and Mathematics, is an educational framework that blends students' imagination and skills with the conventional framework of STEM learning. This holistic model promotes creativity, the development of problem-solving skills, and learning through practice for future careers. Combining arts and science through STEAM design education develops the technical and creative skills needed to solve real-world challenges (Yakman, 2008). Thus, diverse students can collaboratively engage in an interdisciplinary project, integrating artistic principles with STEM disciplines. This hands-on approach fosters critical thinking skills by encouraging students to approach problem-solving from multiple perspectives, while the fusion of creative expression and technical knowledge cultivates a spirit of innovation. (Fig. **2**) visually exemplifies the "Bridging the Barrier between Science and Arts" concept, showcasing the dynamic potential of collaborative learning across traditionally separate fields.

Fig. (2). Diverse students engage in interdisciplinary projects blending arts and STEM, fostering critical thinking and innovation.

Cross-disciplinary collaborations in academic and research

When a complex problem is approached from different fields to create new perspectives, there is cross-disciplinary collaboration. Scholars are able to cultivate originality through creativity and new ideas by integrating approaches from interdisciplinary issues. Such integration fosters innovation across technology, medicine, environmental studies, and other areas, fuels discovery, and enhances existing theoretical frameworks and practical applications. The knowledge from different fields allows researchers to carve new paths that would otherwise be impossible if approached using traditional stratified academic disciplines (Repko, 2012).

ROLE OF DIGITAL AND MULTIMEDIA ART IN SCIENTIFIC VISUALIZATION

In today's science, visual science has profound digital and multimedia artistic applications that help transform complex data into easy and understandable information. Artists aid in explaining abstract concepts through 3D models, animation, and more advanced techniques. This helps in spreading the captured imaginations to the masses. This fusion improves the interpretation of scientific information, supports well-informed decision-making, and connects different scientific communities with the public, thereby promoting awareness and education (McCormack, 2010).

APPLICATIONS IN VARIOUS FIELDS

The combination of art and science has led to the development of new applications in many fields. For example, digital art is used to represent complex medical data in the healthcare industry, which helps in accurate diagnosis and better understanding by the patients. In teaching, the incorporation of art and scientific subjects motivates students to understand more science from a creative perspective (Fig. **3**). Artistic representations have also been introduced by some environmental researchers to help the public understand the deeper implications of climate change. Collaborative interdisciplinary approaches provide new ideas and rational solutions to a variety of issues; that is, the use of artistic and scientific methods gives us information we would never expect (Mäkelä, 2015).

Medicine & Biomedical Imaging – Artistic visualization of complex biological processes

In the context of biomedical imaging and treatment procedures, art has the particular function of further explaining complex biological mechanisms in a stepwise manner, shaping raw information into something meaningful and manageable for people. Improvements in understanding of anatomy, disease manifestations, and treatment outcomes enhance communication between patients and healthcare practitioners, and their involvement leads to better 'healthcare' decisions (Dunn, 2012). In the context of biomedical imaging and medicine, artistic visualization serves as a technique for portraying biological information. Using new technologies like 3D models, animations, and other design objects, these sophisticated aids promote communication in healthcare and understanding of complex biological systems, while augmenting diagnostic accuracy. They enhance life sciences teaching and research to improve informed decisions in life (Ma, Yuan, Sun, Lawson, & Wang, 2024).

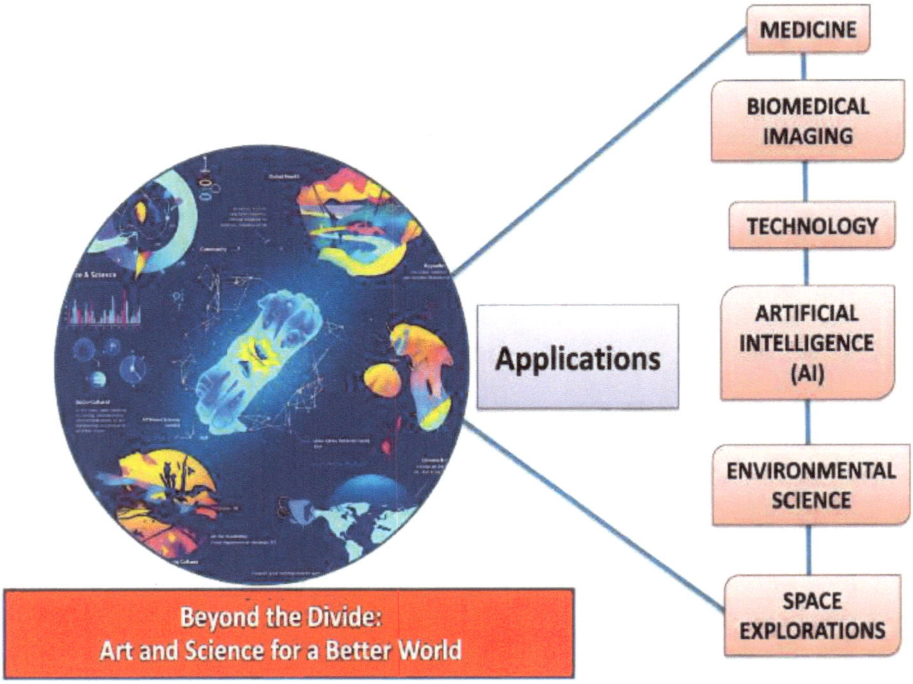

Fig. (3). The interdisciplinary role of art. This digital artwork illustrates the convergence of art and science across various domains.

The use of art in UX/UI design and human-centred AI

Both human-centred artificial intelligence (AI) and UX/UI design are notably affected by artistic application. Designers apply colour schemes with aesthetics and order to structure intuitive user interaction and experience designs for people with differing mental capabilities, such as children and elderly persons. Similarly, human-centred AI combines artistic ideas to make AI systems more humane, accessible, engaging, and user-friendly. By fusing this interdisciplinary approach, not only is overall efficacy and usability improved, but trust in technological systems, and their practical and aesthetic appeal (Norman, 2013), is reinforced. Artistic principles are important in the realms of technology, as they work as a catalyst to improve human-centred AI and UX/UI design. By drawing on design principles such as colour theory, spatial composition, and aesthetic appeal, developers can enhance user engagement and create interfaces that are intuitive and easy to use. In addition to improving user interaction, this integration helps build trust and confidence in AI systems, which, in turn, increases their efficacy and adoption (Brown, 2018).

Environmental Science – Art as a medium to communicate climate change and sustainability

Art is a potent medium for expressing climate change and sustainability in the field of environmental science. In their art, creators translate challenging environmental conditions into the common, especially in terms of feelings and emotions. This approach not only creates awareness but also initiates discussions and encourages actions, making people take up sustainable practices and look forward to environmental issues (Shaw, 2025). Art plays a critical role in environmental science by transforming complex scientific data into compelling visual narratives to enable the public's understanding and participation. The environment becomes more accessible and emotionally resonant because artists' creative endeavours, by their very nature, connect them to the audience in a more intimate and meaningful way. Not only does art and science synergy raise awareness, but it also triggers conversation and motivates action, which lures us towards environmental preservation (Girod, 2017).

Space Exploration – Scientific illustrations and immersive experiences in astrophysics

In the context of space exploration, scientific illustrations and immersive experiences are used to explain complex astrophysical phenomena. Artists facilitate the comprehension of abstract scientific notions through detailed depictions of celestial entities, events, and simulations of space missions. Both scientists and the general public can participate in immersive experiences, such as virtual reality, that deepen their engagement with the universe (Sagan, 1994). Yet scientific visualizations and immersive experiences are central to bringing complex astrophysical concepts to life in the context of space exploration. Artists create intricate animations and visually compelling representations in order to help interpret huge datasets resulting from space missions, making celestial objects like black holes and nebulae better understood. Immersive technologies such as virtual reality (VR) also assist with public engagement and encourage a deep understanding of the cosmos (Hubble, 2011).

CHALLENGES AND BARRIERS

The resolution of epistemological differences between the arts and sciences is a fundamental challenge to bridge the gap between them, as art prioritizes subjective expression and interpretative creativity, while science is based on empirical evidence and objective analysis. In addition, resistance to collaboration between specialists in both domains and a lack of multidisciplinary training are major obstacles to integration. The lack of funding and institutional support for such an interdisciplinary endeavour impedes the easy integration of artistic and scientific methodologies (Drucker, 2015).

Deep institutional and cultural resistance

There is deep institutional and cultural resistance to integrating art and science into the long-standing traditions of academic culture. Disciplinary specialization is often preferred over interdisciplinary collaboration by educational and research institutions, and prevailing cultural norms sustain the dichotomy between creative and empirical domains. In addition to limiting opportunities for cross-disciplinary engagement, this entrenched resistance also impedes the creation of novel solutions arising from the combination of artistic and scientific methodologies (Gibbons, 2000). Art and science are historically separated, and continue to be so; institutional and cultural resistance to their convergence persists. Specialized expertise is often encouraged in many academic institutions at the expense of interdisciplinary collaboration, thereby limiting creative exploration as well as the cross-pollination of ideas among these fields (Ziman, 2000).

Integration of art and science into academic and professional circles

Obtaining funding and being recognized in academic and professional circles is a difficult task to integrate art and science. Interdisciplinary research is often underfunded or unsupported by traditional funding mechanisms that generally support specialized research within established disciplines. Lack of institutional backing can lead to insufficient attention paid to the innovative art-science collaboration (Becker, 2008), thus depriving it of the possibility of professional recognition. (Becker, 2008). In formal and professional contexts, securing funding and gaining appreciation for art-science transdisciplinary integration remain among the greatest challenges. Most of the institutions seem to lean toward more precise and single-discipline research-defined loci of study, making cross-disciplinary partnerships harder for institutional funding and endorsement. This inclination tends to reduce the perceived value and growth potential of such integrative initiatives (Klein, 2010).

Barriers to Interdisciplinary Communication

Navigating interdisciplinary communication is often more difficult due to the unique vocabularies, strategies, and goals of different academic disciplines. For example, scientists focus on objectivity and verification, while artists focus on their interpretation. These differences often result in conflict and considerable difficulties in collaboration (Snow, 1998). There are many different approaches and terms employed in each discipline, which makes collaborative work within interdisciplinary projects more difficult. For example, many artists prioritize imagination or emotionally driven content, while many scientists focus on logical reasoning and evidence. Many fundamental differences affect communication across disciplines. (Barry, 2001).

CASE STUDIES AND SUCCESS STORIES

Case studies and notable success stories of art and science bridge the spaces of interdisciplinary collaboration as a demonstration of the transformative potential of the practice. Perhaps the most prominent example is the work of neuroscientist and artist Heather Barnett, who employs artistic methods of investigation of the human brain. The Neuro-Image is a work of art that gives us the ability to visualize cognitive processes and neural activity in a way that has not been seen in the field of neuroscience previously. The ArtScience Prize, a joint effort of MIT and the Harvard Art Museums, does the same in recognizing and celebrating work that crosses the boundaries of art and science. Such cross-disciplinary partnerships provide not only creative inquiry but also scientific exploration (Cohen & Waite-Stupiansky, 2020). This initiative illustrates the great benefits that such partnerships have to offer.

Some examples of artists and scientists working together (MIT Media Lab, NASA collaborations, BioArt)

Artists and scientists have worked together to merge scientific exploration with artistic production. The MIT Media Lab is a place where engineers and artists work together, developing innovative technology and interactive artworks, and fostering a creative back-and-forth. Likewise, NASA's "Art of Exploration" program invites artists to paint the space missions and astronomical discoveries they depict, bringing complex scientific concepts to the public. Eduardo Kac is an example of a fusion of art and biology through his BioArt movement. Reflection is also induced by Kac's work, including the making of transgenic organisms (such as fluorescent rabbits) and environmental issues related to genetic engineering. Mäkelä, (2015) suggests that these interdisciplinary collaborations demonstrate how bringing together different fields can produce new insights and increase public engagement. But interdisciplinary collaborations between artists and scientists have generated innovative projects that combine scientific inquiry with artistic expression. MIT Media Lab is a place where engineers and artists work together to create cutting-edge technologies and interactive artwork, where ideas are passed among each other. Likewise, NASA's "Art of Exploration" program brings artists to describe space missions and astronomical discoveries in ways that make complex scientific concepts understandable to the public (Packer, 2008).

The role of museums and exhibitions showing science through art

Museums and exhibitions that connect science with art to bridge the gap between creativity and empirical study have a very important role. This includes the Science Gallery in Dublin and the Exploratorium in San Francisco, both of which feature interactive exhibits and art installations to communicate complex scientific principles in an engaging and accessible way. Exhibitions like Art and Science: A New Beginning at the Smithsonian's National Museum of Natural History employ artistic mediums to simplify intricate scientific ideas into more easily understood concepts, to help the general public understand interdisciplinary knowledge (Jones, 2017). The artistic expression seamlessly combined with scientific inquiry is shown in exhibitions at prestigious institutions, such as the Natural History Museum in London and the Museum of Science and Industry in Chicago. For instance, the Art in Science exhibit at the California Academy of Sciences uses artistic imagery along with ecological science to entice audiences and make complex scientific concepts more intellectually accessible and engaging (MacDonald, 2014).

FUTURE DIRECTIONS AND INNOVATIONS:

The future of innovations for the integration of art and science will be founded on Biotech's intersection with Artificial Intelligence and immersive applications of virtual and augmented reality. These innovations will enhance how we visualize science, encouraging new, sophisticated ways to interact with complex scientific phenomena. Collaborations from different fields will be supported, Sullivan predicts, and new educational systems will be developed that integrate and apply art, science, and technology more efficiently (Sullivan, 2018).

Emerging technologies (VR, AR, AI-generated art):

New technologies (VR, AR, AI art): New capable technologies like virtual reality (VR) and augmented reality (AR) are transforming the fusion of science and art, while AI-generated art is taking them to a whole new level. VR & AR provide interactive participation where the audience can see and manipulate scientific concepts like molecules or even galaxies in 3D, which improves understanding and interest. Also, AI art uses machine learning algorithms to create original artworks, unlike traditional artworks, which expands the scope of artistic endeavors. All of these developments from various fields provide new platforms for developing complex scientific concepts for the public. (McCormack, 2019). In the same way, AI-powered art helps develop new ways to portray and explain advanced scientific ideas. It uses VR and AR technologies to provide active, immersive experiences that blend empirical exploration and artistic creativity (Gero, 2020).

The role of communicators in education as a part of focused science communication

Polyvalent education and educated science communicators play an essential role in the harmony of art and science. The science communicators, in particular, assume the pivotal role of a mediator. He or she translates everything in the science domain into text or audio-visual materials that can be readily understood by the public. At the same time, interdisciplinary education promotes collaboration by teaching students to draw from various fields, synthesize ideas, and therefore create out-of-the-box solutions. Such a strategy promotes active understanding and publicization of great scientific inventions and discoveries, and, more importantly, the invention of new ways to address complex issues around the world, and educates people to become more imaginative, creative, and inclusive (Bucchi, 2019). Interdisciplinary education also enables students from different subjects to collaborate on projects spanning multiple subjects or branches of knowledge. Science communicators, on the other hand, use fascinating narratives and other forms of storytelling to paint rich pictures that clarify complex technical concepts (Jacobson, 2017).

Potential for further integration in research and industry

The integration of art disciplines with technology and scientific exploration underscores the need to employ science and modern technology in environmental and healthcare services. Art, as a form of representation of technology and science, helps simplify most scientific ideas. Alongside art, its achievements, such as modern devices and methodologies, broaden the horizons of science. New art techniques associated with creative thinking aid in making user-friendly products and devising smart communication tools. Such advances increase the effectiveness, use, and appeal of technology and science for the general public (Schön, 2017). Art disciplines merging with science and technology are still in the early stages of application. The application of artistic imagination to advanced technology gives rise to aesthetic innovation. For example, technological instruments can be scientifically improved by integrating their design. Environmental science and healthcare are among the many interdisciplinary collaborations that could produce groundbreaking solutions (Velkova & Kaun, 2021).

CONCLUDING REMARKS

This chapter has laid the groundwork for understanding the artificial and limiting divide that has historically separated science and art. While conventional thought often positions these disciplines as contrasting – one rooted in objective structure, the other in subjective intuition – we have argued that at their core, both are driven by a fundamental human desire to understand, interpret, and shape the world around us.

Presenting "Bridging the Barrier between Science and Arts" as a novel approach, this chapter has highlighted the fruitful intersection where these fields converge. Drawing upon the insights of thinkers like Shlain, Barrett, Kelley, and Chang, we have established that the interplay between scientific rigour and artistic imagination is not merely a point of contact but a powerful catalyst for innovation and deeper understanding.

The purpose and significance of bridging this gap are manifold, extending from fostering innovation and problem-solving to enriching cultural understanding, as evidenced by its impact in diverse global contexts. Theoretically, this integration is grounded in the shared cognitive processes of creation, observation, and expression, challenging traditional distinctions between practitioners. Philosophically, both art and science engage in representation and the imaginative questioning of reality, as articulated by Danto. The advocacy of key figures such as Leonardo da Vinci and C.P. Snow further underscores the long-recognized value of their convergence.

Contemporary approaches to integration, such as STEAM education and cross-disciplinary collaborations, demonstrate the practical benefits of this synthesis, fostering creativity and innovation across various fields. The role of digital and multimedia art in scientific visualization, along with compelling applications in medicine, UX/UI design, environmental science, and space exploration, vividly illustrates the tangible outcomes of this interdisciplinary synergy.

While acknowledging the inherent challenges and barriers – including epistemological differences, institutional resistance, funding limitations, and communication hurdles – this chapter has also showcased compelling case studies and success stories that underscore the transformative potential of art-science collaborations. Examples like the work of Heather Barnett, the ArtScience Prize, and initiatives at the MIT Media Lab and NASA demonstrate the innovative insights and enhanced public engagement that arise from bridging these domains.

The crucial role of science communicators and interdisciplinary education will be paramount in fostering a more unified and collaborative landscape. Ultimately, the integration of art and science holds immense potential to drive innovation, deepen understanding, and address complex global challenges in more creative and effective ways. This chapter serves as an introduction to this wonderful amalgamation, setting the stage for a deeper exploration of the methodologies, applications, and transformative power of "Bridging the Barrier between Science and Arts."

REFERENCES

Barrett, T. (2012). Art and science: A study in the intersection of creativity and knowledge. *Journal of Arts and Sciences, 34*(2), 123-138.

Barry, A. (2001). *Political machines: Governing a technological society.* The Athlone Press.

Becker, H.S. (2008). Art worlds. *University of California Press.*

Brown, T. (2018). *Change by design: How design thinking creates new alternatives for business and society.* HarperBusiness. Available form: https://harpercollins.co.uk/products/change-by-design-revised-and-up-ated-how-design-thinking-transforms-organizations-and-inspires-innov-tion-tim-brown?variant=39430167199822

Bucchi, M. (2019). *Science in society: An introduction to the sociology of science.* Routledge.

Cohen, L. E., Waite-Stupiansky, S. (2020). STEM in Early Childhood Education. *Ny: Routledge.*

Csikszentmihalyi, M. (1996). *Creativity: Flow and the psychology of discovery and invention.* HarperCollins.

Danto, A.C. (1981). The transfiguration of the commonplace: A philosophy of art. *Harvard University Press.*

Drucker, J. (2015). *Graphesis: Visual forms of knowledge production.* Harvard University Press.

Dunn, L. (2012). Artistic visualization in biomedical imaging. *J. Med. Imaging (Bellingham), 24*(4), 68-78.

Gero, J. (2020). Technological innovations in art and science collaboration. *Journal of Creative Technologies, 18*(4), 134-145.

Gibbons, M. (2000). Mode 2 knowledge and the new production of knowledge. *Sci. Technol. Human Values, 25*(4), 346-364.

Girod, B. (2017). Art as a catalyst for climate change awareness and action. *Journal of Environmental Communication, 14*(1), 78-90.

Goodman, N. (1976). *Languages of art: An approach to a theory of symbols.* Hackett Publishing. [http://dx.doi.org/10.5040/9781350928541]

Green, A. E., Beaty, R. E., Kenett, Y. N., Kaufman, J. C. (2024). The process definition of creativity. *Creativity Research Journal, 36*(3), 544-572.

Hubble, E. (2011). Exploring the universe: The role of visualizations in astrophysics. *Astrophys. J., 21*(3), 102-115.

Husserl, E. (1999). *The crisis of European sciences and transcendental phenomenology.* Northwestern University Press.

Jacobson, R. (2017). Bridging the gap: The role of science communication in interdisciplinary education. *Science Education Review, 16*(2), 101-112.

Jones, R. (2017). Art meets science: Museums and exhibitions exploring scientific concepts. *Museum Studies Journal, 28*(3), 45-60.

Kelley, T., Chang, J. (2018). Crossing boundaries: How the arts and sciences can collaborate. *International Review of Arts and Technology, 27*(1), 45-60.

Klein, J.T. (2010). *Creating interdisciplinary campus cultures: A model for strength and sustainability.* Jossey-Bass.

Ma, H., Yuan, X., Sun, X., Lawson, G., Wang, Q. (2024). Seeing your stories: Visualization for narrative medicine. *Health Data Science, 4*, 0103.

MacDonald, L. (2014). Art and science in museums: Fostering public engagement. *Mus. Soc., 12*(3), 102-115.

Mäkelä, M. (2015). Art and science collaborations: New models and new perspectives. *J. Interdiscip. Stud., 18*(2), 21-35.

Mäkelä, M. (2015). The intersection of art and science: Bridging the gap in various fields. *J. Interdiscip. Stud., 18*(3), 56-69.

McCormack, D. (2010). Digital art and scientific visualization: Bridging art and science. *Journal of Media Arts, 18*(3), 95-110.

McCormack, D. (2019). Emerging technologies and the intersection of art and science. *Journal of Digital Art and Technology, 22*(3), 55-67.

Norman, D.A. (2013). *The design of everyday things: Revised and expanded edition.* Basic Books.

Okeke, C. (2015). Art, science, and technology: Bridging the gap in Africa. *African Journal of Innovation and Technology, 10*(3), 45-58.

Packer, R. (2008). Multimedia art and scientific collaboration: Exploring new horizons. *Journal of Art and Technology, 10*(4), 98-110.

Repko, A.F. (2012). *Interdisciplinary research: Process and theory.* Sage Publications.

Sagan, C. (1994). *Cosmos..* Penguin Random House.

Sanz-Camarero, R., Ortiz-Revilla, J., Greca, I. M. (2023). The impact of integrated STEAM education on arts education: A systematic review. *Education Sciences, 13*(11), 1139. [http://dx.doi.org/doi.org/10.3390/educsci13111139]

Schön, D.A. (2017). *The reflective practitioner: How professionals think in action.* (1st ed.). Routledge. [http://dx.doi.org/10.4324/9781315237473]

Shaw, K. R. (2025). *Using Visual Art to Facilitate Conservation Education and Communication.* [Doctoral dissertation, University of Wisconsin–Madison]. UW-Madison Libraries.

Shlain, L. (1991). *Art & physics: Parallel visions in space, time, and light.* HarperCollins.

Snow, C.P. (1998). *The two cultures and the scientific revolution.* Cambridge University Press.

Sullivan, G. (2018). Art, science, and technology: The future of interdisciplinary innovation. *Journal of Creative Practice, 15*(2), 123-137.

Velkova, J., Kaun, A. (2021). Algorithmic resistance: Media practices and the politics of repair. *Information, Communication & Society, 24*(4), 523-540.

Yakman, G. (2008). STEAM: A framework for teaching across the disciplines. *International Journal of Arts Education, 6*(2), 1-10.

Yılmaz, M. (2019). Interdisciplinary approaches: Bridging the gap between art and science. *J. Interdiscip. Stud., 7*(2), 112-125.

Ziman, J. (2000). *Real science: What it is, and what it means.* Cambridge University Press. [http://dx.doi.org/10.1017/CBO9780511541391]

CHAPTER 3

Pedagogical Strategies for Interdisciplinary Learning

Waseem Zahra[1,*] and **Manisha Singh**[2]

[1] *Department of B.Ed., Mahila Vidyalaya Degree College, Lucknow, Uttar Pradesh 226018, India*

[2] *Department of Education, Integral University, Lucknow, Uttar Pradesh 226026, India*

Abstract: In modern education, interdisciplinary learning (IDL) has gained popularity as a method to develop students' creativity, critical thinking, and problem-solving abilities. By combining perspectives, knowledge, and skills from various subjects, interdisciplinary learning enables students to comprehend complex topics more effectively. Unlike traditional learning, which is subject-centred and isolates disciplines, IDL develops analytical thinking and problem-solving by connecting various study areas and providing a holistic picture. One of the biggest advantages of this approach is its ability to address real-world challenges, such as those related to artificial intelligence, healthcare, or climate change. This chapter explores the teaching of interdisciplinary strategies that can enhance learning by fostering significant connections between subjects, promoting collaborative inquiry, and integrating real-world applications into the curriculum. Furthermore, it examines the role of technology in interdisciplinary learning, providing access to diverse resources and enabling interactive, multimedia-based teaching. The chapter also highlights the challenges teachers face when implementing multidisciplinary teaching methods and offers practical solutions to address these challenges.

Keywords: Challenges, Collaborative learning, Experiential learning, Interdisciplinary Learning (IDL), Pedagogical strategies, Problem-based learning, Project-based learning, Solutions, Team-teaching, Technologies.

INTRODUCTION

Interdisciplinary learning combines insights from various fields, providing students with a comprehensive and interrelated understanding of complex issues. Unlike traditional education, which limits knowledge to separate academic disciplines, this approach encourages students to explore the connections between subjects, developing critical thinking, creativity, and problem-solving skills. This method is particularly effective in addressing real-life challenges that cannot be

* **Corresponding Author Waseem Zahra:** Department of B.Ed., Mahila Vidyalaya Degree College, Lucknow, Uttar Pradesh 226018, India; E-mail: waseemzahra85@gmail.com

Tahmeena Khan, Manisha Singh, & Saman Raza (Eds.)

adequately addressed within the scope of a single subject. For example, understanding climate change requires knowledge from various subjects, including environmental science (to comprehend its causes and impacts), economics (to assess financial implications and policies), and political science (to examine government initiatives and international agreements) (Serafini, Serafini, Moura, *et al.*, 2022). Similarly, effective healthcare solutions often require an amalgamation of biology, medical science, psychology, and technology. In the field of artificial intelligence, collaboration among computer science, ethics, psychology, and law is necessary for developing responsible and structured systems. By breaking down academic silos, the interdisciplinary approach nurtures students' ability to think outside the box and problem-solving skills. It prepares them for careers that require flexibility and a broad range of knowledge. This approach helps students develop a deeper, more interconnected understanding of the world, equipping them with the skills to address the challenges of the 21st century. A multidisciplinary individual can easily shift between different fields, drawing knowledge and skills from diverse areas to provide new insights into their issues (Barron & Darling-Hammond, 2008). This chapter throws light on the following aspects of interdisciplinary learning:

- Identification and analysis of effective pedagogical strategies that support interdisciplinary learning in educational institutions.
- Evaluation of the role of technology and experiential learning in enhancing interdisciplinary education outcomes.
- Exploration of the challenges teachers face in implementing interdisciplinary teaching methods and proposing solutions. Some benefiFs of interdisciplinary learning are illustrated in Fig. (**1**).
- It encompasses a wide range of content areas, leading to better cognitive development as deeper learning occurs.
- Instead of separated educational experiences, it teaches hands-on learning. When we gain knowledge in everyday life, it is usually interdisciplinary. When trying to cook or bake a new dish, people often research the best recipes and techniques, read instructions carefully, and apply mathematical concepts during the cooking or baking process.
- It allows students to improve their critical thinking and problem-solving skills. Given the nature of interdisciplinary learning, which is deep and project-based, students are encouraged to draw their own conclusions and, in the process, share their insights on what they have learned.
- It is a student-centric approach. The emphasis is on the student rather than the teacher, and it avoids lower levels of Bloom's taxonomy, which are typically associated with rote-learning activities.

- It tends to engage students significantly, as interdisciplinary tasks usually have at least one area of content that particularly interests them, showcasing their strengths.
- It creates opportunities for students to develop an interest in subjects they may not like, as they begin to see interrelationships between topics and are keen to learn about other subjects.

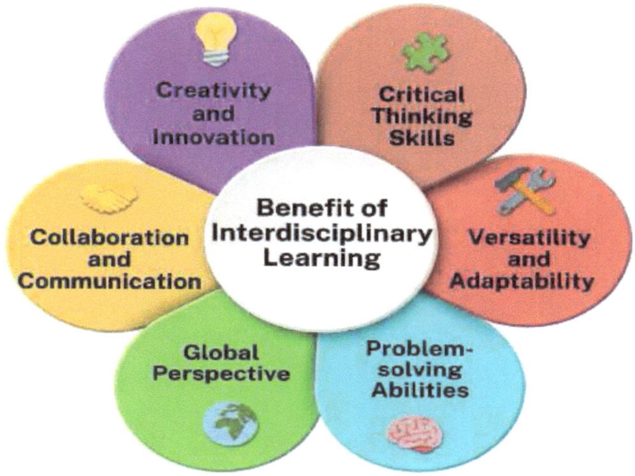

Fig. (1). Benefits of interdisciplinary learning.

Overview of Past Research

Interdisciplinary learning (IDL) has been extensively explored in academic research as a crucial strategy for contemporary education. Scholars argue that conventional disciplinary divisions impede a holistic understanding and limit students' capacity to tackle complex issues (Sánchez-García & Reyes-de-Cózar, 2025). Evidence suggests that interdisciplinary education enhances cognitive flexibility, promotes the amalgamation of knowledge, and cultivates higher-order thinking skills among students (Boix Mansilla, 2016). Various models of interdisciplinary learning have been proposed. Klein presented a classification that distinguishes between multidisciplinary, interdisciplinary, and transdisciplinary approaches. Multidisciplinary learning involves placing different disciplines side by side without integration, while interdisciplinary learning intertwines insights from various fields to achieve a more thorough understanding (Klein, 1990). Transdisciplinary learning goes beyond academic environments by integrating real-world perspectives and societal knowledge. Research also highlights numerous teaching strategies that facilitate interdisciplinary education. Project-based learning (PBL) and problem-based learning (PrBL) have demonstrated their effectiveness in engaging students, fostering collaboration, and enhancing problem-solving skills (Barron & Darling-Hammond, 2008; Dula & Porter, 2021). Collaborative teaching, primarily through co-teaching by instructors from diverse disciplines, further enriches the learning process, though it necessitates careful coordination and institutional backing (Friend & Cook, 2016; Lattuca, Voigt, & Fath, 2004). Technology-enhanced learning, which utilizes digital resources and virtual simulations, along with experiential strategies such as internships and service learning, further bolsters interdisciplinary pedagogy (Zydney, *et al.*, 2018); (Kolb, 2015). Despite these advancements, challenges such as faculty resistance, a lack of interdisciplinary training, and rigid institutional frameworks persist (Holley, 2017). Traditional assessment techniques also fail to accurately represent the depth of interdisciplinary learning outcomes, prompting the exploration of innovative evaluation methods, such as portfolios and reflective journals (Newell, 2001; McGowan, 2013). Considering these conditions, the rapidly changing society, technology, and profession necessitate an educational approach beyond the boundaries of traditional subjects. Interdisciplinary learning is necessary to meet this need because it enables students to integrate knowledge, think critically, and develop adaptability. Therefore, this chapter is crucial for determining and evaluating effective teaching methods that facilitate interdisciplinary learning. This chapter explores various instructional approaches to establish a connection between theory and practice, thereby informing curriculum development, teacher training, and institutional policies that foster interdisciplinary education.

Identification and Analysis of Effective Pedagogical Strategies that Support Interdisciplinary Learning in Educational Institutions

Education has its greatest impact when students not only gain knowledge about the world but also take an active role in shaping it. Interdisciplinary teaching methods break down traditional subject boundaries, converting education into a vibrant, interconnected journey. These methods motivate students to explore diverse areas of study, fostering a deeper understanding of complex issues. Below are several effective interdisciplinary teaching strategies:

Project-Based Learning (PBL)

In a world filled with multiple complex challenges, students require more than basic information; they need the ability to solve real-world problems. Project-based Learning (PBL) is a solution that addresses this need by involving students in extensive, meaningful projects that encourage critical analysis and the application of knowledge in real-world settings. This method naturally encourages interdisciplinary learning by integrating various disciplines, thereby increasing students' understanding of the topic. Through PBL, students acquire essential skills such as a scientific temper, cooperative learning, and problem-solving.

Example: Suppose students receive a project centred on climate change that can encompass multiple disciplines. They are collaborating on a climate change initiative, with each discipline adding depth to their understanding:

- Geography: They examine the effects of global warming on various regions, ranging from melting glaciers to shifts in weather patterns.
- Science: They investigate the greenhouse effect and climate models, uncovering the scientific principles behind global warming.
- Social Studies: They discuss government policies and initiatives, analysing how laws influence climate action.
- Economics: They assess the financial implications of climate change, including disaster recovery costs and the implementation of sustainable energy policies.

By combining their research for the project, students go beyond merely learning facts about climate change, and they suggest practical solutions. Whether designing sustainable projects or suggesting policy modifications, they develop a comprehensive, action-oriented perspective that extends beyond the classroom.

Team Teaching

Team Teaching enhances the educational experience by uniting educators from diverse fields to collaboratively deliver instruction, creating a more vibrant and interconnected learning environment. It goes beyond merely sharing a classroom; it integrates expertise, links diverse concepts, and presents students with a multi-faceted perspective on intricate subjects. This method promotes interdisciplinary learning by making lessons more stimulating and applicable, transforming subjects into a unified experience that animates content. Team teaching not only enhances student learning but also supports the professional development of teachers (Isfanescu-Ivan, 2024). Fostering critical thinking and enabling students to examine topics from different viewpoints encourages more profound discussions. It helps learners cultivate a wider understanding of the world, linking the "how" and "why" behind the information.

Example: A history and literature teacher collaborates to teach a unit on World War II. When in a classroom where history and literature teachers join hands to dig into World War II, rather than studying facts in isolation, students learn from both the political circumstances and personal narratives of that time:

- Literature: They learn through stories such as The Diary of Anne Frank or Night by Elie Wiesel, experiencing the conflict through the perspectives of those who lived it.

- History: They evaluate significant events, political alliances, and the global impact of the war.

By combining historical facts with the personal touch of storytelling, students not only acquire knowledge about history, but they feel it, question it, and connect with it. Team teaching transforms a lesson into an engaging and practical learning experience.

Problem-Based Learning (PBL)

In today's complex environment, students require more than just answers; they need to learn how to ask thought-provoking questions and solve real-world problems. Problem-Based Learning (PBL) addresses this need by making real-world challenges the centrepiece of the learning experience. Rather than being passive absorbers of information, students work collaboratively to analyse problems, formulate hypotheses, assess solutions, and produce well-informed answers.

This strategy naturally generates cross-disciplinary thinking, as real-world problems often require integrating knowledge from diverse fields such as science, economics, and social studies. When students work on realistic problems, they deepen their understanding and develop the critical life skills necessary to lead and influence their world outside the classroom. Problem-oriented teaching is a commonly implemented teaching strategy, particularly in applied sciences and technology (Nikitina, 2006); (Tolsby, 2005). In this technique, teachers play the role of stage-setters, presenting students with a specific problem scenario (Anderson *et al.*, 2021). This helps them relate theoretical concepts to real-world applications by utilizing their technical knowledge to create practical solutions (Mahaffey, Kinard, & Daughrity, 2020).

Example: Suppose students are assigned a project on food insecurity. They approach this topic from various perspectives:

- Biology: They study malnutrition, food composition, and the human body's nutritional needs.
- Economics: They evaluate food costs, supply chains, and economic inequalities that impact access to nutritious food.
- Social Sciences: They critically analyse the influence of governmental policies, cultural beliefs, and community initiatives on food distribution.

By comparing real-world data, generating innovative solutions, and proposing ideas such as community gardens, food bank projects, or policy changes, students not only learn but also become agents of change.

Inquiry-based learning

Through inquiry-based learning, the classroom becomes a place of exploration and discovery, where students transition from passive learning to actively seeking out answers and constructing their knowledge. It provides an opportunity for students to discover and ask questions about topics that truly inspire them, allowing them to take control of their learning. Instead of learning predetermined facts, students delve into questions and challenges, utilizing knowledge from various disciplines to examine issues from multiple academic perspectives and gain a holistic understanding. (Rafiq, Kamran, & Afzal, 2024) state that not only does this method promote interdisciplinary learning, but it also enhances students' critical thinking and problem-solving skills, enabling them to address real-life challenges and make their education both meaningful and empowering.

For example, consider an inquiry-based unit focused on artificial intelligence (AI). A group of students understands AI through the lens of various subjects:

- Computer Science: They decode AI algorithms, experiment with machine learning, and even attempt to program their chatbots.
- Psychology: They study the human decision-making process and compare it with AI-driven choices, analysing how machines replicate cognitive processes.
- Ethics: They engage in group discussions about the implications of AI, including bias, privacy, cyberbullying, and societal impact.

By the end of this unit, students will have developed a deep understanding of AI and be able to comprehend it from multiple perspectives, facilitating meaningful discussions and cultivating a well-rounded, informed viewpoint. Inquiry-based learning focuses not only on finding the correct answers but also on asking the right questions.

STEAM Approach (Science, Technology, Engineering, Arts, Mathematics)

When science, technology, engineering, arts, and mathematics merge, they create STEAM. It is an exciting, interdisciplinary approach to education that makes the process of gaining knowledge entertaining and interactive. STEAM drives innovation by merging analytical thinking with creative expression, encouraging students to address problems from innovative and imaginative perspectives (Wang *et al.*, 2024). This approach helps in connecting theory with practice, making abstract ideas more concrete through practical applications in the real world. This deepens understanding and reflects the collaborative essence of today's work scenarios, where students from diverse backgrounds and skill sets collaborate to address complex challenges. By combining logic and creativity, STEAM teaches students to become flexible, forward-thinking problem solvers (Singh, *et al.*, 2024).

Example: Suppose students have to prepare a project on sustainable building design. In project work, a group of students addresses the challenge of creating eco-friendly structures. Each discipline contributes significantly:

- Physics – They investigate forces and structural integrity to ensure the building remains stable.
- Mathematics – They compute dimensions, material amounts, and energy efficiency to enhance sustainability.
- Engineering – They devise innovative frameworks and select sustainable materials for durability.
- Art – They incorporate aesthetic elements, demonstrating that green buildings can be both functional and visually pleasing.

Through this collaborative, hands-on method, students move beyond learning theories as they self-innovate, create, and design solutions for a better future.

Application of Case Studies

Learning becomes more effective and enjoyable when students learn to apply their knowledge in real-world contexts rather than relying solely on theory. Case studies provide an ideal platform for this change. Students are encouraged to think critically, assess alternative perspectives, and integrate information from multiple disciplines by examining and resolving complex problems. This approach fosters in-depth, interdisciplinary learning as students apply theoretical concepts to real-world situations, whether addressing global challenges or local issues (Gierhart & Brown, 2018). Through case studies, students not only enhance their analytical and problem-solving skills but also gain a clearer understanding of how different disciplines contribute to addressing everyday challenges, making education both relevant and engaging.

Example: Imagine students are required to conduct a case study on pandemic response and tasked with creating a global strategy to address a pandemic. They need to combine several disciplines to address the crisis effectively:

- Biology: Understanding how viruses are produced, evolve, and how vaccines are prepared.
- Statistics: Examining infection rates, modelling outbreak patterns, and interpreting health data.
- Public Policy: Assessing governmental initiatives, from lockdowns to vaccination efforts, and their effectiveness.
- Sociology: Investigating human behaviour, misinformation, and societal reactions to health emergencies.

As students discuss strategies, analyse data, and suggest innovative solutions for future pandemics, they recognize how knowledge from various subjects interconnects to resolve real-life challenges. Case studies do not merely teach; they transform students into critical thinkers, strategic thinkers, and agents of change.

Flipped classroom

The flipped classroom reinvents the traditional education system by delivering information outside the classroom hours and transforming class time into a space for active involvement. Rather than passively attending lectures, students actively engage with videos, articles, and research materials at their own pace before class, coming prepared to collaborate, debate, and address problems (Khan, Egbue, Palkie, & Madden, 2017). It helps students become producers of knowledge and become vested in interdisciplinary subjects. In the flipped classroom, lessons become opportunities for critical thinking and creative exchange, where students draw on knowledge from various fields to explore complex issues and make profound connections. Flipped combines subjects to encourage teamwork, imbuing students with a more integrated and deeper understanding of how multiple disciplines come together in the real world.

Example: A flipped classroom focused on space exploration: Envision a classroom filled with enthusiasm as students arrive, already equipped with knowledge about space exploration. Their research has covered several subjects, setting the groundwork for in-depth group discussions and collaborative problem-solving:

- History: They trace the timeline of the space program, Cold War conflicts, and the development of international cooperation in space.
- Physics: They illustrate Newton's laws, propulsion systems, and the mechanics involved in space travel.
- Economics: They explore the budget involved in space exploration, including government-sponsored missions and the emergence of private space enterprises.

Rather than only receiving information, students actively participate, discuss, and expand their ideas. They ask relevant questions such as: Should governments focus on space travel? How might commercial spaceflight transform the world? Adopting a flipped classroom approach transforms the teaching and learning process into one that fosters curiosity, teamwork, and exploration.

Role-Playing and Simulations

What if learning transcended the classroom and became an experiential journey? This is exactly what role-playing and simulations do: they immerse students in practical situations where they must think critically, make assertive decisions, and communicate effectively, just like professionals do in their respective fields. By giving roles such as administrators, diplomats, scientists, or business leaders, students engage in practical learning that integrates knowledge across various disciplines. This method enhances both engagement and empathy, as learners face the emotional and intellectual complexities of real-world issues. It inculcates crucial skills such as problem-solving, negotiation, and decision-making, turning abstract concepts into a concrete understanding. Through experiential learning, role-playing, and simulations, a deeper and more meaningful comprehension of interdisciplinary topics is cultivated (Homanova, Prextova, & Klubal, 2018).

Example: Students are given a simulation of the United Nations. Imagine a classroom transformed into an international summit, where students portray world leaders, economists, and policymakers, each representing their respective countries. They navigate diplomatic discussions, financial strategies, and public dialogue to address urgent global issues.

- Political Science – Gaining insight into international relations, alliances, and the complex dynamics of diplomacy.
- Economics– Assessing trade policies, foreign assistance, and the economic repercussions of worldwide decisions.
- Public Speaking – Crafting articulate arguments, delivering impactful speeches, and compelling others to form coalitions.

By learning how to debate and negotiate, and through collaborative efforts, students develop a broader picture of the topic that connects theoretical knowledge to real-world applications. Role-playing and simulations not only impart knowledge; they turn students into proactive thinkers, decision-makers, and informed global citizens.

Integration of Technology

We live in a fast-changing world, and technology has become more than just a tool; it opens a gateway to unlimited interdisciplinary learning. By integrating digital resources into education, students break free from traditional methods and engage in interactive, engaging experiences that mirror real-world connections. Technology converts abstract concepts into concrete, hands-on learning with the help of virtual reality, simulations, AI, coding, and data analysis. It fosters creativity and innovation, empowering students to utilize and create with technology, developing apps, designing digital prototypes, and engineering solutions (Exploring the Interdisciplinary Nature of Economics, 2023). Moreover, digital platforms widen global viewpoints, connecting students to diverse cultures, real-time information, and worldwide knowledge, ultimately enhancing their understanding of multiple disciplines.

Example: Students are preparing a virtual reality (VR) module on ancient civilizations. Imagine students travelling back in time, not through textbooks, but via a virtual reality experience that allows them to explore the cities of the Indus Valley Civilization or uncover the elaborate tombs of Egyptian pharaohs. This interdisciplinary approach includes:

- History– Discovering the narratives behind ancient civilizations, including their architecture, language, and societal dynamics.
- Art– Examining artistic styles, symbols, and the cultural relevance of murals, sculptures, and artefacts.
- Computer Science – Understanding how VR applications are designed, programmed, and developed to create realistic, 3-D experiences.

By integrating technology into education, students don't merely learn about history; they experience it firsthand. They don't just read about innovation, they become innovators. Technology enhances the traditional teaching-learning process.

Final Thoughts

By utilizing these innovative strategies, students significantly enhance their ability to draw connections between various disciplines, engage in critical analysis, and apply their knowledge to address real-world challenges. Interdisciplinary learning not only nurtures creativity and collaboration but also develops essential problem-solving skills, effectively preparing students for a range of diverse and ever-evolving career paths. This holistic approach to education prepares students to flourish in this interconnected and complex world. Economics as an Interdisciplinary Approach is depicted in Fig. (**2**).

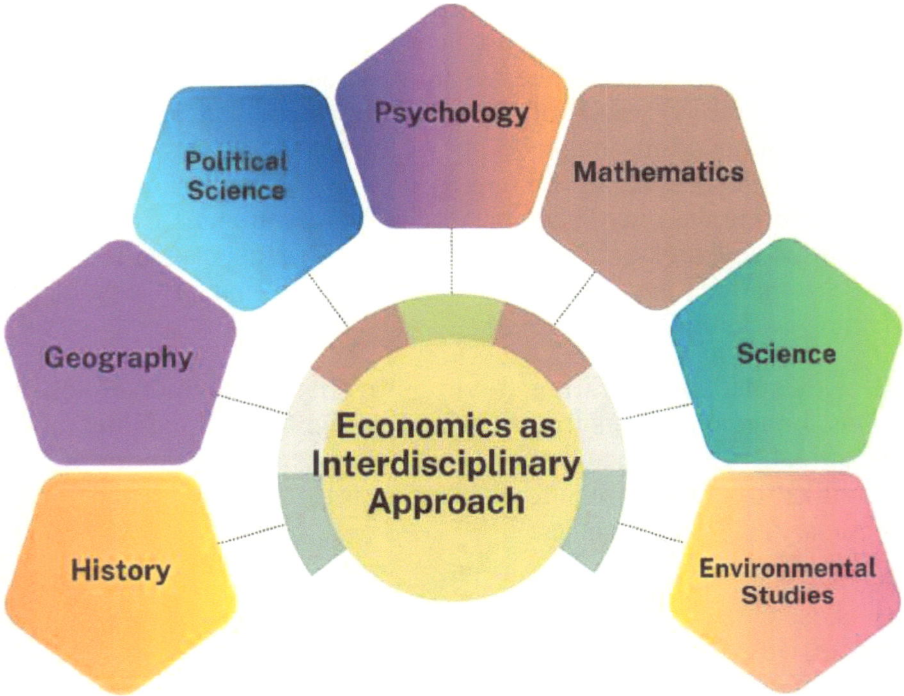

Fig. (2). Economics as an interdisciplinary approach.

Economics is a subject that blends remarkably well with other subjects because it primarily deals with real-life situations, such as decision-making, resource allocation, and human behaviour (Tan, Nicholas, Scribner, & Francis, 2019). It does not stand alone but is associated with history, geography, politics, psychology, and science. Let us try to understand this with some examples:

Economics and History: Economics and history are inseparable. History teaches 'why,' and economics explains 'because. For example, economic factors such as poverty, unemployment, or rising food prices have been significant in the study of revolutions or wars. The French Revolution was not primarily a victory of political ideas, but rather a response to widespread suffering caused by hunger and high prices.

Economics and Geography: Economics also gels with geography. Geography teaches us about a place's physical features and natural resources, but economics shows us how to use them effectively. For example, countries rich in oil, such as Saudi Arabia, use this resource to grow wealth. This is where geography provides us with 'where' and economics provides us with 'how and why'.

Economics and Political Science: Political science is an essential field of study because it relates to economics in terms of understanding the government's policies and initiatives. Raising taxes, giving, or reducing subsidies is a political decision that directly impacts the economy. For example, suppose the government increases the prices of fuel, which affects transportation. In that case, it will result in a price hike for food products, which will ultimately impact household budgets —a topic addressed in economics.

Economics and Psychology: Economics also complements psychology well. Psychology informs us about how people think, feel, and behave. Economics utilizes this knowledge to examine how individuals make choices and decisions, what they purchase, when to save, and why they take risks. For instance, impulsive shopping during the sale season, as people will buy things they do not need when they're on sale. This is both psychological and economic thinking of a kind.

Economics and Mathematics: Another subject that supports economics is mathematics. Economists use math to determine how much profit will be generated, predict traffic trends, analyse data, and produce graphs. The cost and revenue of businesses are calculated using simple formulas, and statistics are utilized in economics to understand concepts such as inflation, unemployment, and population growth.

Economics and Science: Science also works with economics. Innovation is the result of science, and economics determines the most effective way to apply science. For example, science invents solar panels, but economics determines whether and how they can be made affordable and fairly distributed to different locations.

Economics and Environmental Science: In the modern world, environmental studies are closely intermingled with economics. Besides concentrating on development, we need to conserve our environment. Economics teaches us how to utilize natural resources without harming the environment. For instance, economics can seek methods to reduce pollution and use technology-friendly machinery to promote a healthy environment for the future.

Economics is closely related to many subjects because it affects all aspects of our lives. It teaches us that not everything is about money (or markets), people, behaviour, policies, history, or the planet. However, economics is a robust tool for making better decisions and leading a better life.

Evaluation of the Role of Technology and Experiential Learning in Enhancing Interdisciplinary Education Outcomes

Students take advantage of interdisciplinary learning when they can apply knowledge from different fields, and this can be facilitated by the addition of two approaches: technology integration and experiential learning. Technology serves as a bridge across disciplines, enabling students to grasp concepts, collaborate globally, and apply theoretical knowledge in real-world settings.

The Role of Technology in Interdisciplinary Education

Technology is necessary for interdisciplinary education, as it connects various academic disciplines and facilitates a more integrated, comprehensive educational environment. Digital tools, such as collaborative platforms, virtual labs, and simulations, can help students investigate complex issues across multiple disciplines in ways that foster creativity and critical thinking (Khan *et al.*, 2024). Furthermore, technology enables the integration of unlimited resources, such as interactive and multimedia models, together with online databases, making learning dynamic and accessible. This helps teachers and students from diverse educational backgrounds gain a better understanding of different fields in real-world settings. Overall, technology enhances interdisciplinary learning by making it more dynamic, engaging, and relevant to current realities of the world (Singh, *et al.*, 2024). The role of technology in interdisciplinary education is depicted in Fig. (**3**).

Virtual Reality (VR) for Medical and Engineering Integration

Virtual reality is transforming the relationship between medical and engineering education by offering numerous hands-on learning opportunities. Medical students can now conduct virtual surgeries through platforms such as Osso VR, applying knowledge from biology, physics, and engineering in a realistic, risk-free setting. Simultaneously, engineering students study biomechanics to design prosthetic limbs and devices, utilizing advanced medical imaging software and 3D modelling tools such as AutoCAD. This interdisciplinary strategy not only enhances practical skills but also fosters collaboration among aspiring doctors and engineers. The outcome is a significant improvement in precision, reduced chances of errors, and the development of innovative solutions in both the medical and engineering sectors.

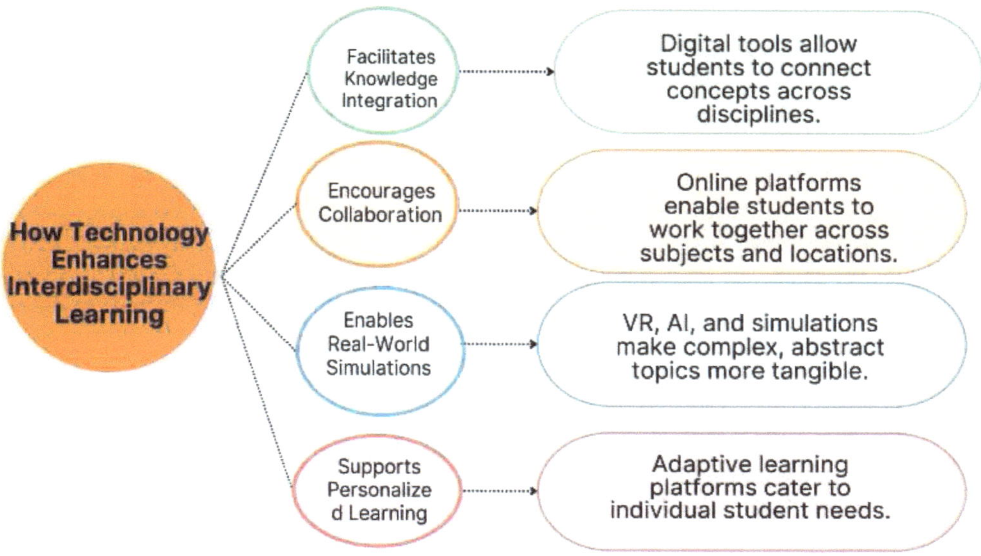

Fig. (3). Examples of technology in interdisciplinary education.

AI and Big Data for Environmental Science & Economics

Artificial Intelligence (AI) and Big Data are significantly altering the landscape of Environmental Science and Economics. Students utilize AI-powered data analytics to study climate change patterns, employing tools like Google Earth Engine to analyse satellite images and environmental data on a large scale (Tandon *et al.*, 2024). By incorporating economic concepts, they assess the cost-effectiveness of various sustainable initiatives, such as renewable energy implementation or conservation efforts. Programming languages like Python are utilized for climate and economic modelling, enabling students to simulate scenarios and forecast results. This interdisciplinary approach equips students with essential skills, including critical thinking, data literacy, and policy-making, preparing them to address complex global issues with informed, data-driven decisions.

Coding and Robotics in STEM and Humanities

The amalgamation of coding and robotics in both STEM and humanities subjects is nurturing creativity and innovation across various fields. History students use Python to create interactive timelines, making historical events more engaging and easier to understand. In literature, students can explore Natural Language

Processing (NLP) tools to develop chatbots that blend technology with creative writing, analyse texts, and generate AI-driven narratives. Similarly, engineering students apply their technical expertise to design robots that replicate biological movements, such as robotic arms inspired by human anatomy (Markauskaite *et al.*, 2024). This interdisciplinary approach not only enhances students' problem-solving and creativity skills but also cultivates technical skills and collaboration across academic disciplines (Singh *et al.*, 2024; Khan *et al.*, 2024). The role of experiential learning in interdisciplinary education is presented in Fig. (**4**).

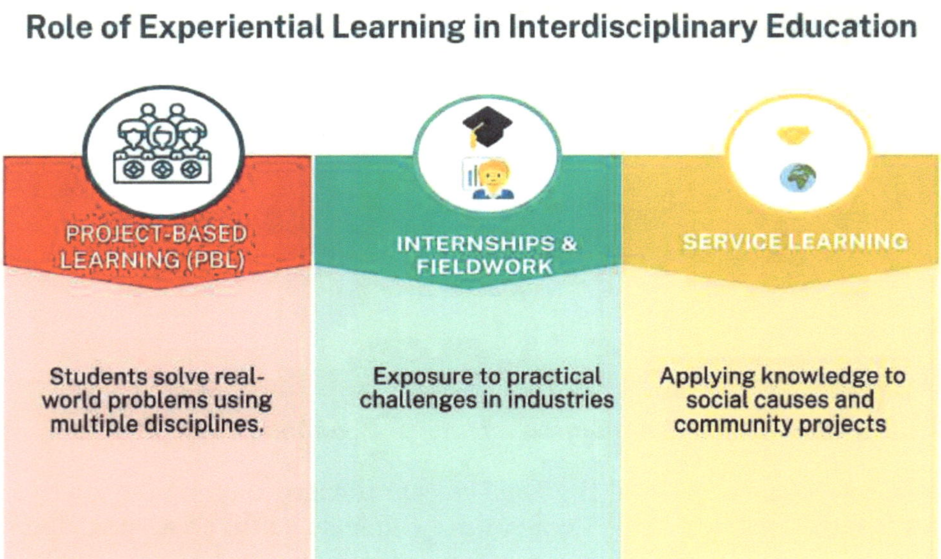

Fig. (4). The role of experiential learning in interdisciplinary education.

Examples of Experiential Learning in Interdisciplinary Education

- ***Smart Cities Development (Engineering, Economics, and Urban Planning)***

The development of Smart Cities is an excellent example of interdisciplinary learning that combines Engineering, Economics, and Urban Planning. In this method, students collaborate to design sustainable urban spaces by integrating concepts from architecture, eco-friendly engineering, and environmental planning. Economics students assess the financial implications of infrastructure projects to ensure that developments are both innovative and financially feasible. Simultaneously, sociology students also examine how such initiatives impact communities socially, for example, on issues of accessibility, equity, and quality of life. Experiential techniques, such as site visits to key smart city projects in

Singapore, building a physical model, and conducting digital simulations, provide valuable opportunities to gain practical experience. This in-depth learning process prepares students for careers in urban planning and sustainable development policy-making.

- ***Food Science and Business Entrepreneurship***

Food science and business entrepreneurship presents an engaging interdisciplinary perspective that collaborates scientific temper with practical business skills. Biology students learn food preservation methods to extend the shelf life of edibles and minimize waste, while chemistry students focus on checking nutritional content and ensuring food safety. At the same time, business students develop marketing strategies to promote healthy food products, positioning them effectively in the marketplace. The experiential activities include managing a food startup within the school, including acquiring real customers and learning about costs from the ground up. Collaboration with professional food scientists and local businesses offers valuable mentorship and insights into industry practices. Innovation is facilitated, scientific understanding is deepened, and business skills are developed among students through this model of study.

- ***Environmental Conservation and Policy Making***

Environmental conservation and policy making is a critical interdisciplinary learning experience that links science, law, and communication, providing a solid foundation for a rich learning experience. For example, it first gathers data by conducting field research to measure pollution levels in a local river, assessing the ecological impact of polluted water. Then, law students evaluate specific environmental laws and propose policy solutions to address the problems identified. At the same time, journalism students serve an essential role in reporting on the outcomes and developing advocacy and awareness efforts to inform and educate the public. Through experiential activities such as community engagement projects and collaborations with environmental NGOs, students attain practical knowledge in facilitating change. This collaborative approach not only enriches students' understanding of the environmental problem but also fosters awareness and responsibility for the issue, providing them with the tools to solve real-world problems.

- ***Synergy of Technology and Experiential Learning in Interdisciplinary Education***

Combining technology and experiential learning significantly enhances interdisciplinary education by creating immersive and impactful learning opportunities, for instance, the application of Augmented Reality (AR) in subjects such as archaeology and history. Through AR, History and Anthropology students can virtually visit historical sites, gaining access to places they might not be able to see physically. Advanced AI technologies are used to reconstruct 3D models of extinct civilizations, allowing students to visualize and interact with history dynamically. Students also participate in developing interactive historical storytelling projects that merge research with digital creativity. This amalgam of technologies fosters a deeper understanding, enhances student engagement, and facilitates interactive, hands-on learning that seamlessly integrates various subjects.

Overall, technology and hands-on education are the future of interdisciplinary learning, as they are becoming more engaging, efficient, and aligned with the needs of the future. VR simulations and AI models help make abstract concepts more tangible, enabling students to understand complicated ideas in a simplified manner (Kanmaz, 2022). This method also encourages students to collaborate across fields such as STEM, humanities, and business, while thinking creatively. In addition to developing critical thinking, problem-solving skills, creativity, and adaptability, students also learn the real-life applications of science and environmental conservation, which helps them adapt to the problems of the modern world.

Exploration of the Challenges Faced by Educators in Implementing Interdisciplinary Teaching Methods and Proposed Solutions

Interdisciplinary teaching emerges from a combination of diverse courses; however, implementing these methods poses a challenge for teachers. The demand for interdisciplinary education arises from the recognition that many real-world problems are complex and require solutions that integrate insights from multiple knowledge domains (Sailsman, 2022). These obstacles include limitations in the curriculum and challenges in evaluating interdisciplinary skills. In the following sections, we will delve into these problems in greater detail and offer practical solutions, accompanied by relevant examples.

Curriculum Constraints and Rigid Subject Boundaries

Most educational systems adhere to a strict curriculum with distinct subjects, making it challenging to incorporate interdisciplinary learning. Teachers often find themselves under pressure to complete the syllabus within the allotted time, which restricts the opportunity for cross-disciplinary exploration. Although teachers acknowledge the importance of interdisciplinary methods, they often struggle to implement them effectively in the classroom, as these approaches are not sufficiently integrated into the curriculum (Boix *et al.*, 2007).

To address this problem, schools and policymakers should encourage schools to develop interdisciplinary modules that align with the existing curriculum. By teaching within a broader, connected framework, teachers can address multiple topics while meeting curriculum requirements. For example, "Sustainable Development" can be studied in science (climate change and biodiversity), economics (the cost of renewable energy), social studies (government policy), and mathematics (analysing carbon footprint data) to ensure that students gain a holistic understanding of the subject while meeting educational standards.

Lack of Training and Expertise in Multiple Disciplines

While many teachers focus on a single subject, they may lack the training to successfully integrate other disciplines into their lessons, thereby limiting the establishment of rich interdisciplinary connections. To counter this, schools must offer workshops and courses to train teachers in interdisciplinary teaching methods, equipping them with the skills and confidence to work across disciplines. Learning from one another is possible as teachers from other subjects can co-teach units. For example, a physics teacher might collaborate with an art teacher to explore the angle of light, establishing an optical connection between how a physicist would describe optics and Renaissance art techniques. Similar to mathematics and economics, a teacher could jointly teach financial literacy by addressing concepts such as interest rates, budgeting, and economic policies. Such collaborative work in learning allows every teacher to have a lighter workload and enhances students' ability to relate to different branches of study.

Time Constraints and Increased Workload

Involving multiple disciplines is very demanding to achieve and therefore often adds to teachers' workloads and time pressures. Collaboration among faculty members of different subjects is one of the main challenges for interdisciplinary teaching. A consistent interdisciplinary curriculum requires aligning teaching methods and learning outcomes across various subjects, a process that demands time and energy (Pihlajamaa *et al.*, 2016). Schools can collaborate with teachers to establish a repository of ready-to-use, easily modifiable interdisciplinary lesson plans, thereby reducing the need for teachers to create lessons from scratch. Schools should also set aside time for interdisciplinary collaborative planning sessions, where educators can work together to develop effective lesson plans. Schools could create modules such as Space Exploration (combining physical science, history, and economics) or Food Science (integrating chemical science, nutrition, and agriculture). If teachers use these pre-constructed modules, they would save time, adapt the content for their classrooms more quickly, and implement an interdisciplinary approach to learning more effectively.

Challenges in Evaluating Student Performance

Nowadays, traditional assessment methods tend to focus on individual subjects, making it challenging to evaluate students effectively. Many standardized tests only measure students' ability to synthesize knowledge across multiple disciplines, rather than holistic skills necessary for learning across several disciplines. Thus, to address this problem, teachers may choose interdisciplinary projects as a more adequate assessment strategy. Instead of providing students with a traditional exam, teachers could create a project-based assignment that requires students to demonstrate how the knowledge acquired from different fields can be applied. It should aim to assess skills such as critical thinking and problem-solving. For example, in a project called 'Design a Smart City', students would explore engineering (infrastructure and transportation design), environmental science (sustainable energy solutions), mathematics (budget calculations and data analysis), and social studies (government policies and urban planning).

Resistance to Conventional Teaching Methods

Many teachers are familiar with traditional subject-centred teaching and may be hesitant to adopt an interdisciplinary approach due to unfamiliarity or doubts about its success. This unease is frequently due to conventional academic frameworks and a lack of faculty training in interdisciplinary teaching (Fulton *et al.*, 2020; Mishra & Koehler, 2006). This can be addressed through a gradual integration strategy that introduces a bit of interdisciplinary activity before fully adopting multidisciplinary courses. Successful interdisciplinary projects and their outcomes can be highlighted to illustrate the value of this approach. Additionally, experienced interdisciplinary teachers can mentor newly recruited teachers to guide and support those who are not familiar with using it. For instance, a school could start creating learning days across subjects or projects; these events would serve as groundwork for more extensive integration in the future. This method allows both teachers and students to adopt interdisciplinary learning more comfortably and effectively.

Absence of Administrative and Institutional Support

Interdisciplinary learning is often hindered by school policies, schedules, and resources, and without strong administrative support, teachers struggle to coordinate cross-subject initiatives. The limited supply of resources presents a challenge that aligns with the broader notion that interdisciplinary work requires institutional support and resources (Applebee *et al.*, 2007). However, administrators need to recognize the necessity of Interdisciplinary Education and, in return, should be provided with the necessary structural support. Specific time slots in the school schedule could be made flexible to accommodate interdisciplinary learning and facilitate cross-subject collaboration. Investments in shared materials, technology, and teacher training are crucial for cultivating a culture of interdisciplinary teaching. Imagine a school designating one day each month as "Interdisciplinary Learning Day," during which regular classes are replaced with hands-on projects that incorporate multiple disciplines. This strategy would not only foster interdisciplinary collaboration but also help to create a sustainable, supportive framework for interdisciplinary learning.

Challenges in Engaging Students with Varied Learning Backgrounds

Some students struggle with interdisciplinary education, particularly if they are used to traditional methods that emphasize a single subject (Siemens, 2005). To address this issue, teachers can employ various instructional strategies, including the use of audio-visual aids, interactive activities, and cooperative learning, to engage learners effectively. Starting with simple interdisciplinary links and gradually increasing complexity can help build students' confidence. For example, using practical budgeting scenarios in economics can make math concepts more relatable.

The lack of appropriate teaching materials for interdisciplinary education is a challenge, as most textbooks focus on individual subjects (Rafiq *et al.*, 2024). Schools can leverage Open Educational Resources (OERs) from platforms like Khan Academy and TED-Ed videos that combine subjects, such as "The Physics of Music."

Despite its many benefits, interdisciplinary learning faces challenges, including curriculum limitations, training gaps, resource constraints, time constraints, and resistance to change. Solutions include adaptable curriculum design, collaborative planning, project-based assessments, and strong administrative support. By implementing these approaches, schools can effectively adopt interdisciplinary teaching to facilitate real-life problem-solving, critical thinking, and enrich student learning.

CONCLUDING REMARKS

This chapter emphasizes the importance of employing effective pedagogical strategies to promote interdisciplinary learning in educational institutions. By helping students relate concepts across disciplines, teachers can increase understanding, creativity, and critical thinking skills. The integration of technology and experiential learning, including digital tools, simulations, and real-life experiences, enhances the teaching and learning process, making it more relevant and impactful.

Despite challenges such as curriculum restrictions and limited resources, solutions like interdisciplinary curriculum design and collaborative teaching can address these problems. Ultimately, promoting interdisciplinary learning requires a systemic approach that supports teachers and empowers students, aligning institutional goals with the requirements of an interrelated world. Such efforts will better prepare students to address future challenges.

ACKNOWLEDGEMENT

The corresponding author acknowledges the management of Mahila Vidyalaya Degree College, Lucknow, and Integral University for their support. This chapter acknowledges the use of AI-assisted tools, including Perplexity for research and fact-checking, Grammarly for writing enhancement and grammar correction, and Zotero for efficient reference management. These tools were selected to improve workflow efficiency, accuracy, and clarity while maintaining rigorous academic standards. Their integration complemented traditional research and writing methods by providing reliable source verification, streamlined citation, and polished language, thereby enhancing the overall quality and transparency of the research process.

REFERENCES

Applebee, A.N., Adler, M., Flihan, S. (2007). Interdisciplinary curricula in middle and high school classrooms: Case studies of approaches to curriculum and instruction. *Am. Educ. Res. J., 44*(4), 1002-1039. [http://dx.doi.org/10.3102/0002831207308219]

Anderson, R., Ruef, J., Reigh, E., Chavez, R., Williamson, P., Villa, A. (2021). "Math is so much more": The design, implementation, and outcomes of an elective mathematics methods course. *Teach. Educ. Q., 48*(4), 28-51.

Barron, B., Darling-Hammond, L. (2008). *Teaching for meaningful learning: A review of research on inquiry-based and cooperative learning..* George Lucas Educational Foundation.

Boix Mansilla, V. (2017). Interdisciplinary learning: A cognitive-epistemological foundation. In R. Frodeman (Ed.) In: Frodeman, R., (Ed.), *The Oxford Handbook of Interdisciplinarity* 2nd ed., Oxford University Press. [http://dx.doi.org/10.1093/oxfordhb/9780198733522.013.22]

Dula, C.A.C., Porter, A.L. (2021). Addressing challenges in skills-based education through innovation and collaboration. *Am. J. Pharm. Educ., 85*(7), 8788. [http://dx.doi.org/10.5688/ajpe8788] [PMID: 34544750]

Peak Potentia. (2024, February 19). (2023). *Exploring the interdisciplinary nature of economics: Its harmonious blend across subjects.* Available at https://www.linkedin.com/pulse/exploring-interdisciplinar--nature-economics-its-harmonious-rbiyc/

Friend, M., Cook, L. (2016). *Interactions: Collaboration skills for school professionals..* Pearson.

Fulton, T., Edelen, D., Kebreab, L., Greer, M., Caton, J., Brewer, J., Bush, S., Mohr-Schroeder, M. (2020). Integrative virtual STEM: Ensuring access in distance learning. *Elementary STEM Journal, 25*(2), 22-25.

Gierhart, A., Brown, R. (2018). Developing technological literacy: Third graders use Google Classroom. *Elementary STEM Journal, 22*(4), 22-25.

Holley, K.A. (2017). *Interdisciplinary education and the future of higher education..* Rutgers University

Press.

Homanova, Z., Prextova, T., Klubal, L. (2018). Connectivism in elementary school instruction. In: *Proceedings of the 17th European Conference on e-Learning (ECEL 2018)* Academic Conferences and Publishing International Limited.

Ivan, R., Zambori, A-E. (2024). Interdisciplinarity team teaching: Opportunities and challenges. Romanian. *Romanian Review of Geographical Education, 13*(1-2), 22-34. [http://dx.doi.org/10.23741/RRGE20242]

Kanmaz, A. (2022). A study on interdisciplinary teaching practices: Primary and secondary education curricula. *Am. Educ. Res. J., 10*(2), 200-210.

Khan, A., Egbue, O., Palkie, B., Madden, J. (2017). Active learning: Engaging students to maximize learning in an online course. *Electron. J. e-Learn., 15*(2), 107-115.

Kolb, D.A. (2015). *Experiential learning: Experience as the source of learning and development..* Pearson Education.

Lattuca, L.R., Voigt, L.J., Fath, K.Q. (2004). Does interdisciplinarity promote learning? *J. Higher Educ., 75*(1), 66-88.

McGowan, M. (2013). Assessing interdisciplinary learning outcomes. *J. Gen. Educ., 62*(4), 249-263.

Markauskaite, L., Schwarz, B., Damşa, C., & Muukkonen, H. (2024). Beyond disciplinary engagement: Researching the ecologies of interdisciplinary learning. *Journal of the Learning Sciences, 33*(2), 213-241..

Mahaffey, F., Kinard, W., Daughrity, L. (2020). Google Docs in elementary gifted education. *Delta Kappa Gamma Bulletin, 86*(5), 23-29.

Mishra, P., Koehler, M.J. (2006). Technological pedagogical content knowledge: A framework for teacher knowledge. *Teach. Coll. Rec., 108*(6), 1017-1054. [http://dx.doi.org/10.1111/j.1467-9620.2006.00684.x]

Nikitina, S. (2006). Three strategies for interdisciplinary teaching: contextualizing, conceptualizing, and problem-centring. *J. Curric. Stud., 38*(3), 251-271. [http://dx.doi.org/10.1080/00220270500422632]

Newell, W.H. (2001). A theory of interdisciplinary studies. *Issues in Integrative Studies, 19*, 1-25.

Pihlajamaa, J., Karukka, M., Alander, H. (2016). Comparison of higher education student and teacher perceptions of e-learning. In: Novotná, J., (Ed.), *Proceedings of the 15th European Conference on e-Learning,* (pp. 782-785). Reading, UK: Academic Conferences and Publishing International.

Rafiq, S., Kamran, F., Afzal, A. (2024). Investigating the benefits and challenges of interdisciplinary education in higher education settings. *Journal of Social Research Development, 5*(1), 87-100. [http://dx.doi.org/10.53664/JSRD/05-01-2024-08-87-100]

Sánchez-García, R., & Reyes-de-Cózar, S. (2025). Enhancing project-based learning: A framework for optimizing structural design and implementation—A systematic review with a sustainable focus. *Sustainability, 17*(11), 4978.

Sailsman, S. (2022). Utilizing online collaborative technology to promote active learning in nursing students. *Q. Rev. Distance Educ., 23*(2), 91-93, 100.

Serafini, P.G., Moura, J.M., Almeida, M.R., Rezende, J.F.D., Rezende, D. (2022). Sustainable development goals in higher education institutions: A systematic literature review. *J. Clean. Prod., 370*, 133473. [http://dx.doi.org/10.1016/j.jclepro.2022.133473]

Siemens, G. (2005). Connectivism: A learning theory for the digital age. *International Journal of Instructional Technology and Distance Learning, 2*(1), 3-10.

Singh, M., Azad, I., Qayyoom, M.A., Khan, T. (2024). A study on perceptions and practices of STEAM-based education with university students. *Soc. Sci. Humanit. Open, 10*, 101162.

[http://dx.doi.org/10.1016/j.ssaho.2024.101162]

Tan, V., Nicholas, C., Scribner, J., Francis, D. (2019). Enhancing STEM learning through an interdisciplinary, industry-generated project. *Technology and Engineering Teacher, 79*(1), 26-31.

Tandon, U., Tandon, A., Mehrotra, T. (2024). Transformation in the world of commerce and economics through AI. *Artificial Intelligence: A Multidisciplinary Approach towards Teaching and Learning.,* Bentham Science Publishers.(Pp. 194-215).

Tolsby, J. (2005). Bridging hand and heart to stimulate learning in the organization. *Dev. Learn. Organ., 19*(2), 8-10.
[http://dx.doi.org/10.1108/14777280510580663]

Wang, X., Yuan, L., Li, S., Yu, H., Tuo, P., Du, Y., Zhang, B. (2024). Developing and validating an Interdisciplinary Teaching Readiness Scale (ITRS) for pre-service teachers in China. *PLoS One, 19*(12), e0315723.
[http://dx.doi.org/10.1371/journal.pone.0315723] [PMID: 39739698]

Zydney, J.M., DeJonckheere, M., Karp, T. (2018). Technology-supported interdisciplinary learning. *Educ. Technol. Res. Dev., 66*(5), 1119-1142.

CHAPTER 4

The STEAM Paradigm: Integrating Disciplines for a Holistic Learning Experience

Priya Mishra[1], **Satya** [1], **Kulsum Hashmi**[1], **Sakshi Gupta**[1] and **Seema Joshi**[1,*]

[1] *Department of Chemistry, Isabella Thoburn College, Lucknow, Uttar Pradesh 226007, India*

Abstract: The STEAM paradigm, which combines Science, Technology, Engineering, Arts, and Mathematics into a single educational framework, represents a major transition from conventional subject-centred teaching to an interdisciplinary model that emphasizes experiential learning. STEAM education promotes creativity, emotional intelligence, and innovative thinking alongside analytical and technical skills by integrating the Arts into the established STEM disciplines. By allowing students to tackle problems from various angles, this broadened scope fosters creativity, empathy, and adaptability, skills that are ever more essential in our rapidly evolving world. STEAM, in contrast to traditional educational models that frequently divide learning into separate categories, emphasizes real-world applications and holistic growth. It fosters cognitive involvement as well as emotional and social development, resulting in a balanced learning environment where both logic and imagination flourish. Students actively engage in designing, creating, and collaborating across disciplines rather than merely being passive recipients of information. This chapter investigates the theoretical foundations of the STEAM approach and explores its educational benefits, including heightened student motivation, improved critical thinking, and enhanced problem-solving skills. The discussion demonstrates how STEAM connects academic learning with life skills through detailed case studies, curriculum models, and classroom practices. The model's significance in fostering well-rounded individuals who can tackle intricate global challenges is highlighted by these examples. With the evolution of industries and societies, education needs to adjust accordingly. With its alignment with educational objectives and societal needs, the STEAM paradigm offers a dynamic model that prepares students for life beyond exams.

Keywords: Interdisciplinary education, Innovative learning, Arts integration, Pedagogy shift, Creative thinking, Problem-solving approach, Artistic engineer.

* **Corresponding Author Seema Joshi:** Department of Chemistry, Isabella Thoburn College, Lucknow, Uttar Pradesh 226007, India; E-mail: sjoshiitc@gamil.com

Tahmeena Khan, Manisha Singh, & Saman Raza (Eds.)

INTRODUCTION

In recent years, the growing complexity of global challenges and the evolving demands of the 21st-century workforce have necessitated a shift from traditional disciplinary silos toward more integrated models of education. One prominent development in this direction is the evolution of the STEM (Science, Technology, Engineering, and Mathematics) framework into STEAM, which incorporates the Arts as a vital component. The inclusion of the "A" signifies more than just artistic disciplines—it reflects a broader emphasis on creativity, empathy, cultural awareness, and innovative thinking in educational systems. This paradigm shift recognizes that solving multifaceted problems requires not only analytical and technical proficiency but also the imaginative, interpretive, and expressive capacities often fostered through the arts (Land, 2013). While the STEM model has long been celebrated for promoting rigorous scientific inquiry and technological advancement, it has also faced criticism for neglecting the humanistic aspects of learning. STEAM education addresses this gap by advocating for a balanced integration of the arts—encompassing visual arts, music, drama, dance, and literary studies—into science and engineering curricula. This integration fosters emotional intelligence, critical reflection, and communication skills, which are increasingly essential in diverse professional contexts, from design and architecture to AI ethics and environmental sustainability (Larson & Miller, 2011) (Henriksen *et al.*, 2019) . Moreover, the arts serve as a bridge to engage underrepresented groups in STEM fields, particularly by appealing to different learning styles and cultural backgrounds.

The STEAM model emphasizes interconnected and interdisciplinary learning, encouraging students to engage with concepts through multiple perspectives-scientific, technological, mathematical, and artistic. This approach cultivates metacognitive awareness, enabling learners to draw connections across domains and apply knowledge in novel contexts. According to research, STEAM-based classrooms often feature project-based learning (PBL), real-world problem solving, and collaborative inquiry, where students design experiments, create multimedia projects, and prototype innovative solutions. These activities not only deepen conceptual understanding but also build transferable skills such as adaptability, teamwork, and creative risk-taking (Georgette, 2008). Implementing STEAM education, however, demands a reevaluation of curriculum structures, pedagogical strategies, and assessment practices. A successful transition involves aligning learning objectives with interdisciplinary goals, integrating arts-based assessment methods, and fostering collaborative learning environments. Schools and institutions that have adopted STEAM report a host of benefits, including increased student motivation, higher retention in STEM fields, and improved performance on both standardized and non-traditional assessments (Perales & Aróstegui, 2021). Importantly, STEAM encourages inclusivity by validating diverse talents and nurturing learners' unique identities. For educators, the shift to

STEAM necessitates targeted professional development that goes beyond content knowledge. Teachers must be equipped to design cross-curricular projects, facilitate open-ended inquiry, and assess both the creative and cognitive dimensions of learning. Programs offering hands-on training, mentorship, and collaborative planning are vital in building teachers' capacity to implement STEAM effectively. Moreover, creating a school culture that values experimentation, resilience, and the interplay between imagination and logic is key to sustaining STEAM practices (Watson & Watson, 2013). Despite its transformative potential, the STEAM paradigm also faces considerable challenges. These include limited funding for arts programs, entrenched curricular rigidities, a lack of interdisciplinary planning time, and insufficient training opportunities for teachers. Additionally, standardization pressures often prioritize rote learning over exploratory practices. Addressing these barriers requires systemic support from policymakers, school leaders, and curriculum developers who recognize the long-term benefits of nurturing whole-child education. In conclusion, the STEAM approach embodies a dynamic reimagining of education—one that prepares learners not only to consume knowledge but to co-create solutions for an interconnected and uncertain world. This chapter delves into the philosophical foundations of STEAM, outlines practical strategies for classroom integration, highlights effective models of teacher preparation, and maps the future trajectory of STEAM as a transformative educational movement.

STEM AND A SHIFT TOWARDS STEAM

In the 1990s, the U.S. National Science Foundation introduced the STEM acronym to highlight the importance of education in crucial areas for economic competitiveness (Galama & Hosek, 2008). In the early 2000s, concerns about U.S. students falling behind globally, fuelled by The World Is Flat and similar reports, spurred major STEM initiatives (Teitelbaum, 2014). In 2001, the NSF temporarily used the term SMET (Science, Math, Engineering, Technology) before adopting STEM. Subsequently, states and the federal government started funding STEM-focused initiatives, such as Educate to Innovate in 2009 (Mohr-Schroeder *et al.*, 2015). STEM professions expanded quickly, doubling the rate of other occupations and highlighting STEM's contribution to economic growth. As time passed, educators observed that an exclusive emphasis on STEM neglected vital aspects of creativity and design (Rothwell, 2013 ; Deming & Noray, 2018). In 2006, educator Georgette Yakman established a STEAM framework that explicitly incorporated the Arts into STEM, connecting art and design with science and math (Yakman, 2010) (Kamienski & Radziwill, 2018). In the early 2010s, figures such as John Maeda, the president of the Rhode Island School of Design, began to popularize STEAM. He contended that STEM subjects by themselves would not foster the type of innovation required in the 21st century

(Somerson & L. Hermano, 2013). Several models and initiatives employ STEAM to combine arts with science (Sousa & Pilecki, 2018), such as:

1. Project-based Learning: Lessons frequently incorporate practical projects that demand design thinking and creativity (Ubben, 2019). A UK case study outlines STEAM projects in which students explored marine science through photography, sculpture, and design, and employed literature like The Hunger Games as a basis for physics and engineering challenges (Z. Wang *et al.*, 2024).
2. Cross-disciplinary Partnerships: Initiatives that connect educational institutions with organizations in the arts and sciences (Wu, 2022). In 2017, Arts Council England (ACE) funded seven regional STEAM networks that united schools, cultural organizations, and STEM partners to create curriculum and teacher resources collaboratively (Rimmer, 2018 ; Butterworth, 2020). The networks organized themed activities such as art exhibits on beach pollution and code-breaking art inspired by Alan Turing to enhance STEM learning (Hodges, 2009), (Lee, 2023).
3. Professional Development: Teacher training is a vital element. Research indicates that educators frequently begin with a limited grasp of STEAM, but through ongoing collaborative support, they can develop the ability to create STEAM lessons suited to their specific context. One initiative demonstrated that teachers significantly enhanced their STEAM integration and utilized it for their specific grade level and discipline when they were provided with continuous mentorship and resources (Herro & Quigley, 2016),(Conradty & Bogner, 2020),(Martins & Baptista, 2024).
4. Institutional Initiatives: Certain ministries and schools offer formal STEAM programs. In 2011, South Korea's education ministry allocated funds for STEAM teacher research councils aimed at helping 12 teachers create STEAM curricula and resources (Y. Kim *et al.*, 2015). In Malaysia, arts components were incorporated into STEM subjects in the national curriculum to promote creativity and innovation (Bahrum *et al.*, 2017).

STEAM offers a more inclusive platform by engaging a broader range of student interests and intelligences, particularly those who may not see themselves represented in purely technical disciplines (Sousa & Pilecki, 2018). Hardiman and co-workers showed that when students are exposed to STEAM, they exhibit greater engagement and motivation, improved long-term content retention, enhanced spatial reasoning and visual literacy, and increased collaboration and communication skills (Hardiman *et al.*, 2014). The shift represents a more expansive educational vision that places equal value on creativity, emotional intelligence, cultural awareness, and communication, along with analytical and scientific proficiency. Incorporating the arts into STEM creates a more comprehensive learning environment. It motivates students to think creatively, cultivate innovative ideas, and tackle problems with compassion and creativity. For example, a student working on the design of a sustainable building incorporates engineering principles and also takes into account factors such as aesthetic appeal, cultural relevance, and human experience, abilities that are developed through the arts (Moore & Smith, 2014 ; Helms *et al.*, 2016). Moreover, STEAM supports diverse learning styles and opens the door for broader student engagement, particularly for those who may not resonate with traditional STEM instruction (Boy, 2013). By bridging the gap between technical skills and human-centred design, STEAM education equips learners with the tools they need to navigate and shape a dynamic, interconnected world. The shift from STEM to STEAM is not about replacing science with art but about creating synergy between disciplines. It reflects an educational philosophy that values both rigour and creativity, making students not only problem solvers but also solution creators and innovators.

KEY ELEMENTS OF THE STEAM MODEL

Interdisciplinary Learning

Interdisciplinary learning in the STEAM model emphasizes the seamless integration of science, technology, engineering, arts, and mathematics to provide a unified learning experience. This approach breaks down traditional subject boundaries and encourages students to apply knowledge from multiple domains (Thuneberg *et al.*, 2017). By integrating multiple disciplines, learners can gain a more comprehensive knowledge and learn to use diverse perspectives to address complex problems (Quigley *et al.*, 2020). Guimeráns-Sánchez and co-workers highlighted the crucial importance of the art within the STEAM framework. They contend that incorporating artistic viewpoints enhances creativity and critical thinking, allowing students to engage with scientific and mathematical concepts more thoughtfully. This interdisciplinary approach nurtures a deeper comprehension of fields like environmental science and engineering by encouraging both analytical and imaginative thinking (Guimeráns-Sánchez *et al.*, 2023). Zhan and co-workers investigated how a design-based learning model incorporating computational thinking improved interdisciplinary skills in students. By integrating graphical programming with science and math, the study highlighted significant enhancements in student engagement and problem-solving capabilities (Zhan *et al.*, 2023). To foster interdisciplinary learning, educators should design curriculum units that purposefully blend science, technology, engineering, arts, and mathematics. Collaborative teaching across subjects, as well as real-world problem scenarios, should be used to encourage students to draw upon diverse knowledge areas to develop creative and well-rounded solutions. A practical example of interdisciplinary learning in STEAM comes from the Erasmus+ STEAM-Connect project, as reported by Pocalana and co-workers. In this study, teachers from art, music, technology, and mathematics collaborated to design classroom activities based on symmetry. Each teacher contributed their subject's perspective such as modular art, geometric patterns, musical canons and mathematical formalization using Cartesian coordinates. Though it began with a focus on math, the activity developed into a balanced integration of all disciplines. This case highlighted how joint planning and shared teaching goals can help educators create meaningful, cross-disciplinary STEAM experiences (Pocalana *et al.*, 2024).

Critical and Creative Thinking

Critical and creative thinking are foundational cognitive processes (Fig. **1**) that empower students to analyze, evaluate, and innovate. It enables students to evaluate information critically and produce imaginative solutions to complex problems. Within the STEAM framework, the integration of both thinking styles encourages students to approach problems from diverse perspectives, challenge assumptions, and devise original solutions (Zhan *et al.*, 2022). Integrating the arts into the STEAM framework is essential for nurturing creativity and innovation. The inclusion of artistic practices allows students to convey their ideas in varied and imaginative forms, which is vital for crafting unique solutions to challenges (Bartholomew *et al.*, 2023). Developing critical and creative thinking by engaging students in open-ended tasks that require deep analysis and the generation of innovative ideas. Teaching strategies such as brainstorming, reflective journals, and design thinking frameworks can cultivate these abilities across STEAM subjects. Filipe and co-workers studied the iSTEAM approach, where high school students created soundtracks for animations using physics principles and digital tools. The interdisciplinary process significantly boosted students' creativity and critical thinking (Filipe *et al.*, 2024).

Practical and Problem-Solving Approaches

STEAM education emphasizes hands-on, experiential engagement with real-world challenges. Students are encouraged to translate theoretical knowledge into practical applications, thereby strengthening their understanding and skill development. Through activities such as design challenges, experiments, and simulations, learners cultivate resilience, resourcefulness, and informed decision-making abilities (Ellison, 2023). Following implementation, the evaluation phase requires students to critically reflect on their solutions, which is essential for enhancing analytical, evaluative, and problem-solving skills. Technology further enhances the problem-solving process (Fig. **2**) by offering dynamic and interactive tools for modelling, testing, and refining solutions (Lage-Gómez & Ros, 2021).

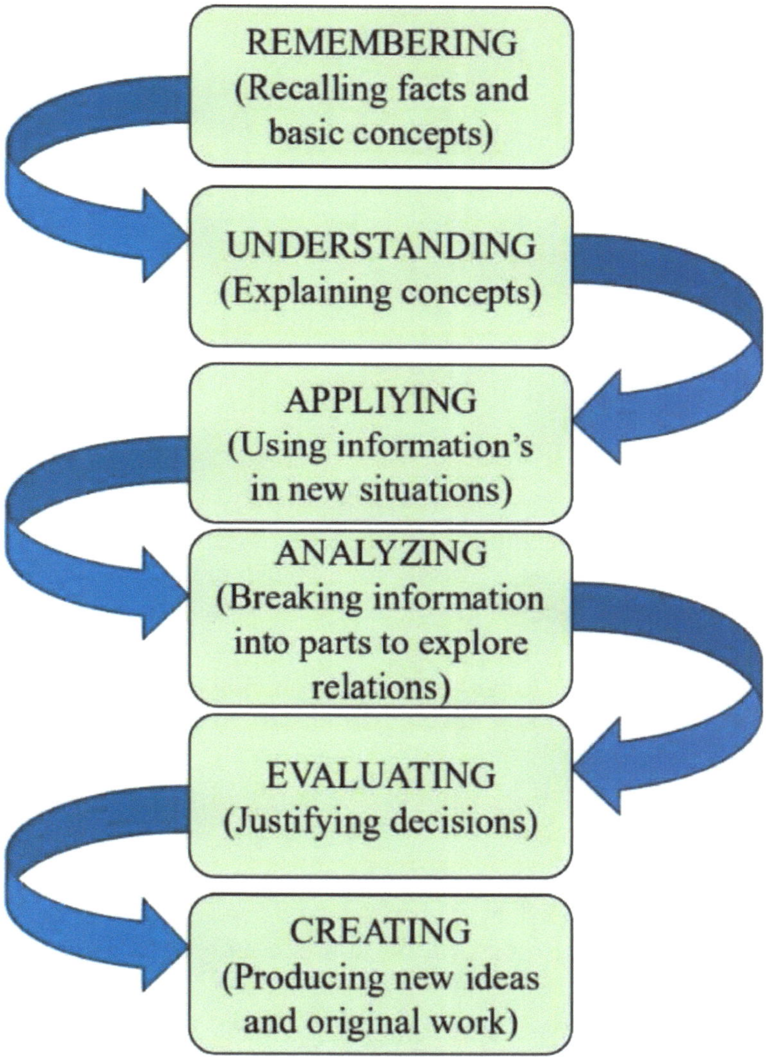

Fig. (1). Critical and creative thinking process.

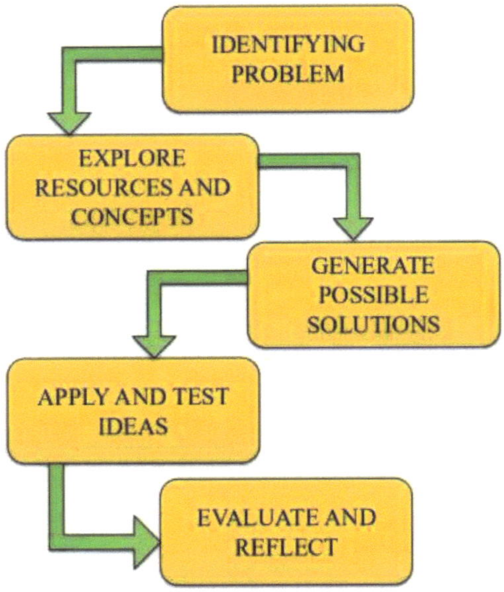

Fig. (2). STEAM problem-solving approach.

To support practical and problem-solving skills, classrooms should be equipped with maker spaces and access to tools for experimentation. Real-life scenarios, simulations, and problem-based assignments allow learners to apply theoretical knowledge in meaningful ways. Arpaci and co-workers demonstrated how the STEAM learning module in science education improved students' conceptual understanding and engagement through problem-solving and practical inquiry (Arpaci *et al.*, 2023).

Personalized and Inclusive Approach

An essential principle of effective STEAM education is the ability to adapt to the diverse needs, backgrounds, and abilities of all learners. This strategy actively engages students in the learning process and encourages them to take ownership of their educational experiences (Hidayat & Wardat, 2024). A personalized and inclusive approach helps to create equitable learning environments where every student gets respected and supported. Through the use of differentiated instruction, adaptive technologies, and culturally responsive teaching practices, educators can develop individual strengths while addressing diverse learning needs (Koch *et al.*, 2016). Such an inclusive approach increases learners' motivation and confidence and promotes equitable educational outcomes. A study in Chile revealed that early childhood and primary teachers, although initially unfamiliar with STEAM, embraced inclusive strategies when supported through training. The findings underscore the potential of inclusive, personalized STEAM frameworks in early education (Silva-Hormazábal & Alsina, 2023). Educators can support personalized and inclusive environments by differentiating instruction, adapting assessments, and integrating culturally responsive pedagogy. Using assistive technologies and flexible teaching methods can further support diverse learners (Kayyali, 2025). Wang and co-workers described in their study how STEAM-based instruction can strengthen students' problem-solving abilities. In this research, students aged 10 to 14 participated in integrated activities involving robotics, science, art, and engineering. The learning design focused on active engagement, encouraging learners to work through real-life scenarios by applying knowledge from multiple disciplines. Results showed that older students (13–14 years) were more capable in areas such as reflection and solution development than their younger peers. The study highlighted how thoughtfully designed STEAM tasks can nurture analytical thinking and practical problem-solving, making it an effective approach for developing these essential skills in diverse age groups (Wang *et al.*, 2024).

Collaborative and Project-Based Learning

STEAM education emphasizes the development of interpersonal skills through collaborative and project-based learning. Research has consistently shown that engaging in group projects improves students' collaborative abilities, which are crucial for effective communication and conflict resolution. In team-based activities, students learn to express themselves clearly, exchange information, and respect different viewpoints. These interactions are crucial for successfully planning, executing, and evaluating projects, building leadership, accountability, and the ability to engage in thoughtful, productive discussions (Papadopoulou, 2024 ; Triwoelandari *et al.*, 2023). López Carrillo and co-workers found that project-based STEAM learning improved students' self-regulation and collaboration. Group activities prompted learners to plan, monitor, and evaluate their learning effectively (Khaled & Alghfeli, 2025). Zhao and co-workers found that the students involved in collaborative STEAM projects showed significant enhancement in their ability to communicate, particularly in explaining intricate scientific ideas to their classmates. The research highlighted the role of collaboration and presentation opportunities in improving students' communication skills (Nguyen, 2025). Project-based learning in STEAM education often tackles real-world problems, encouraging students to apply theoretical knowledge in practical settings (Caplan & Oropeza, n.d.). This approach not only deepens their understanding by highlighting the relevance of what they have learned but also strengthens essential collaboration and communication skills through teamwork (Baidal-Bustamante *et al.*, 2023). Studies have emphasized that such learning methods significantly increase student engagement, making the connection between theory and practice more tangible and impactful (Chistyakov *et al.*, 2023). Collaborative and project-based learning can be developed by assigning interdisciplinary group tasks and providing platforms for shared learning and peer evaluation. Teachers can facilitate roles within groups to foster responsibility and leadership skills. Project-based learning in STEAM can be conceptualized through six essential features shown in Fig. (**3**) (Pérez Torres *et al.*, 2023).

Flexibility for Real-World Challenges

Flexibility means cultivating adaptability, critical awareness, and the ability to apply knowledge in uncertain or evolving contexts. The dynamic nature of today's world demands that learners be adaptable, solution-oriented, and prepared to engage with complex global issues. Contemporary challenges such as sustainability, public health, and digital transformation are explored by students through the flexibility provided by STEAM education in both content and pedagogy. This adaptability not only ensures the relevance of learning but also empowers students to contribute actively to society, cultivating a deep sense of community engagement (Cheng *et al.*, 2022). Flexibility can be nurtured through dynamic curriculum design that reflects current global challenges and technological advances. Encouraging experimentation, iterative feedback, and real-time learning adaptations equip students to thrive in unpredictable environments. Delgado-Rodríguez and co-workers implemented immersive Augmented Reality (AR) in STEAM learning. The approach offered dynamic, interactive experiences, enhanced adaptability in learning new digital tools, strengthened problem-solving and innovation through design and coding, and collaborative engagement in real-world application development (Hidayat & Wardat, 2024). The success of the AR-based workshop underlines the potential of integrating emerging technologies into STEAM education to better prepare students for modern challenges. Manikutty and co-workers described how STEAM learning adapts to real-world problems through the "STEAM for Social Good" program in India. In this project, pre-university students took part in a 3.5-month mentoring program where they worked on solving real social issues connected to the United Nations Sustainable Development Goals (UN SDGs). Even during the COVID-19 pandemic, students worked together online, using both digital tools and hands-on skills to build creative solutions. They turned their ideas into real projects and shared their work through video presentations. This program showed how STEAM education can stay flexible and meaningful, helping students learn to solve problems in changing and challenging situations (Manikutty *et al.*, 2022).

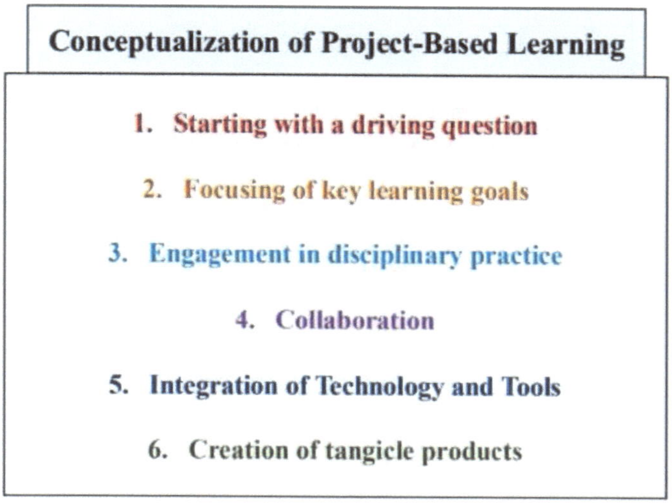

Fig. (3). Conceptualization of project-based learning.

Implementation and Impact

The recent introduction of STEAM into the educational agenda, due to its novelty or other structural factors like school logistics and teacher training (Kim & Kim, 2016), has yet to produce initiatives with considerable quality and quantity. Based on their literature review, Perignat and Katz-Buonincontro concluded that learning outcomes in areas such as creativity, problem-solving, and the arts are insufficiently reported in existing research (Perignat & Katz-Buonincontro, 2019). Simultaneously, research indicated that STEAM increases children's self-confidence and makes them more engaged and capable of taking initiative using their knowledge (Wahyuningsih *et al.*, 2020). Kim and Kim established three learning objectives for STEAM education: situation, creative design, and emotional touch. The situational context stimulates students' awareness of the necessity to solve problems, creative design encourages self-directed problem-solving, and emotional touch inspires enthusiasm to embrace new challenges through motivation, interest, and joy of success. These initiatives involve activities like hands-on work, visiting museums, reading interactive books, conducting fun and online experiments, gamification, and creating simulations and videos (Li & Wong, 2020), along with using drama to design rhyming poems and songs that help in memorizing information (Colucci-Gray *et al.*, 2019). Gamification, a process of transforming traditional educational content and activities into game-like formats, is becoming more widely embraced (Boytchev & Boytcheva, 2020).

Some notable STEAM implementation projects include:

i. Global Science Opera. Every year, the science education network Scientix in Europe (www.scientix.eu) introduces its STEAM project for the development of a science opera with a designated theme. Schools from various countries around the world create their two-and-a-half-minute scenes—complete with music, libretto, costumes, and set—coordinated from a central location and streamed globally online. Furthermore, the groups engage in metacognitive activities related to the chosen theme, including visits to scientific institutions or museums, video conferences between participants, and even activities focused on inquiry or design thinking (Urbaniak *et al.*, 2021).
ii. The UK CREATIONS projects, Programs such as "The Imagineerium", Sonic Pi, and similar projects, promote a collaborative approach involving teachers, artists, and science experts to foster a more holistic understanding of both science and art. Students have the opportunity to scientifically engage with traditional problem-solving methods, digital creation, and design through a variety of art forms, including dance, opera, visual arts, sound, and theatre.
iii. In the U.S., Costantino discusses a project developed alongside RISD's Lab, Museum, and Academic Affairs, where art and natural history objects serve as a basis for transdisciplinary learning. Students explore abstract concepts through various forms of inquiry in science, art, and design, gaining experience in tackling problems from multiple angles (Costantino, 2017).
iv. The GetWet project, which involves a water pumping station at a heritage site, encourages students to explore real-life scenarios where science and technological progress play a crucial role in community life, aligning with science education on social and environmental issues and the concept of social-ecological justice (Colucci-Gray *et al.*, 2019).

Integrating STEAM education into curricula from an interdisciplinary perspective has prompted educators to seek methods that foster deep learning and transferable skills. The integration of STEAM education into curricula from an interdisciplinary perspective has motivated educators to explore methods that foster deep learning and transferable abilities (Kelley & Knowles, 2016) (Leavy *et al.*, 2023). The rapid technological advancements have also led researchers to explore the influence of STEAM education on the essential skills learners need to navigate a digital and interconnected world (Supriyadi *et al.*, 2023). The impact of a STEAM-based assistive technology curriculum was examined by Hung and co-workers. Their findings demonstrated that improvements in student satisfaction and professionalism were achieved through a structured educational framework. These studies show that STEAM-based instruction significantly enhances learners' affective factors like motivation, interest, self-confidence, self-regulation skills, attitude, satisfaction, engagement, and teamwork. According to the studies reviewed, STEAM education integration has positively impacted learners' creative abilities, educational experiences, motivation, and satisfaction levels (Hung *et al.*, 2024).

The application of STEAM education in teaching should emphasize the core aspects of interdisciplinary learning, problem- and project-based activities, the development of students' problem-solving skills, fostering their innovation, and promoting the use of creative and divergent thinking to address real-world problems.

i. **Developing Students' Creative Skills from an Interdisciplinary Perspective**

STEAM education promotes the integration of various disciplines such as science, mathematics, arts, and engineering. To foster students' creativity, schools may promote STEAM education through modern digital technologies, interdisciplinary curriculum design, and the creation of relevant teaching resources. Firstly, the application of modern information network technology enables teachers to present lessons to students through a variety of teaching approaches, helping students better comprehend the content and learn more effectively within the STEAM education framework. Moreover, students can use online resources to collect, organize, and analyze information to identify efficient problem-solving strategies (Nan *et al.*, 2025). In addition, schools should provide a diverse selection of courses to address the different needs of students while ensuring the courses meet professional standards. Students should then be allowed to select their projects freely and be guided to apply project-based and problem-based approaches for effective learning. Implementing interdisciplinary education, as previously discussed, integrating multiple disciplines enables students to employ various innovative thinking approaches to identify appropriate materials and tools, effectively addressing specific problems. This approach significantly contributes to the development of innovative talents.

ii. **Implementing Practical Projects to Develop Students' Innovative Abilities**

STEAM education emphasizes hands-on learning and inquiry-based approaches, which enable students to develop critical problem-solving skills. Creativity is like human body muscles. Regular exercise can help muscles become more flexible; learning is also similar. Encouraging students to regularly test their knowledge through real-world projects and consistently promoting project-based practice helps foster their technological creativity and problem-solving abilities. Projects should not only be seen as a basic verification of acquired knowledge, but also as an opportunity for students to be motivated to solve problems in various creative ways. To fulfil these educational objectives, schools and educators need to focus on nurturing students' enthusiasm for learning when applying STEAM education. To achieve a high level of creativity, individuals often need a persistent learning attitude. To encourage students to learn, it is crucial to start with their favourite subjects and gradually expand their knowledge to other subjects. Students can enhance their creativity by interacting with peers or exploring resources available on the internet (Aguilera & Ortiz-Revilla, Leavy *et al.*, 2023).

iii. **Cultivating students' problem-solving skills by focusing on real-life situations**

STEAM Education encourages students to engage in solving real-life problems and emphasizes learning through projects, with a focus on problem-based, project-centred approaches. The project's authenticity is based on the problem's comprehensiveness and authenticity. Students face complex and diverse problems in their future life and work, often more challenging than those encountered in classes. As a result, problems should involve multiple fields of knowledge, prompting students to utilize interdisciplinary learning and helping them apply their knowledge to practical challenges. The "Pulley labour-saving device" project is a good example. This knowledge has been taught to junior high students in their classes. STEAM education encourages students first to grasp scientific theories and concepts, then design and create labour-saving devices, and ultimately take part in group competitions. First of all, the teacher introduces the relevant scientific theories and then instructs students on the proper and safe handling of tools such as laser cutting devices, CNC milling machines, and 3D printers. Furthermore, students are encouraged to collaborate freely to design, produce, and debug devices.

Students learn the principles and uses of the labour-saving device in this project, which reflects the "science" subject knowledge in STEAM education, and using digital devices reflects "technology" subject knowledge. The components used in the students' devices' designs reflect the knowledge of the "engineering" field, and the arrangement of parts shows the knowledge of the "arts" discipline. The knowledge of the "mathematics" discipline is essential in adjusting the pulley labour-saving device, including selecting suitable fixed and moving pulleys, pulley direction, and rope length. The project provides students with a comprehensive, problem-based STEAM education that enhances their innovation skills through real-life experiences (W. Li *et al.*, 2018). A well-developed educational platform is necessary for schools to implement problem-based and project-based learning successfully. The school is encouraged to establish an open

STEAM education platform that integrates virtual learning with hands-on lab activities. Students can explore STEAM learning freely and access resources through the internet, while laboratories are equipped with a variety of digital tools like laser cutting machines, 3D printers, CNC milling machines, *etc.*, and open-source hardware. Under the guidance of the instructors, students can conduct project-based learning in the laboratories. STEAM education aims to help students combine academic knowledge with practical problem-solving in real-world scenarios, stimulate their interest in learning, and improve their creativity and teamwork abilities, thereby fostering the development of innovative talents.

TRAINING OF TEACHERS TO ADOPT THE STEAM MODEL: KNOW-HOW

Teachers' preparedness and adaptability are key factors in successfully integrating STEAM education into classrooms. Teachers, as enablers of interdisciplinary learning, need to possess not just expertise in their subjects but also the ability to promote creativity, collaboration, and critical thinking across different fields (Mpuangnan, 2024).

Professional Development Programs play a crucial role in helping educators grasp the fundamental concepts of STEAM (Herro & Quigley, 2016). These programs ought to provide hands-on training in project-based learning, design thinking, and assessment methods customized for integrated instruction. Teachers can be immersed in the methodologies they are expected to implement through workshops, certification courses, and hands-on STEAM labs. Collaborative Teaching Models are advantageous as well, involving educators from various subject areas in co-planning and co-teaching lessons (Pratt *et al.*, 2017). This method contributes to the breakdown of academic silos and fosters a cohesive teaching strategy that is in harmony with the STEAM philosophy. Planning lessons in an interdisciplinary way enhances both the quality of instruction and the richness of students' learning experiences. Access to resources and technology is essential, as teachers need well-designed lesson plans, digital tools, and innovative materials to facilitate STEAM activities; therefore, schools and educational institutions should invest in equipping classrooms with maker spaces, art supplies, science kits, and exploratory software to support hands-on learning and experimentation (Lavicza *et al.*, 2022). Fostering a growth mindset and creative confidence among educators is vital, as teachers must be encouraged to experiment, reflect, and adapt their practices without fear of failure; ongoing mentorship and peer support networks can cultivate a culture of innovation and continuous learning, empowering educators to model these values for their students (Seaton, 2017) (Brooks & Goldstein, 2008). Wynn and Harris proposed that institutions such as the STEAM Education Centre (Canada) and the Teaching

Institute for Excellence in STEM (TIES) provide workshops and modules aimed at helping educators grasp how to link arts and design with fundamental STEM content via project-based learning (Wynn & Harris, 2012). By investing in comprehensive training and supportive infrastructures, educational systems can equip teachers to become effective stewards of the STEAM movement, ultimately nurturing a generation of imaginative, inquisitive learners prepared to navigate the complexities of the modern world (Y. H. Kim & Na, 2022).

CHALLENGES AND PROSPECTS

STEAM education faces various challenges. These challenges include the preparation of teachers to carry out integrated lessons and the professional development necessary to incorporate the transdisciplinary method and philosophy into their teaching practices (Herro *et al.*, 2018). This initiative needs sufficient time for planning, along with the support of skilled professionals. Inadequate cooperation between departments and colleges remains a significant challenge for implementing arts-based reform. One of the main obstacles in integrating arts into the science and math curriculum is the pre-existing idea among faculty and students that engineering and the arts are incompatible. This perspective indicates a lack of mutual respect and trust among faculty members from different disciplines. Some faculty members in science and engineering question whether the arts can be used effectively to teach an abstract concept in engineering, science, math, and vice versa (Rabalais, 2014). Another obstacle to STEAM pedagogy is the lack of resources and funding, especially in developing countries, where schools struggle with insufficient books, desk benches, large class sizes, and technological tools. For instance, by the end of 2003, around 113 million children globally were unable to attend school due to poverty, according to the United Nations Development Program. In this context, it would be very difficult to implement STEAM integration in many schools in developing countries that are underfunded. Departments and faculties display disciplinary egocentrism, showing an unwillingness to exchange ideas with members of other departments. The integration of math, science, the arts, and engineering with technology faces challenges due to students' limited readiness to embrace, accept, and engage in transdisciplinary education. Several factors may be responsible for this. Some staff members in different departments are either unwilling or unable to adopt alternative methods in their discipline, as they believe it is unnecessary due to negative attitudes or perspectives. The primary challenge is the failure to acknowledge interconnections between disciplines, which obstructs the integration of innovative ideas and practices into teaching. Another reason is having a negative attitude towards other perspectives beyond one's discipline. These perspectives illustrate the psychological difficulties associated with STEAM education. Other challenges in STEAM education are related to semesterization and modularization, which can hinder its implementation due to limited time for preparation or training, and insufficient resources to incorporate effective classroom practices (Connor *et al.*, 2015). The specific components that should be included in STEAM curricula remain unclear and disputed. Should the emphasis be on integrating the curriculum, teaching methods, and learning approaches, or fostering interdisciplinary faculty collaboration for projects and teaching and learning? Such confusion arises because the definition and meaning of a STEAM-integrated curriculum and pedagogy are unclear in substance, not just in its acronym (Falls, 2020). This perspective confuses the integration of STEAM discipline content. It is impossible to implement process integration without integrating content. Transdisciplinary or interdisciplinary collaboration presents a significant challenge due to intellectual and logistical factors. It is logistically difficult to establish a common area that includes all the required support resources. Even a regular planning time is a challenge. It was reported by various faculties that the schedule was complicated, especially when co-teaching is required within the same class. This challenge also affects teaching loads across various departments and colleges. It is also essential to find a suitable class location that promotes interdisciplinary learning, as well as coordinate times and days that fit the diverse schedules of faculty and students across various disciplines. Another key challenge is the comprehension of the basic interdisciplinary knowledge and skills requirement, as STEAM pedagogy expects teachers to have an extensive knowledge range, which is particularly difficult for educators teaching at the middle or high school levels and beyond. Departments across schools and universities have been designed with a focus on discipline-based curricula, teaching methods, projects and research, and assessments. Traditional institutional models can present obstacles to fostering collaboration among faculty members. The disparity in knowledge, philosophy, and paradigms across different disciplines may pose a significant challenge. Conflicting areas of knowledge across disciplines may lead to confrontations, where faculty members from different departments challenge each other's values and beliefs, making collaboration more difficult or even unfeasible (Costantino, 2017). Partnerships at lower grade levels where disciplinary boundaries are not clearly defined can be more effective than those at higher levels. The axiological, epistemological, and ontological differences in disciplinary paradigms may create a conflict between advocates of STEM education and those promoting STEAM lessons (Belbase *et al.*, 2022). Why would scholars and experts in the individual disciplines of Science, Technology, Engineering, Arts, and Mathematics support and encourage the implementation of STEAM education? Two views were expressed by NCTM's past President Matt Larson that make sense in this context. Firstly, several professions require STEAM/STEM skills, and the starting salaries in these fields are usually higher than in other sectors. Secondly, he believes that promoting STEM/STEAM can also benefit mathematics education, even though implementing it effectively presents challenges for meaningful learning in science, mathematics, and other subjects ("Message From Julius H. Hlavaty, President, NCTM," 1970). Also, there are differences between STEAM teachers' beliefs and what they implement in their teaching. For instance, Shin reported that although around 65% of South Korean teachers supported the need for STEAM education, only 18% applied it in their classrooms (Chu *et al.*, 2019). Conversely, the artistic aspect of STEAM may not adequately promote social justice issues in education, either broadly or within the context of STEAM education. It is challenging to comprehend how STEAM provides a feasible solution to the social and financial exclusion of underrepresented populations (Bennett, 2016). The integration of disciplines in STEAM is a concerted effort to consolidate various aspects without allowing for adjustments, relocations, or remediation of each other's shortcomings. Some of the challenges of STEAM education integration in Dubai were highlighted by Shaer and co-workers. These issues include maintaining a balance between curriculum focus on STEAM learning outcomes, teacher workload, resources in schools, time and curriculum constraints, and integration of different disciplines into the curriculum (Al Murshidi, 2019). The specific requirements of Science, Technology, Engineering, Arts, and Mathematics, along with these challenges, are recognized as significant concerns by stakeholders globally and within the education sector. STEAM education is a complex endeavour with numerous challenges, including teacher preparation, faculty members, school leadership, university departments, integration costs, and resource management. This study has significant theoretical and practical implications, necessitating further research to address the existing research gaps.

Future studies should focus on observing students over multiple semesters to assess the outcomes of STEAM education. Researchers should also focus on analyzing the skills developed through STEAM teaching. Future research should explore the efficacy of artificial intelligence in STEAM education. For instance, future research could compare the effectiveness of traditional teaching methods with AI in terms of student skills and learning outcomes. Future research should explore the impact of mentoring programs on teacher education within the context of STEAM instruction. The 15 articles on STEAM implementation suggest that there is a growing need for more thorough research on STEAM education in the coming years. Research indicates that interdisciplinary course instructors can enhance the design of appropriate STEAM activities for their higher-education students, leading to positive learning outcomes. For the development of interdisciplinary skills in students through STEAM education, faculty and researchers must have the proper teaching skills to implement it effectively. To enhance the effectiveness of teaching and provide more effective examples of STEAM education, future research should focus on a more systematic approach to designing and implementing STEAM activities in schools. Although STEAM instruction plays a vital role in schools, research on its implementation remains limited. Future research should focus on implementing STEAM education in teacher education.

CONCLUDING REMARKS

The transition from STEM to STEAM represents a pivotal shift in educational philosophy, emphasizing the integration of the arts with traditional science, technology, engineering, and mathematics disciplines. This transformation enriches the learning experience by fostering creativity, innovation, and holistic problem-solving approaches. The elements of STEAM, critical and creative thinking, interdisciplinary learning, practical problem-solving, collaboration, personalization, and real-world flexibility empower students to address complex challenges in an increasingly interconnected world. As we continue to explore and implement STEAM-based curricula, we see the profound impact on students' ability to innovate, communicate, and adapt across multiple domains. However, the journey towards widespread STEAM adoption is not without challenges. Teacher preparedness and structural limitations within educational systems emphasize the need for a collective effort to redefine teaching practices within the STEAM framework. Despite these challenges, there are abundant opportunities to address these issues. Further research is essential to explore the intersection of arts and STEAM education in practice. Focused studies on professional development programs for art teachers tailored to STEAM integration, as well as successful models of collaboration between STEM and art educators, can help identify best practices and encourage reciprocal teaching methods. The potential for STEAM to cultivate future-ready learners remains vast, and its integration into education systems holds promise for nurturing the next generation of thinkers, creators, and leaders.

ACKNOWLEDGEMENT

This chapter acknowledges the use of AI-assisted tools, including Perplexity and Elicit AI for research and fact-checking, Grammarly for writing enhancement and grammar correction, and Zotero for efficient reference management. These tools were selected to improve workflow efficiency, accuracy, and clarity while maintaining rigorous academic standards. Their integration complemented traditional research and writing methods by providing reliable source verification, streamlined citation, and polished language, thereby enhancing the overall quality and transparency of the research process.

REFERENCES

Arpaci, I., Dogru, M.S., Kanj, H., Ali, N., Bahari, M. (2023). An Experimental Study on the Implementation of a STEAM-Based Learning Module in Science Education. *Sustainability (Basel), 15*(8), 6807. [http://dx.doi.org/10.3390/su15086807]

Bahrum, S., Wahid, N., Ibrahim, N. (2017). Integration of STEM Education in Malaysia and Why to STEAM. *Int. J. Acad. Res. Bus. Soc. Sci., 7*(6), 645-654.

Baidal-Bustamante, E., Mora, C., Alvarez-Alvarado, M.S. (2023). STEAM Project-Based Learning Approach to Enhance Teaching-Learning Process in the Topic of Pascal's Principle. *IEEE Trans. Educ., 66*(6), 632-641. [http://dx.doi.org/10.1109/TE.2023.3283850]

Bartholomew, S., Yauney, J., Wuthrich, V., Wolfley, K., Elya, E., Rich, P., Shumway, S., Wright, G. (2023). Digital Storyboards: Making CS Elementary. In: *23 Intermountain Engineering, Technology and Computing (IETC),* Provo, UT, USA. [http://dx.doi.org/10.1109/IETC57902.2023.10152087]

Belbase, S., Mainali, B.R., Kasemsukpipat, W., Tairab, H., Gochoo, M., Jarrah, A. (2022). At the dawn of science, technology, engineering, arts, and mathematics (STEAM) education: prospects, priorities, processes, and problems. *Int. J. Math. Educ. Sci. Technol., 53*(11), 2919-2955. [http://dx.doi.org/10.1080/0020739X.2021.1922943]

Bennett, A.G. (2016). Ethnocomputational creativity in STEAM education: a cultural framework for generative justice. *Teknokultura. Revista de Cultura Digital y Movimientos Sociales, 13*(2), 587-612. [http://dx.doi.org/10.5209/rev_TEKN.2016.v13.n2.52843]

Boy, G. A. (2013). From STEM to STEAM: Toward a human-centred education, creativity & learning thinking. In: *Proceedings of the 31st European Conference on Cognitive Ergonomics,* Association for Computing Machinery.1-7. [http://dx.doi.org/10.1145/2501907.2501934]

Boytchev, P., Boytcheva, S. (2020). Gamified Evaluation in STEAM for Higher Education: A Case Study. *Information (Basel), 11*(6), 316. [http://dx.doi.org/10.3390/info11060316]

Brooks, R., Goldstein, S. (2008). The Mindset of Teachers Capable of Fostering Resilience in Students. *Can. J. Sch. Psychol., 23*(1), 114-126. [http://dx.doi.org/10.1177/0829573508316597]

Butterworth, C. (2020). *Data-centric policies and practice: Implications for publicly funded arts and cultural organisations in England*[Doctoral dissertation]. Lancaster University. England. Available from: https://search.proquest.com/openview/3a1acf35b6ab175e0118696b23e0c3d9/1?pq-origsite=gscholar&cbl=18750&diss=y

Caplan, M., Oropeza, E. (2019). The STEAM Conference: An Event to Promote Youth to Explore STEAM-related Fields and Potential Careers. *ASEE Annual Conference & Exposition Proceedings.* ASEE Conferences. Tampa, Florida. [http://dx.doi.org/10.18260/1-2--33427]

Cheng, L., Wang, M., Chen, Y., Niu, W., Hong, M., Zhu, Y. (2022). Design My Music Instrument: A Project-Based Science, Technology, Engineering, Arts, and Mathematics Program on The Development of Creativity. *Front. Psychol., 12*, 763948. [http://dx.doi.org/10.3389/fpsyg.2021.763948] [PMID: 35069336]

Chistyakov, A.A., Zhdanov, S.P., Avdeeva, E.L., Dyadichenko, E.A., Kunitsyna, M.L., Yagudina, R.I. (2023). Exploring the characteristics and effectiveness of project-based learning for science and STEAM education. *Eurasia J. Math. Sci. Technol. Educ., 19*(5), em2256. [http://dx.doi.org/10.29333/ejmste/13128]

Chu, H.E., Martin, S.N., Park, J. (2019). A Theoretical Framework for Developing an Intercultural STEAM

Program for Australian and Korean Students to Enhance Science Teaching and Learning. *Int. J. Sci. Math. Educ., 17*(7), 1251-1266.
[http://dx.doi.org/10.1007/s10763-018-9922-y]

Colucci-Gray, L., Burnard, P., Gray, D., Cooke, C. (2019). A Critical Review of STEAM (Science, Technology, Engineering, Arts, and Mathematics). In: *Oxford Research Encyclopedia of Education.*. Oxford University Press.
[http://dx.doi.org/10.1093/acrefore/9780190264093.013.398]

Connor, A.M., Karmokar, S., Whittington, C. (2015). From STEM to STEAM: Strategies for Enhancing Engineering & Technology Education. *International Journal of Engineering Pedagogy (iJEP), 5*(2), 37-47.
[http://dx.doi.org/10.3991/ijep.v5i2.4458]

Conradty, C., Bogner, F.X. (2020). STEAM teaching professional development works: effects on students' creativity and motivation. *Smart Learning Environments, 7*(1), 26.
[http://dx.doi.org/10.1186/s40561-020-00132-9]

Costantino, T. (2018). STEAM by another name: Transdisciplinary practice in art and design education. *Arts Educ. Policy Rev., 119*(2), 100-106.
[http://dx.doi.org/10.1080/10632913.2017.1292973]

Deming, D., Noray, K. (2018). *STEM Careers and the Changing Skill Requirements of Work.*. National Bureau of Economic Research.
[http://dx.doi.org/10.3386/w25065]

Falls, Z. (2020). Beyond Boundaries: Pre-Service Teachers' Experiences of Transdisciplinary Education via STEAM Making Projects. *J. Comput. Math. Sci. Teach., 39*(1), 19-31.
[http://dx.doi.org/10.70725/844963cgsfxa]

Galama, T., Hosek, J. (2008). *Is the United States Losing Its Edge in Science and Technology?*. RAND Corporation.
[http://dx.doi.org/10.7249/RB9347]

Georgette, Y. (2008). STEAM Education: An Overview of Creating a Model of STEAM in the Classroom. *J. Res. Educ., 3*(21), 1-28.

Guimeráns-Sánchez, P., Alonso-Ferreiro, A., Zabalza-Cerdeiriña, M.A., Monreal-Guerrero, I.M. (2023). E-textiles para la educación STEAM en educación primaria: una revisión sistemática. *RIED-Revista Iberoamericana de Educación a Distancia, 27*(1), 417-448.
[http://dx.doi.org/10.5944/ried.27.1.37645]

Hardiman, M., Rinne, L., Yarmolinskaya, J. (2014). The Effects of Arts Integration on Long-Term Retention of Academic Content. *Mind Brain Educ., 8*(3), 144-148.
[http://dx.doi.org/10.1111/mbe.12053]

Helms, M., Moore, R., Edwards, D., Freeman, J. (2016). STEAM-based interventions: Why student engagement is only part of the story. *2016 Research on Equity and Sustained Participation in Engineering, Computing, and Technology (RESPECT)* Atlanta, GA, USA.
[http://dx.doi.org/10.1109/RESPECT.2016.7836171]

Henriksen, D., Mehta, R., Mehta, S. (2019). Design Thinking Gives STEAM to Teaching: A Framework That Breaks Disciplinary Boundaries. In: Khine, M.S., Areepattamannil, S., (Eds.), *STEAM Education.* Cham: Springer.
[http://dx.doi.org/10.1007/978-3-030-04003-1_4]

Herro, D., Quigley, C. (2017). Exploring teachers' perceptions of STEAM teaching through professional development: implications for teacher educators. *Prof. Dev. Educ., 43*(3), 416-438.
[http://dx.doi.org/10.1080/19415257.2016.1205507]

Herro, D., Quigley, C., Cian, H. (2019). The Challenges of STEAM Instruction: Lessons from the Field. *Action Teach. Educ., 41*(2), 172-190.
[http://dx.doi.org/10.1080/01626620.2018.1551159]

Hodges, A. (2009). Alan Turing and the Turing Test. In: Epstein, R., Roberts, G., Beber, G., (Eds.), *Parsing the Turing Test.* Dordrecht: Springer.
[http://dx.doi.org/10.1007/978-1-4020-6710-5_2]

Hung, C.H., Chen, M.H., Fan, S.C. (2024). Enhancing occupational therapy education: Evaluating the impact of a STEAM-based assistive technology curriculum using Kirkpatrick's four-level model. *Br. J. Occup. Ther., 87*(8), 512-523.
[http://dx.doi.org/10.1177/03080226241239563] [PMID: 40337475]

Hidayat, R., & Wardat, Y. (2024). A systematic review of augmented reality in science, technology, engineering and mathematics education. *Education and Information Technologies, 29*(8), 9257-9282.

Kamienski, N., Radziwill, N. (2018). Design for STEAM: Creating Participatory Art with Purpose. *STEAM, 3*(2), 1-17.
[http://dx.doi.org/10.5642/steam.20180302.08]

Kayyali, M. (2025). The Impact of Personalized Learning on Student Engagement and Achievement in STEAM. In N. Son (Ed.), *Integrating Personalized Learning Methods Into STEAM Education.* IGI Global Scientific Publishing.
[http://dx.doi.org/10.4018/979-8-3693-7718-5.ch003]

Kelley, T.R., Knowles, J.G. (2016). A conceptual framework for integrated STEM education. *Int. J. STEM Educ., 3*(1), 11.
[http://dx.doi.org/10.1186/s40594-016-0046-z]

Kim, B., Kim, J. (2016). Development and Validation of Evaluation Indicators for Teaching Competency in STEAM Education in Korea. *Eurasia J. Math. Sci. Technol. Educ., 12*(7), 1909-1924.
[http://dx.doi.org/10.12973/eurasia.2016.1537a]

Kim, Y., Chu, H-E., Lim, G. (2015). Science Curriculum Changes and STEM Education in East Asia. In: *Science Education in East Asia.* Springer International Publishing.
[http://dx.doi.org/10.1007/978-3-319-16390-1_7]

Kim, Y.H., Na, S.I. (2022). Using structural equation modelling for understanding relationships influencing the middle school technology teacher's attitudes toward STEAM education in Korea. *Int. J. Technol. Des. Educ., 32*(5), 2495-2526.
[http://dx.doi.org/10.1007/s10798-021-09708-z]

Koch, F.D., Dirsch-Weigand, A., Awolin, M., Pinkelman, R.J., Hampe, M.J. (2017). Motivating first-year university students by interdisciplinary study projects. *Eur. J. Eng. Educ., 42*(1), 17-31.
[http://dx.doi.org/10.1080/03043797.2016.1193126]

Lage-Gómez, C., Ros, G. (2021). Transdisciplinary integration and its implementation in primary education through two STEAM projects (*La integración transdisciplinar y su aplicación en Educación Primaria a través de dos proyectos STEAM*). *Infanc. Aprendiz., 44*(4), 801-837.
[http://dx.doi.org/10.1080/02103702.2021.1925474]

Leavy, A., Dick, L., Meletiou-Mavrotheris, M., Paparistodemou, E., & Stylianou, E. (2023). The prevalence and use of emerging technologies in STEAM education: A systematic review of the literature. *Journal of Computer Assisted Learning, 39*(4), 1061-1082.

Land, M.H. (2013). Full STEAM Ahead: The Benefits of Integrating the Arts Into STEM. *Procedia Comput. Sci., 20*, 547-552.
[http://dx.doi.org/10.1016/j.procs.2013.09.317]

Larson, L.C., Miller, T.N. (2011). 21st Century Skills: Prepare Students for the Future. *Kappa Delta Pi Rec., 47*(3), 121-123.
[http://dx.doi.org/10.1080/00228958.2011.10516575]

Lavicza, Z., Weinhandl, R., Prodromou, T., Anđić, B., Lieban, D., Hohenwarter, M., Fenyvesi, K., Brownell, C., Diego-Mantecón, J.M. (2022). Developing and Evaluating Educational Innovations for STEAM Education in Rapidly Changing Digital Technology Environments. *Sustainability (Basel), 14*(12), 7237.

[http://dx.doi.org/10.3390/su14127237]

Leavy, A., Dick, L., Meletiou-Mavrotheris, M., Paparistodemou, E., Stylianou, E. (2023). The prevalence and use of emerging technologies in STEAM education: A systematic review of the literature. *J. Comput. Assist. Learn., 39*(4), 1061-1082.
[http://dx.doi.org/10.1111/jcal.12806]

Lee, C. (2023). Documenting Children's Spatial Reasoning through Art: A Case Study on Play-Based STEAM Education. *Sustainability (Basel), 15*(19), 14051.
[http://dx.doi.org/10.3390/su151914051]

Lewis Ellison, T. (2023). Normalizing Black Students/Youth and their Families' Digital and STEAM Literacies. *Read. Teach., 76*(5), 594-600.
[http://dx.doi.org/10.1002/trtr.2182]

Li, K.C., Wong, B.T.M. (2020). Trends of learning analytics in STE(A)M education: a review of case studies. *Interact. Technol. Smart Educ., 17*(3), 323-335.
[http://dx.doi.org/10.1108/ITSE-11-2019-0073]

Li, W., Li, G., Mo, W., Li, J. (2018). The Influence of STEAM Education on the Improvment of Studentsr Creative Thinking. *Proceedings of the 4th International Conference on Arts, Design and Contemporary Education (ICADCE 2018)* Atlantis Press.
[http://dx.doi.org/10.2991/icadce-18.2018.200]

López Carrillo, M.D., Calonge García, A., Lebrón Moreno, J.A. (2024). Self-Regulation of Student Learning in a STEAM Project. *Educ. Sci. (Basel), 14*(6), 579.
[http://dx.doi.org/10.3390/educsci14060579]

Manikutty, G., Sasidharan, S., Rao, B. (2022). Driving innovation through project based learning: A pre-university STEAM for Social Good initiative. *IEEE Frontiers in Education Conference (FIE).* Uppsala, Sweden.
[http://dx.doi.org/10.1109/FIE56618.2022.9962420]

Murshidi, G.A. (2019). STEM education in the United Arab Emirates: Challenges and possibilities. *International Journal of Learning, Teaching and Educational Research, 18*(12), 316-332.

Martins, I., Baptista, M. (2024). Teacher Professional Development in Integrated STEAM Education: A Study on Its Contribution to the Development of the PCK of Physics Teachers. *Educ. Sci. (Basel), 14*(2), 164.
[http://dx.doi.org/10.3390/educsci14020164]

Message From Julius, H. (1970). Message From Julius H. Hlavaty, President, NCTM. *J. Res. Math. Educ., 1*(1), 7-8.
[http://dx.doi.org/10.5951/jresematheduc.1.1.0007]

Mohr-Schroeder, M.J., Cavalcanti, M., Blyman, K. (2015). Stem education: Understanding the changing landscape. In: Sahin, A. (eds), *A Practice-based Model of STEM Teaching: STEM Students on the Stage (SOS)* SensePublishers, Rotterdam.
[http://dx.doi.org/10.1007/978-94-6300-019-2_1]

Moore, T.J., Smith, K.A. (2014). Advancing the State of the Art of STEM Integration. *J. STEM Educ., 15*(1), 5. Available from: https://karlsmithmn.org/wp-content/uploads/2017/08/Moore-Smith-JSTE-Ed-GuestEditorialF.pdf

Mpuangnan, K.N. (2024). Teacher preparedness and professional development needs for successful technology integration in teacher education. *Cogent Education, 11*(1), 2408837.
[http://dx.doi.org/10.1080/2331186X.2024.2408837]

Nan, X., An, C., Zhu, W., Jin, J., Xu, H. (2025). Consolidating the Digital Foundation to Strengthen Public Support for Education. In: Zeng, H., Pan, J., Xu, J., Sun, Q. (eds), *Lecture Notes in Educational Technology.* Springer, Singapore.
[http://dx.doi.org/10.1007/978-981-97-8144-7_4]

Nguyen, A.Q. (2025). Pedagogical Approaches in STEAM Education. In S. Behera, A. Sorayyaei Azar, S. Curle, & J. Dials (Eds.), *Transformative Approaches to STEAM Integration in Modern Education* IGI Global Scientific Publishing.
[http://dx.doi.org/10.4018/979-8-3693-7408-5.ch003]

Papadopoulou, E.A. (2024). Advancements in STEAM Education for 21st Century Learners. *Int. J. Educ., 16*(4), 39.
[http://dx.doi.org/10.5296/ije.v16i4.22270]

Perales, F.J., Aróstegui, J.L. (2024). The STEAM approach: Implementation and educational, social and economic consequences. *Arts Educ. Policy Rev., 125*(2), 59-67.
[http://dx.doi.org/10.1080/10632913.2021.1974997]

Pérez Torres, M., Couso Lagarón, D., Marquez Bargalló, C. (2023). Evaluation of STEAM Project-Based Learning (STEAM PBL) Instructional Designs from the STEM Practices Perspective. *Educ. Sci. (Basel), 14*(1), 53.
[http://dx.doi.org/10.3390/educsci14010053]

Perignat, E., Katz-Buonincontro, J. (2019). STEAM in practice and research: An integrative literature review. *Think. Skills Creativity, 31*, 31-43.
[http://dx.doi.org/10.1016/j.tsc.2018.10.002]

Pocalana, G., Robutti, O., Ciartano, E. (2024). Resources and Praxeologies Involved in Teachers' Design of an Interdisciplinary STEAM Activity. *Educ. Sci. (Basel), 14*(3), 333.
[http://dx.doi.org/10.3390/educsci14030333]

Pratt, S.M., Imbody, S.M., Wolf, L.D., Patterson, A.L. (2017). Co-planning in Co-teaching. *Intervention Sch. Clin., 52*(4), 243-249.
[http://dx.doi.org/10.1177/1053451216659474]

Quigley, C.F., Herro, D., King, E., Plank, H. (2020). STEAM Designed and Enacted: Understanding the Process of Design and Implementation of STEAM Curriculum in an Elementary School. *J. Sci. Educ. Technol., 29*(4), 499-518.
[http://dx.doi.org/10.1007/s10956-020-09832-w]

Rabalais, M.E. (2014). *STEAM : A National Study of the Integration of the Arts Into STEM Instruction and its Impact on Student Achievement A.* available at https://search.proquest.com/openview/d4dadbab54198740e20fd2bf013ed5d4/1?pq-origsite=gscholar&cbl=18750

Rimmer, M. (2020). The art of survival: Community-based arts organisations in times of austerity. *Community Dev. J., 55*(2), 295-312.
[http://dx.doi.org/10.1093/cdj/bsy036]

Rothwell, J. (2013). *The hidden STEM Economy.* (pp. 1-38). Washington, DC: Brookings.

Seaton, F.S. (2018). Empowering teachers to implement a growth mindset. *Educ. Psychol. Pract., 34*(1), 41-57.
[http://dx.doi.org/10.1080/02667363.2017.1382333]

Silva-Hormazábal, M., Alsina, Á. (2023). Exploring the Impact of Integrated STEAM Education in Early Childhood and Primary Education Teachers. *Educ. Sci. (Basel), 13*(8), 842.
[http://dx.doi.org/10.3390/educsci13080842]

Somerson, R., Hermano, L. (2013). *The Art of Critical Making: Rhode Island School of Design on Creative Practice.* Wiley.

Sousa, D.A., Pilecki, T. (2018). *From STEM to STEAM: Brain-Compatible Strategies and Lessons That Integrate the Arts.* Corwin.
[http://dx.doi.org/10.4135/9781544357393]

Supriyadi, E., Turmudi, Dahlan, J. A., Juandi, D. (2023). Publication Trends from STEAM in Education from

Scopus Database: Bibliometric Analysis. *Jurnal Penelitian Pendidikan IPA, 9*(6), 104-111.

Teitelbaum, M.S. (2014). *Falling Behind?*. Princeton University Press.
[http://dx.doi.org/10.1515/9781400850143]

Thuneberg, H., Salmi, H., Fenyvesi, K. (2017). Hands-On Math and Art Exhibition Promoting Science Attitudes and Educational Plans. *Education Research International, 2017,* 9132791.
[http://dx.doi.org/10.1155/2017/9132791]

Triwoelandari, R., Handayani, E.W., Arif, S. (2023). Development of Project-Based Learning Science E-Module To Improve Collaboration Skills Of Elementary School Students. *Jurnal Cakrawala Pendas, 9*(4), 762-774.
[http://dx.doi.org/10.31949/jcp.v9i4.6611]

Urbaniak, K, Venkatesh, V, Ben-Horin, O (2021). A creative global science classroom: crafting the Global Science Opera. In: Holdhus, K, Murphy, R, Espeland, MI, (Eds.), *Music education as craft* Springer.
[http://dx.doi.org/10.1007/978-3-030-67704-6_12]

Ubben, G. (2019). Using Project-Based Learning to Teach STEAM. In: Stewart, A.J., Mueller, M.P., Tippins, D.J. (eds), *Converting STEM into STEAM Programs.* Environmental Discourses in Science Education, Springer, Cham.
[http://dx.doi.org/10.1007/978-3-030-25101-7_6]

Wahyuningsih, S., Nurjanah, N.E., Rasmani, U.E.E., Hafidah, R., Pudyaningtyas, A.R., Syamsuddin, M.M. (2020). STEAM Learning in Early Childhood Education: A Literature Review. *International Journal of Pedagogy and Teacher Education, 4*(1), 33.
[http://dx.doi.org/10.20961/ijpte.v4i1.39855]

Wang, Y., Qin, M., Hu, Y., Huang, G. (2024). A Practical Study of STEAM Instructional Activities Directed Toward Students' Problem-Solving Skill. *Proceedings of the 2024 15th International Conference on E-Education, E-Business, E-Management and E-Learning.* (pp. 160–169), York, NY, USA.
[http://dx.doi.org/10.1145/3670013.3670024]

Wang, Z., Zhao, D., Heidari, A.A., Chen, H. (2024). Hunger games search algorithm based on stochastic individual information for engineering design optimization problems. *J. Comput. Des. Eng., 11*(3), 280-307.
[http://dx.doi.org/10.1093/jcde/qwae050]

Watson, A.D., Watson, G.H. (2013). Transitioning STEM to STEAM Reformation. *J. Qual. Particip., 36*, 1-4. Availabe from: https://www.academia.edu/download/35120086/TransitioningSTEMToSTEAM_Watson_-_JQP_-_October_2013-libre.pdf

Wu, Z. (2022). Understanding teachers' cross-disciplinary collaboration for STEAM education: Building a digital community of practice. *Think. Skills Creativity, 46*, 101178.
[http://dx.doi.org/10.1016/j.tsc.2022.101178]

Wynn, T., Harris, J. (2012). Toward A Stem + Arts Curriculum: Creating the Teacher Team. *Art Educ., 65*(5), 42-47.
[http://dx.doi.org/10.1080/00043125.2012.11519191]

Yakman, G. (2010). What is the Point of STEAM - A Brief Overview. *Steam: A Framework for Teaching Across the Disciplines. STEAM Education, 7*(9), 1-9.

Zhan, Z., Hu, Q., Liu, X., Wang, S. (2023). STEAM Education and the Innovative Pedagogies in the Intelligence Era. *Appl. Sci. (Basel), 13*(9), 5381.
[http://dx.doi.org/10.3390/app13095381]

Zhan, Z., Yao, X., Li, T. (2023). Effects of association interventions on students' creative thinking, aptitude, empathy, and design scheme in a STEAM course: considering remote and close association. *Int. J. Technol. Des. Educ., 33*(5), 1773-1795.
[http://dx.doi.org/10.1007/s10798-022-09801-x] [PMID: 36531976]

CHAPTER 5

Interrelationship of Commerce, Management, and ICT

Martha Sharma[1,*] and **Ekta Pal**[2]

[1] *Department of Management, St.John's College, Agra, Uttar Pradesh 282002, India*

[2] *Department of Commerce, St.John's College, Agra, Uttar Pradesh 282002, India*

Abstract: This chapter aims to uncover the essential elements of the relationship between commerce and management and their profound influence on the global market. This chapter delves into the intricate and vital relationship between commerce and management, emphasizing the indispensable role that effective management plays in shaping global economic outcomes. The artful governance of commerce is not merely advantageous; it serves as a cornerstone for enhancing the quality of life within nations and elevating their stature on the global stage. In the absence of proper regulation, unchecked commerce can enable large corporations and conglomerates to amass disproportionate power, rendering them effectively impervious to failure. This disturbing imbalance gives rise to significant negative externalities for the broader population while disproportionately benefiting a select group of business owners. Consequently, it becomes imperative for nations to prioritize the meticulous management of commerce, fostering prosperity that is inclusive of all citizens, particularly those who directly contribute to job creation, the supply of goods, and essential services. Management encompasses a dynamic process of organizing and coordinating business activities with the clear aim of achieving well-defined objectives. Often regarded as a fundamental factor of production alongside machines, materials, and financial resources (Drucker, Peter), management is the backbone of any thriving enterprise. As noted by management guru Peter Drucker, the primary responsibilities of management extend to critical areas such as marketing and innovation. This paper endeavours to illuminate the essential elements of the interplay between commerce and management, dissecting their profound influence on the complexities of the global market.

Keywords: Commerce, Correlation, Effects, Global Market, Management.

INTRODUCTION

The integration of Information and Communications Technology (ICT) in business education has a significant impact on the fields of commerce and

* **Corresponding Author Martha Sharma:** Department of Management, St.John's College, Agra, Uttar Pradesh 282002, India; E-mail: marthawilliam1978@gmail.com

management. As commerce increasingly relies on digital platforms and data analytics, incorporating ICT into teaching methods prepares students to navigate this evolving landscape effectively. In commerce, ICT provides real-time access to financial markets, consumer data, and global trade information. This access is essential for students, as it helps them understand and utilize these tools. Gaining hands-on experience fosters a deeper understanding of market dynamics and consumer behaviour, which is critical for future professionals in the field.

Similarly, in management, ICT enhances collaboration through tools that facilitate communication and project management (Plumb & Zamfir, 2011). It allows students to simulate real-world business scenarios, allowing them to experiment with decision-making and strategic planning in a safe environment. Developing critical thinking and problem-solving skills is vital for aspiring managers who must make informed decisions in fast-paced business settings.

Moreover, the flexibility of ICT challenges traditional teaching methodologies that often leave students passive. By promoting active learning through interactive presentations and online resources, students in commerce and management can engage more deeply with the material, thereby enhancing their creativity and innovation. Effectively integrating ICT into commerce and management education paves the way for a more dynamic curriculum that reflects the complexities of the modern business world, equipping students with the necessary skills to thrive as future leaders. This chapter seeks to explore the key aspects of the relationship between commerce and management and their alignment with ICT (Arjang *et al.*, 2023).

INTERCONNECTED ASPECTS OF MODERN BUSINESS AND SOCIETY

Commerce, management, and ICT (Information and Communications Technology) are interconnected aspects of modern business and society (Nwanmuoh *et al.*, 2024). ICT enables businesses to manage resources effectively, improve communication, and facilitate electronic commerce, while management principles guide the application of these technologies for organizational success. Here is a more detailed breakdown.

Commerce

This refers to the activity of buying and selling goods and services, including the traditional brick-and-mortar stores and the newer realm of e-commerce.

Commerce serves as the heartbeat of economic activity, representing the intricate web of buying and selling goods and services that fuels our modern society. It spans both traditional brick-and-mortar establishments, where face-to-face interactions create personal connections, and the rapidly expanding realm of e-commerce, which offers unparalleled convenience and global reach. The significance of commerce cannot be overstated; it drives economic growth, creates jobs, and shapes consumer experiences by providing access to a vast array of products and services. As technology continues to evolve, businesses must adapt a blend of traditional values and innovative strategies to remain competitive and meet the ever-changing demands of consumers.

Management

This encompasses the planning, organization, direction, and control of resources, including human resources, financial resources, and technology resources, to achieve organizational goals (Deep, 2023). Management is more than just a function; it is a vital leadership discipline that involves the strategic orchestration of resources to achieve an organization's goals and vision. Effective management cultivates a productive organizational culture, empowers teams, and drives innovation, ensuring that goals are not only met but exceeded. In today's fast-paced and competitive landscape, strong management practices are essential for navigating challenges, seizing opportunities, and achieving sustainable success. Leaders who master these skills can inspire their teams, optimize operations, and propel their organizations toward long-term growth.

ICT (Information and Communications Technology)

This refers to the technologies used for creating, storing, processing, and distributing information, including computers, networks, and communication devices.

ICT is at the forefront of the digital revolution, encompassing a wide array of cutting-edge technologies that transform how we create, store, process, and share information. This field includes powerful computers, extensive networks, and sophisticated communication devices that work collaboratively to enable seamless connectivity and access to information (Kumar *et al.*, 2024). With ICT, organizations can enhance their operational efficiency, improve decision-making processes, and foster collaboration across geographical boundaries. As we move further into the digital age, the role of ICT becomes increasingly critical, driving innovation, enhancing productivity, and enabling individuals and businesses to thrive in an information-driven world. Embracing ICT not only positions organizations to respond to market demands but also empowers them to shape the future of their industries.

ICT in E-Commerce

ICT: The Technological Backbone of Modern Business

ICT revolutionizes how businesses operate and how commerce and management are taught. In commerce, ICT enables (Igboeli & Bisallah, 2020):

- Real-time financial market access
- Automated trade and logistics systems
- Consumer behaviour analytics
- E-commerce and digital payment solutions (Fig. **1**).

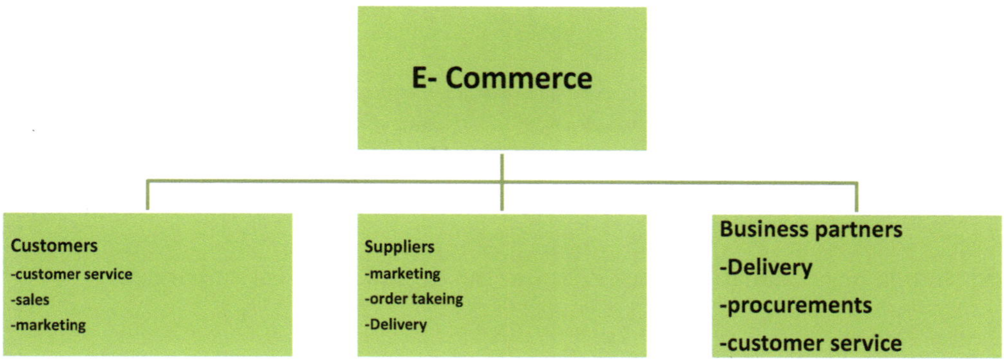

Fig. (1). Business activities in E-Commerce (Behl, 2021).

In management, ICT empowers:

- Cloud-based collaboration
- Project management platforms (*e.g.*, Trello, Asana)
- Business simulation tools for educational purposes
- Data-driven decision-making

Moreover, ICT tools like ERP systems (*e.g.*, SAP, Oracle) streamline business processes and allow managers to forecast, plan, and control operations more effectively.

ICT in Commerce and Management Education

The role of ICT in commerce and management teaching is illustrated in Fig. **(2)**

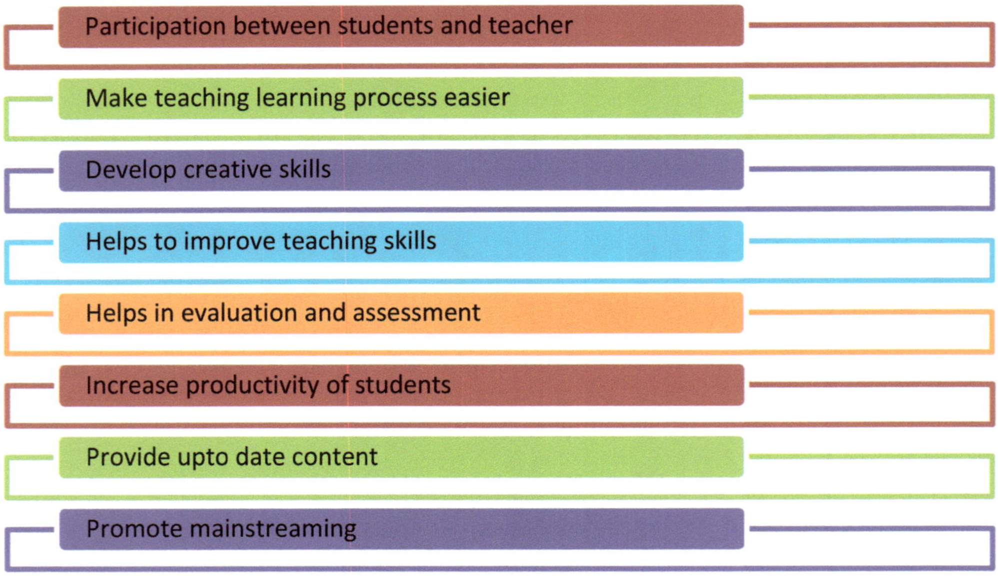

Fig. (2). ICT in commerce and management education (Szymczyk & El Emary, 2021).

Modern business education must adapt to the digital age. ICT enhances:

- Active learning through simulations, case studies, and virtual labs
- Critical thinking via scenario-based analysis
- Innovation by exposing students to the latest tech-driven business models

ICT-rich environments empower students to: (Singh *et al*., 2020)

- Engage in hands-on digital problem-solving
- Understand global commerce operations
- Gain insights into data-driven management strategies

This integration encourages creativity, adaptability, and technical fluency skills that are invaluable in today's dynamic market environment.

Understanding Commerce Management

Commerce management refers to the planning, organizing, directing, and controlling of commercial activities such as trade, sales, marketing, finance, and customer service. It involves strategic decision-making to ensure the effective movement of goods and services from producers to consumers.

Commerce and management are essential to profitability and sustainability. It is to the benefit of nations to establish governmental agencies with policies responsible for promoting and managing commerce to enhance economic growth and the standard of living for their citizens. It is fundamentally important to see the correlation of commerce and management, as without proper management of commerce, the world is left in a state of global meltdown (Jere & Ngidi, 2020).

Commerce management is important because it is imperative to maintain good relationships with other nations. It allows the growth of globalization and produces positive economic. Commerce and management are important to the economic activity of all human beings within their nations.

The Role of Commerce and Management in the Global Economy

Commerce refers to the activities involved in the buying and selling of goods and services, especially on a large scale. Management involves planning, organizing, directing, and controlling these activities to ensure efficiency, profitability, and sustainability.

Effective management

- Aligns organizational goals with market dynamics
- Minimizes negative externalities such as monopolistic practices
- Promotes fair trade and ethical operations
- Ensures resilience against market volatility

Without sound management, commercial activities risk falling into exploitative patterns, creating wealth gaps and systemic imbalances. As Peter Drucker noted, management is not only a resource—it is a *driving force behind* innovation and marketing, which are essential for any enterprise to survive and grow.

Commerce Management: A Framework for Prosperity

Commerce management encompasses:

- Trade coordination
- Sales and marketing strategy
- Financial oversight
- Customer relationship management

Nations that foster robust commerce management systems often experience:

- Economic growth
- Job creation

- Enhanced standard of living
- International trade relationships. Government policies play a vital role in supporting commercial regulation, reducing the risks associated with market monopolies, and ensuring equitable economic participation. An overview of ICT in business is presented in Fig. (**3**).

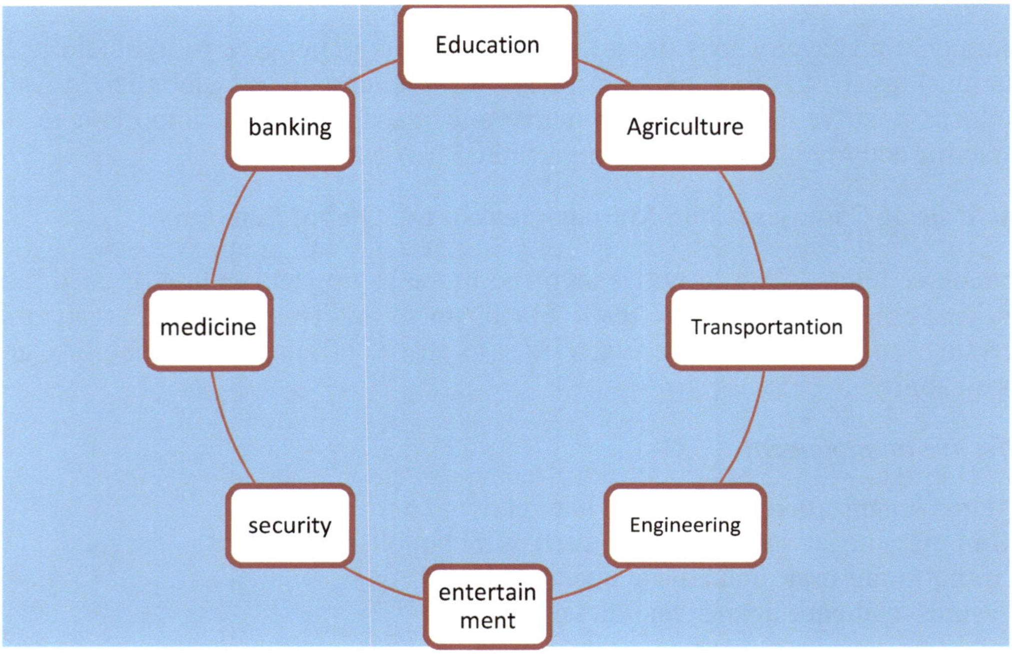

Fig. (3). Overview of ICT in business (Jaganathan *et al.*, 2018).

ICT plays a central role in streamlining business operations, enabling digital communication, and driving innovation. ICT serves as a crucial catalyst for transforming business operations in today's digital age. By streamlining processes, ICT enhances efficiency, allowing organizations to operate more smoothly and reduce costs.

Furthermore, ICT facilitates seamless digital communication, enabling teams to collaborate effortlessly regardless of geographical barriers. This connectivity enhances teamwork and accelerates decision-making, providing a competitive edge in a fast-paced market.

Moreover, ICT drives innovation by providing businesses with the tools and data necessary to develop new products and services. It empowers companies to analyse market trends, customer preferences, and operational performance, leading to informed strategies and creative solutions. (Fig. **4**) depicts ICT key functions in business.

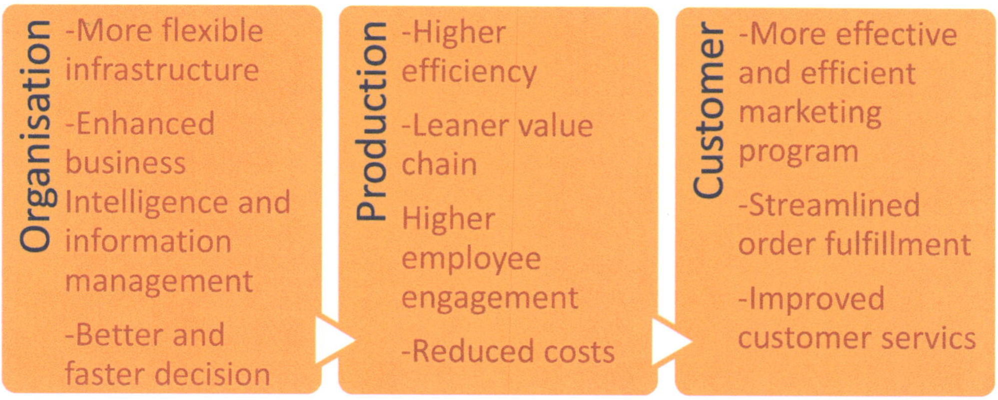

Fig. (4). Benefits of ICT in business (Almasradi *et al.*, 2022).

In essence, embracing ICT is not just about keeping up with technology; it is about leveraging the full potential of digital tools to foster growth, efficiency, and innovation, ultimately positioning businesses for success in an ever-evolving landscape.

KEY FUNCTIONS OF ICT IN BUSINESS

Automation

By automating repetitive tasks such as inventory updates and email marketing campaigns, businesses can significantly enhance efficiency. Automation reduces the time and effort required for these routine activities, allowing employees to focus on more strategic and creative tasks. This not only leads to improved productivity but also minimizes the risk of human error, ensuring more accurate outcomes. For instance, automated inventory management systems can provide real-time stock updates, helping businesses manage their supply chain more effectively and reducing both excess stock and stockouts (Davenport & Ronanki, 2018).

Communication

In today's fast-paced business environment, effective communication is crucial. ICT facilitates instant messaging, video conferencing, and collaborative platforms that enhance teamwork and connectivity, regardless of geographical barriers. Tools like Slack, Zoom, and Microsoft Teams enable teams to interact in real time, fostering a culture of transparency and collaboration (Garrison, 2019). This ensures that all members are aligned on goals, projects, and deadlines, ultimately driving innovation and responsiveness to market changes.

Data Management

Businesses generate and collect vast amounts of data daily. ICT provides robust solutions for secure storage, efficient retrieval, and sophisticated analysis of this data (Mahmood *et al.*, 2024). Advanced data management systems not only protect sensitive information but also allow companies to make better-informed decisions. By employing data analytics tools, organizations can track performance metrics, identify trends, and gain actionable insights that drive operational and strategic improvements. This data-centric approach empowers businesses to respond quickly to market demands and customer preferences.

E-commerce Enablement

With the growing importance of online shopping, ICT plays a vital role in transforming how businesses engage with customers (Kumar & Gupta, 2021). E-commerce platforms support seamless online transactions, allowing businesses to expand their reach beyond traditional storefronts. Features such as customer portals and digital storefronts create personalized shopping experiences, increasing customer satisfaction and loyalty. Additionally, tools like CRM systems help businesses understand customer behaviour, enabling them to tailor marketing strategies effectively and drive sales growth.

Decision Support

In an era where data drives successful businesses, decision support systems leverage advanced technologies, including AI and machine learning, to provide profound insights. Dashboards and analytical tools present complex data in a user-friendly format, making it easier for decision-makers to evaluate key performance indicators (KPIs) and trends. This empowers businesses to make proactive, evidence-based decisions, reduce risks, and seize opportunities more swiftly. Ultimately, a strong decision support framework leads to better strategic planning and long-term success (Tuan *et al.*, 2024).

Remote Work Capabilities

The shift towards remote work has made it essential for organizations to embrace ICT solutions that facilitate workforce flexibility. Cloud services and mobile tools enable employees to access information and collaborate from anywhere, promoting a more adaptable and resilient work environment. These technologies not only support employee productivity but also help organizations reduce overhead costs associated with physical office spaces. By fostering remote work capabilities, companies can attract a diverse talent pool and improve employee satisfaction and retention.

In summary, the integration of Information and Communication Technology (ICT) in business operations is indispensable for achieving efficiency, enhancing communication, managing data securely, enabling e-commerce, supporting informed decision-making, and accommodating modern work styles. By leveraging these key functions, businesses can drive growth and maintain a competitive advantage in an increasingly digital world (Choudhury *et al.*, 2021).

SOME REAL-WORLD EXAMPLES:

- **ERP systems** like SAP and Oracle integrate finance, HR, inventory, and logistics.
- **CRM platforms** like Salesforce centralize customer data and enhance personalization.
- **Marketing automation tools** like HubSpot and Mailchimp improve lead generation.
- **AI and machine learning** optimize forecasting, personalization, and risk assessment.

In short, ICT enhances efficiency, scalability, and competitiveness across all business functions.

Commerce and management are the twin pillars that uphold the structure of modern economic systems. Commerce serves as the mechanism by which goods and services are exchanged, both domestically and internationally, enabling wealth creation, market expansion, and cultural exchange (Islam *et al.*, 2024). Management, in turn, provides the framework necessary to direct, regulate, and optimize these commercial activities, ensuring sustainability and strategic development.

COMMERCE AS A CATALYST FOR ECONOMIC ACTIVITY

At its core, commerce is the lifeblood of any economy. It encompasses all the activities involved in the transfer of goods and services from producers to consumers, including wholesale and retail trade, transportation, logistics, and finance (Islam *et al.*, 2024). This exchange system facilitates job creation, tax revenue generation, and infrastructure development.

In a globalized world, commerce is no longer confined within national borders. International trade agreements, digital platforms, and supply chain integrations have expanded the reach of businesses and consumers alike. Countries that actively engage in global commerce benefit from foreign investment, access to new technologies, and diversified markets.

Management: The Strategic Engine behind Commerce

While commerce drives economic interaction, effective management transforms this activity into a structured, goal-oriented system. Management involves the continuous process of planning, organizing, leading, and controlling business resources to achieve desired objectives. Key managerial roles in the global economy include:

Resource Allocation: Ensuring optimal use of financial, human, and technological capital. We will strategically optimize our financial, human, and technological resources to achieve unparalleled efficiency and productivity across the organization.

Strategic Planning: Anticipating market trends and aligning business goals accordingly.

Risk Management: Mitigating financial, operational, and environmental risks. We are dedicated to proactively identifying, evaluating, and mitigating financial, operational, and environmental risks. This commitment safeguards our assets and paves the way for sustainable growth.

Ethical Governance: Promoting fair trade practices and corporate social responsibility. We passionately champion fair-trade practices and corporate social responsibility, establishing a robust framework for ethical conduct that not only enhances our integrity but also makes a positive impact on society. Ethical Governance: We firmly advocate for fair trade practices and corporate social responsibility, establishing a framework for ethical behaviour that enhances our integrity and positively impacts society.

The Synergy between Commerce and Management

The relationship between commerce and management is a beautifully orchestrated dance of interdependence, where each element enhances the other. Commerce, with its dynamic nature, generates an insatiable demand for organized, meticulous oversight. In response, management emerges as a guiding force, skilfully implementing strategic control and direction to navigate the complexities of the market (Broccardo *et al.*, 2024).

This collaboration nurtures a vibrant succession of growth, efficiency, and innovation, which not only drives both sectors forward but also fosters resilience in the face of changing economic landscapes.

This dynamic synergy is particularly striking in the operations of multinational corporations. These global giants deftly manage intricate and sprawling commercial networks that stretch across continents (Kumar *et al.*, 2024). They rely on advanced management systems that incorporate cutting-edge technology and data analytics, allowing them to optimize performance, anticipate market shifts, and maintain a competitive edge. Such sophisticated management practices ensure that these corporations can harmonize diverse operations, cultures, and consumer behaviours into a cohesive strategy. Simultaneously, small businesses experience a unique yet equally impactful relationship with management principles. In their quest for success, these enterprises prioritize effective management strategies that enable them to adapt to local market demands and stand out amid fierce competition. By fostering strong leadership, cultivating employee engagement, and focusing on customer satisfaction, small businesses can achieve remarkable growth and sustainability despite their limited resources compared to larger counterparts (Igboeli & Bisallah, 2020).

Ultimately, the interplay between commerce and management not only enhances organizational efficiency but also sparks innovation and creativity, paving the way for new ideas, products, and services to emerge in the marketplace. This ongoing collaboration ensures that both large corporations and small businesses can thrive and contribute to a vibrant and evolving economic landscape and those that adapt to these technological advancements will be best positioned for success in the future landscape of commerce and management (Igboeli & Bisallah, 2020).

Impact on National and Global Prosperity through Commerce, Management, and Information and Communication Technology (ICT)

In the current era of the digital economy, the roles of information and communication technology (ICT) and economic complexity are important for controlling environmental unsustainability and formulating policies to address ecological concerns (Arjang *et al.*, 2023). Fig. **5** depicts the conceptual framework of ICT and the digital economy.

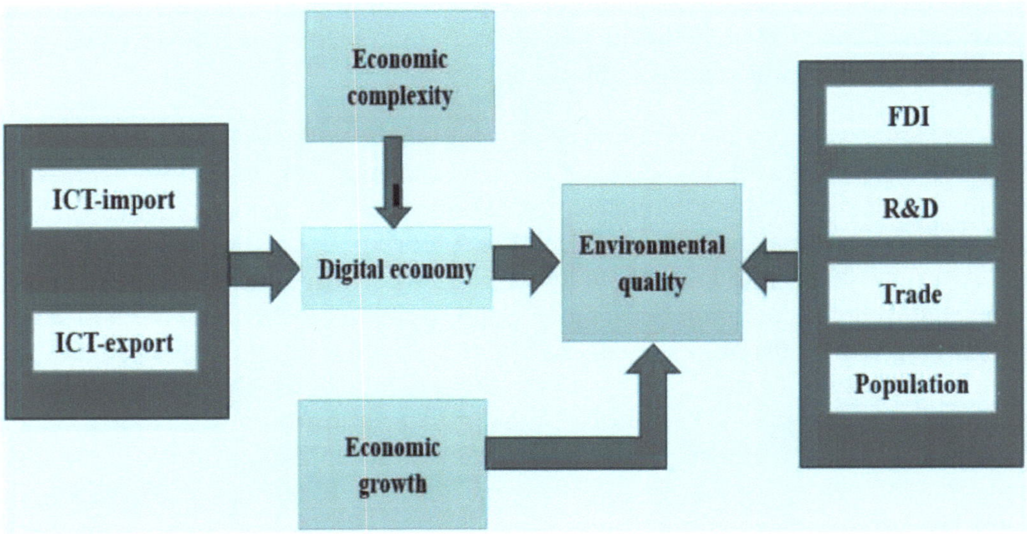

Fig. (5). The Conceptual framework of the model (Arjang *et al.*, 2023).

Countries that prioritize both commerce and effective management, particularly through the integration of Information and Communication Technology (ICT), enjoy multifaceted benefits that contribute to national and global prosperity.

Nations that prioritize both commerce and management enjoy numerous advantages:

Economic Stability: A diverse and well-regulated commercial sector supports GDP growth and job creation. The integration of ICT plays a crucial role in this process by facilitating efficient business operations, enhancing supply chain management, and promoting e-commerce. By leveraging ICT, businesses can reach broader markets, optimize operations, and respond more effectively to consumer needs, thereby contributing to economic resilience and stability (Mousavian, 2012).

Social Development: Effective management systems ensure the equitable distribution of resources and access to essential services. ICT can enhance public management by improving transparency and accountability in governance. For instance, digital platforms can streamline the delivery of public services, making them more accessible to underprivileged communities (Qureshi, 2019). Moreover, ICT enables educational initiatives that foster digital literacy, ensuring that more people can participate in and benefit from socio-economic development.

Global Influence: Nations with strong economic governance often have greater diplomatic and financial influence on the world stage. The adoption of ICT not only strengthens domestic industries but also positions countries as leaders in global technological advancements. Countries that invest in ICT infrastructure and innovation can attract foreign investment, collaborate on international projects, and shape global standards in technology and commerce (Ezeoha *et al.*, 2020). This elevation in status enhances their ability to influence global policies and economic trends.

In conclusion, the symbiotic relationship between commerce, management, and ICT is vital for fostering both national and global prosperity. By embracing these elements, countries can build a robust economic framework, promote social equity, and increase their influence in a rapidly changing global environment.

In essence, the global economy is not merely a network of trade—it reflects how effectively commerce is managed across different regions. This reinforces the need for education systems, government policies, and corporate frameworks to emphasize the integration of commerce and management at all levels.

Core Functions of Management in Modern Commerce

Effective management is essential to the success and sustainability of commercial operations. In today's highly competitive and globalized business environment, the complexity of managing commerce has increased significantly. Organizations must now rely on integrated management functions that support operational efficiency, customer satisfaction, and long-term growth. This section highlights five core functions that play a pivotal role in the effective governance of modern commerce (Bello, 2019). Fig. **6** shows various components of sales management.

Key functions of commerce management include:

- Inventory control
- Marketing and sales management
- Financial planning and analysis
- Customer relationship management (CRM)
- Supply chain and logistics management

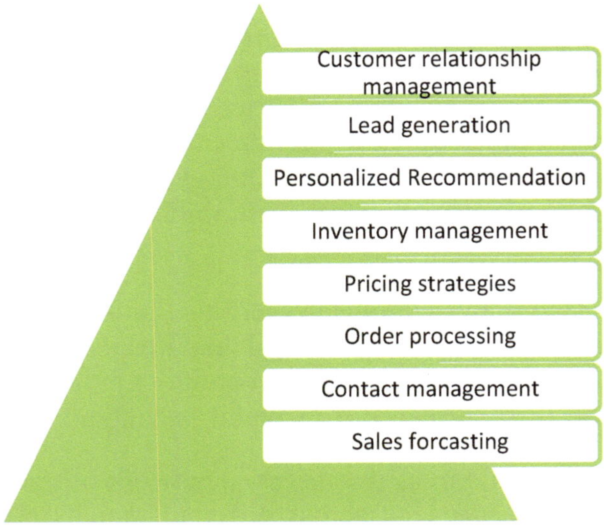

Fig. (6). Components of sales management (Song *et al.*, 2025).

Inventory Control

Inventory control is the process of managing and overseeing the ordering, storage, and use of a company's inventory. Efficient inventory management ensures that a company has the right amount of stock at the right time, reducing costs while maximizing customer satisfaction (Jacobs & Chase, 2018). Technologies like RFID, automated tracking systems, and AI-based forecasting tools have revolutionized this area, allowing real-time insights and data-driven decision-making.

Key Benefits:

- Reduces overstock and under stock situations
- Improves cash flow and profitability
- Enhances responsiveness to market demand

Marketing and Sales Management

Marketing and sales management involves the strategic planning, execution, and monitoring of campaigns and sales tactics to attract, convert, and retain customers. This function encompasses market research, branding, advertising, and sales channel development. Digital transformation has empowered businesses with tools such as social media analytics, SEO, content marketing, and customer segmentation algorithms to reach global audiences efficiently (Kotler & Keller, 2016).

Key Benefits:

- Builds brand equity and market presence
- Drives revenue through targeted outreach
- Enables agile adaptation to market trends

Financial Planning and Analysis

Financial planning and analysis (FP&A) are crucial for aligning business strategies with financial performance. This involves budgeting, forecasting, financial reporting, and strategic decision-making based on quantitative data (Eloff, 2018).

In global commerce, where fluctuations in currency exchange rates, inflation, and trade tariffs can affect profitability, robust FP&A capabilities help businesses remain resilient and adaptive.

Key Benefits:

- Informs capital allocation and investment decisions
- Enhances financial risk mitigation
- Supports strategic growth through informed insights

Customer Relationship Management (CRM)

CRM is the practice of managing a company's interactions with current and potential customers using data-driven systems (Fernando *et al.*, 2023). These systems track customer preferences, purchasing behaviour, and service interactions to enhance satisfaction and loyalty.

Modern CRM platforms like Salesforce, Zoho, and HubSpot offer integrated features including AI recommendations, omnichannel communication, and customer journey mapping.

Key Benefits:

- Improves customer retention and lifetime value
- Enhances personalization of services
- Increases efficiency of marketing and support teams

Supply Chain and Logistics Management

Supply chain and logistics management involve the coordination of production, transportation, warehousing, and distribution of goods. Effective supply chain strategies reduce delays, lower operational costs, and improve customer service levels (Christopher, 2022).

With globalization, supply chains have become more complex, involving international sourcing and multi-tier logistics networks. Technologies such as blockchain, IoT, and predictive analytics are now vital in ensuring transparency, traceability, and efficiency.

Key Benefits:

- Ensures timely delivery and quality assurance
- Minimizes operational bottlenecks
- Enhances competitive advantage in the global market

IMPACT OF ICT IN BUSINESS

There is a diverse impact of ICT on business. A few of the powerful impacts are highlighted in Fig. (7).

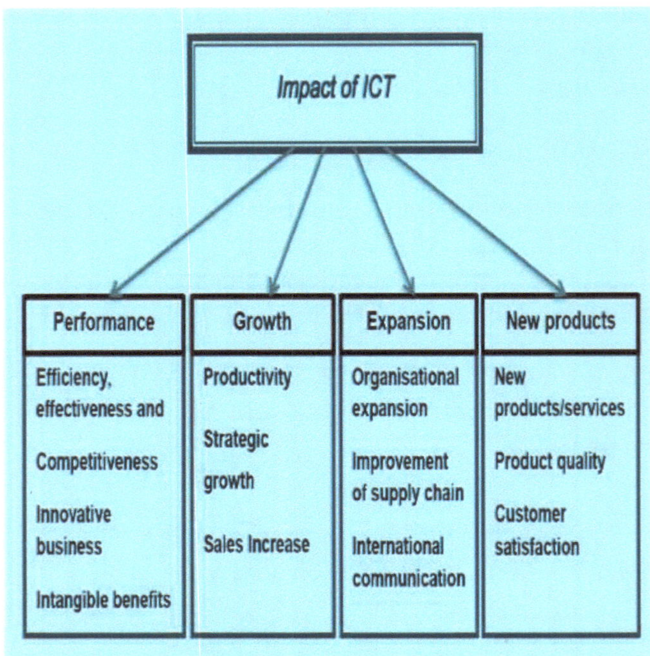

Fig. (7). Impact of ICT (Rainer & Prince, 2021).

In today's digitally driven economy, the convergence of commerce, management, and ICT defines the efficiency and competitiveness of both businesses and nations. Commerce, representing the exchange of goods and services, and management, the structured control of this process, form the bedrock of economic systems (Liu *et al.*, 2021). However, their full potential is only unlocked when complemented by the strategic application of ICT.

As digital platforms reshape market operations, ICT integration in business education and practice becomes indispensable. For students and professionals alike, the mastery of ICT tools in commerce and management is not merely an advantage—it is a necessity.

Digital Tool of ICT

Information and Communication Technology encompasses all digital tools and platforms that facilitate the storage, processing, and exchange of information. The various ICT tools are shown in Fig. (**8**).

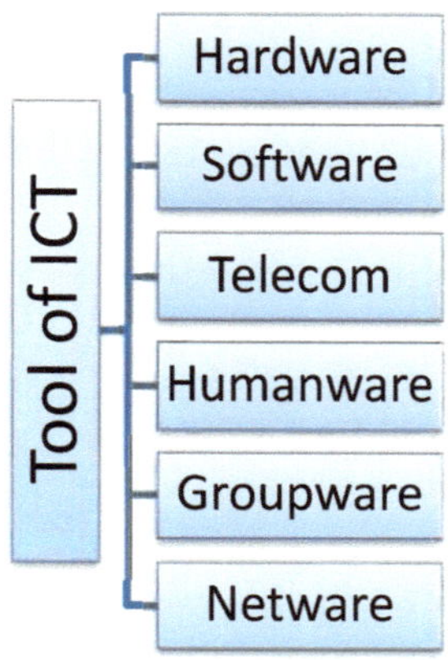

Fig. (8). Digital tool of ICT (Liu *et al.*, 2021).

- **Hardware** (computers, servers, networking devices)
- **Software** (ERP systems, CRM tools, accounting applications)
- **Communication platforms** (email, video conferencing, social media

ICT is not just a support function—it is an enabler of business strategy and operations.

The Symbiotic Relationship: ICT, Commerce, and Management

The digital revolution has irrevocably transformed the landscape of business, and at its heart lies Information and Communication Technology (ICT). No longer merely a support function, ICT has become the very lifeblood of modern commerce and a critical tool for effective management. This chapter delves into the intricate interrelation between ICT, commerce, and management, exploring how technological advancements have reshaped business operations, strategies, and decision-making processes (Ateljević *et al.*, 2023).

The Digital Transformation of Commerce

Commerce, at its core, involves the exchange of goods and services. ICT has fundamentally altered every aspect of this exchange, from the initial sourcing of materials to the final delivery of products and beyond. Consider the following:

- **E-commerce: Redefining Marketplaces:** The rise of the internet and related technologies has given birth to e-commerce, enabling businesses to transcend geographical boundaries and reach a global customer base. Online marketplaces, virtual storefronts, and digital payment systems have streamlined transactions, reduced overhead costs, and offered consumers unparalleled convenience (Laudon & Traver, 2021).
- Think of the vast online retail platforms that connect buyers and sellers across continents, facilitating everything from purchasing everyday necessities to acquiring specialized goods.
- **Supply Chain Optimization:** ICT tools, such as Enterprise Resource Planning (ERP) systems, Radio-Frequency Identification (RFID) tags, and sophisticated logistics software, have revolutionized supply chain management (Pontoh *et al.*, 2024). These technologies enable businesses to track inventory in real-time, optimize transportation routes, predict demand fluctuations, and foster seamless collaboration with suppliers and distributors. This leads to increased efficiency, reduced waste, and faster delivery times, ultimately enhancing customer satisfaction.

Digital Marketing and Customer Relationship Management (CRM):
ICT has empowered businesses to engage with customers in more personalized and effective ways. Digital marketing strategies, leveraging social media, search engine optimization, email marketing, and targeted advertising, allow companies to reach specific demographics and build brand awareness. CRM systems, on the other hand, enable businesses to collect, analyse, and utilize customer data to understand their needs, personalize interactions, and build long-term loyalty (Kotler & Keller, 2016).

- **Financial Technology (FinTech):** The intersection of ICT and finance has spawned a plethora of innovative solutions. Online banking, digital payment gateways, mobile wallets, and blockchain technologies have revolutionised the way financial transactions are conducted, providing speed, security, and accessibility. FinTech innovations are also impacting areas like lending, investment, and insurance, creating new opportunities and challenges for traditional financial institutions (Arner, Barberis, & Buckley, 2016).

- **Management in the Digital Age:** Effective management is crucial for navigating the complexities of the modern business environment, and ICT provides managers with powerful tools to enhance their capabilities across various functions (Kuloğlu, 2024).
- **Data-Driven Decision Making:** ICT enables the collection, storage, and analysis of vast amounts of data. Business intelligence (BI) tools and data analytics platforms empower managers to identify trends, gain insights, and make informed decisions based on evidence rather than intuition (Evans & Lin, 2017). From sales forecasting to risk assessment, data-driven decision-making leads to more strategic and effective outcomes.
- **Enhanced Communication and Collaboration:** ICT facilitates seamless communication and collaboration within and between organizations. Email, instant messaging, video conferencing, and collaborative software platforms enable teams to work together effectively, regardless of their physical location (Jones, 2019). This fosters better information sharing, faster problem-solving, and increased productivity.
- **Process Automation and Efficiency:** ICT allows for the automation of repetitive and routine tasks, freeing up human resources for more strategic and creative endeavours. Robotic Process Automation (RPA), workflow management systems, and artificial intelligence (AI) powered tools can streamline processes, reduce errors, and improve overall efficiency across various departments, from accounting to human resources (Kumar & Singh, 2019)
- **Strategic Planning and Innovation:** ICT provides managers with the tools to analyse market trends, monitor competitor activities, and identify emerging opportunities (Mintzberg *et al.*, 2020). Simulation software and predictive analytics can help in scenario planning and risk management. Furthermore, ICT fosters innovation by enabling rapid prototyping, facilitating knowledge sharing, and connecting businesses with global networks of expertise.

THE FUTURE LANDSCAPE

The interrelation between ICT, commerce, and management will continue to evolve at an accelerating pace. Emerging technologies like artificial intelligence, the IoT, blockchain, and augmented/virtual reality are poised to further transform business models, create new opportunities, and present novel challenges (Kuloğlu, 2024). Businesses that strategically embrace and adapt to these technological advancements will be best positioned for success in the future landscape of commerce and management.

Points of Integration between Commerce Management and ICT

Decision-Making Support

ICT tools such as data analytics and business intelligence systems provide managers with real-time data for informed decision-making. Sales forecasting, financial modelling, and market trend analysis become more accurate with digital tools.

Automation and Efficiency

ERP (Enterprise Resource Planning) systems integrate functions like accounting, HR, and inventory, reducing manual errors and improving operational efficiency.

E-Commerce and Online Transactions

ICT facilitates online business platforms, enabling businesses to reach global markets. Tools like secure payment gateways, digital marketing, and e-commerce websites have revolutionized the retail and wholesale sectors (Talafha, 2024).

Customer Relationship Management

CRM software helps businesses track customer behaviour, preferences, and feedback. This data is crucial for tailoring products and services, enhancing customer satisfaction, and boosting loyalty (Tapscott & Tapscott, 2016)

Supply Chain Optimization

With technologies like RFID, GPS, and IoT, ICT enables real-time tracking of goods and improves supply chain transparency, reducing delays and costs (Manyika *et al.*, 2017).

THE FUTURE OF ICT IN COMMERCE

The future of Information and Communication Technology (ICT) in commerce is set to be significantly transformed by emerging technologies such as AI, blockchain, and cloud computing (Kuloğlu, 2024). These advancements promise to deepen the integration between commerce management and ICT, leading to more efficient and innovative business practices.

Artificial intelligence, for instance, will enable businesses to leverage predictive analytics. This involves analysing vast amounts of data to anticipate customer behaviour and market trends, allowing organizations to make informed decisions and optimize their strategies. Additionally, AI-driven personalized marketing will enable companies to tailor their messaging and offerings to individual customers, enhancing engagement and driving sales (Manyika *et al*., 2017).

Blockchain technology will enhance transparency and security in transactions. By providing a decentralized and immutable ledger, blockchain can help businesses ensure the integrity of their supply chains and manage their operations with greater trust and accountability.

Furthermore, cloud computing offers the scalability and flexibility that modern businesses need to adapt to rapidly changing market conditions. With cloud-based solutions, companies can access tools and resources from anywhere, facilitating collaboration and innovation across teams and locations (Berisha *et al*., 2022).

Finally, autonomous logistics—powered by advancements in robotics and AI—will revolutionize the way products are delivered. Automated systems can optimize routes, manage inventory, and streamline fulfilment processes, ultimately reducing costs and improving customer satisfaction (Tapscott & Tapscott, 2016).

In summary, the integration of these emerging technologies will not only enhance commerce management but also pave the way for new business models and opportunities in the digital economy (Taleb & Mohamed, 2020). As these developments unfold, businesses that embrace these innovations will be better positioned to thrive in a competitive landscape.

CONCLUDING REMARKS

The interrelationship between commerce management and ICT is dynamic and transformative. ICT enhances the efficiency, accuracy, and scalability of commerce operations, while commerce management guides the strategic application of technology. Together, they create a responsive, data-driven business environment essential for modern enterprise success.

The relationship between commerce, management, and ICT is no longer linear—it is symbiotic and multi-dimensional (Srivastava, 2019). Each domain reinforces the others:

- Commerce thrives with strategic management
- Management is empowered by ICT
- ICT revolutionizes both commerce and management

As global markets become increasingly complex, the need for this integration becomes ever more urgent (Ariansyah *et al*., 2025). Educators, policymakers, and business leaders must work together to align these elements, ensuring that economic systems remain resilient, inclusive, and forward-looking.

In conclusion, the interconnected relationship between commerce, management, and ICT is essential for promoting both national and global prosperity. By embracing these elements, countries can establish a strong economic framework, encourage social equity, and increase their influence in a rapidly changing global environment.

ACKNOWLEDGEMENT

This chapter acknowledges the use of AI-assisted tools, including Perplexity for research and fact-checking, Grammarly for writing enhancement and grammar correction and Zotero for efficient reference management. These tools were selected to improve workflow efficiency, accuracy, and clarity while maintaining rigorous academic standards. Their integration complemented traditional research and writing methods by providing reliable source verification, streamlined citation, and polished language, thereby enhancing the overall quality and transparency of the research process.

REFERENCES

Almasradi, R.B., Al-Otaibi, S.A., Akram, M.W., Farheen, N., Mahar, S. (2022). Mediating role of information and communication technology between e-HRM and organizational performance in Pakistan. *Pak. J. Commer. Soc. Sci., 16*(4), 660-681.

Ariansyah, K., Hermawati, I., Wahab, R.A.A., Nugroho, A.C., Dirgahayu, D., Takariani, C.S.D., Nupikso, D. (2025). The role of mobile broadband in poverty alleviation: a comparison of the effects of 3G and 4G network expansion in underdeveloped regions of Indonesia. *Inf. Technol. Dev., 31*(1), 206-231. [http://dx.doi.org/10.1080/02681102.2024.2361482]

Arjang, A., Sutrisno, S., Permana, R.M., Kusumastuti, R., Ausat, A.M. A. (2023). Strategies for improving the competitiveness of MSMEs through the utilisation of information and communication technology. *Al-Buhuts, 19*(1), 462-478.

Ateljević, J., Kulović, D., Đoković, F., Bavčić, M. (2023). *Business strategy and competitive advantage: A reinterpretation of Michael Porter's work.*New York: Routledge. [http://dx.doi.org/10.4324/9781003359173]

Behl, R. (2021). Information technology for management. *Mcgraw Hill India.*

Bello, R-W. (2019). Essential ICT for informatics and management students. *Lap Lambert Academic Publishing.*

Berisha, B, Mëziu, E, Shabani, I (2022). Big data analytics in Cloud computing: an overview. *Journal of Cloud Computing, 11*, 24. [http://dx.doi.org/10.1186/s13677-022-00301-w]

Broccardo, L., Vola, P., Alshibani, S.M., Tiscini, R. (2024). Business processes management as a tool to enhance intellectual capital in the digitalization era: the new challenges to face. *J. Intellect. Cap., 25*(1), 60-91. [http://dx.doi.org/10.1108/JIC-04-2023-0070]

Choudhury, P., Foroughi, C., Larson, B. (2021). Work-from-anywhere: The productivity effects of geographic flexibility. *Strategic Management Journal, 42*(4), 655-683. [http://dx.doi.org/10.1002/smj.3251]

Christopher, M. (2022). *Logistics and supply chain management.*Pearson Uk.

Davenport, T. H., Ronanki, R. (2018). Artificial intelligence for the real world. *Harvard Business Review.* Available form: https://hbr.org/2018/01/artificial-intelligence-for-the-real-world

Deep, A. (2023). Role of ICT in business management: A conceptual review. *International Journal of Creative Research Thoughts, 8*(8), 749-756.

Eloff, M. (2018). Financial planning & analysis and performance management. *John Wiley & Sons.*

Evans, J.R., Lin, J. (2017). Business analytics: Methods, models, and decisions. *Pearson.*

Emma, L (2024). Data-driven decision-making and business intelligence in modern organizations. *ResearchGate.* Available from: https://www.researchgate.net/publication/386385015_Data-Driven_Decisio--Making_and_Business_Intelligence_in_Modern_Organizations

Ezeoha, A.E., Obi, A., Igwe, A., Ezeruigbo, C. (2020). The mobile phone revolution, the Internet and rural electricity: What are the implications for food security in Africa?. *Inf. Dev., 36*(4), 603-622. [http://dx.doi.org/10.1177/0266666919884991]

Fernando, E., Sutomo, R., Prabowo, Y.D., Gatc, J., Winanti, W. (2023). Exploring customer relationship management: Trends, challenges, and innovations. *J. Inf. Syst. Informatics, 5*(3), 984-1001. [http://dx.doi.org/10.51519/journalisi.v5i3.541]

Garrison, M. (2019). Enhancing team communication with ICT: A practical guide. *J. Bus. Commun., 56*(1), 30-45.

Igboeli, U.H., Bisallah, H.I. (2020). Information and communication technology in managing small and medium enterprises in Nigeria. *Open J. Manag. Sci., 1*(2), 1-11. [http://dx.doi.org/10.52417/ojms.v1i2.152]

Islam, M.S., Haque, M.Z., Islam, S.N., Hassan, A., Alam, M.M. (2024). Role of education human capital in ICT-trade relationship. *Int. J. Educ. Econ. Dev., 15*(3), 299-316. [http://dx.doi.org/10.1504/IJEED.2024.139300]

Jacobs, F.R., Chase, R.B. (2018). Operations and supply chain management. *McGraw-Hill Education.*

Jaganathan, M., Ahmad, S., Ishak, K.A., Mohd Nafi, S.N., Uthamaputhran, L. (2018). Determinants for ICT adoption and problems: Evidence from rural based small and medium enterprises in Malaysia. *Int. J. Entrep., 22*(4), 1-13.

Jere, J.N., Ngidi, N. (2020). A technology, organisation and environment framework analysis of information and communication technology adoption by small and medium enterprises in Pietermaritzburg. *S. Afr. J. Inf. Manag., 22*(1), 1-9.

Kotler, P., Keller, K.L. (2016). *Marketing management (15th ed.)* Pearson Education.

Kuloğlu, N. (2024). The impacts of technological transformation on labor and employment: The future of the business world and economic growth. *International Research in Social, Human and Administrative Sciences XVII,* Eğitim Yayınevi.21-42.

Kumar, A., Gupta, S. (2021). E-commerce performance: A research agenda. *J. Retailing Consum. Serv., 58*, 102199.

Kumar, A., Singh, P. (2019). The role of robotic process automation in digital transformation: A study of various industries. *Int. J. Comput. Appl., 178*(44), 1-7.

Kumar, S., Goel, U., Joshi, P., Johri, A. (2024). Factors affecting information & communication technology (ICT) adoption among MSMEs. *J. Open Innov. Techn. Mark. Complex., 10*(1), 100205. [http://dx.doi.org/10.1016/j.joitmc.2023.100205]

Laudon, K. C., & Traver, C. G. (2021). *E-commerce 2020-2021: Business, technology, society* (16th ed.).

Pearson Education.

Liu, H., Luo, Y., Geng, J., Yao, P. (2021). Research hotspots and frontiers of product R&D management under the background of the digital intelligence era—bibliometrics based on citespace and histcite. *Appl. Sci., 11*(15), 6759.

Mahmood, H.S., Abdulqader, D.M., Abdullah, R.M., Rasheed, H., Ismael, Z.N.R., Sami, T.M.G. (2024). Conducting in-depth analysis of AI, IoT, web technology, cloud computing, and enterprise systems integration for enhancing data security and governance to promote sustainable business practices. *J. Inf. Technol. Informatics, 3*(2), 297-332.

Manyika, J., Remes, J., Mischke, J., Krishnan, M. (2017). *The productivity puzzle: A closer look at the United States (Discussion Paper).*(Discussion Paper). McKinsey Global Institute.. Available form: https://www.mckinsey.com/~/media/mckinsey/featured%20insights/employment%20and%20growth/new%20insights%20into%20the%20slowdown%20in%20us%20productivity%20growth/mgi-the-productivi-y-puzzle-discussion-paper.ashx

Mintzberg, H., Ahlstrand, B., Lampel, J.B. (2020). *Strategy safari: The complete guide through the wilds of strategic management. Uk: Pearson.*

Nwanmuoh, E.E., Dibua, E.C., Onyegbuna, C.M., Friday, E.C. (2024). Information Communication Technology (ICT) and the Effective Growth of SMEs in Anambra State, Nigeria. *Int. J. Cap. Build. Educ. Manag., 6*(5), 19-30.

Pontoh, G, Indrijawati, A, Selvi, F, Ningsih, L, Putri, D (2024). A systematic literature review of ERP and RFID implementation in supply chain management. *WSB Journal of Business and Finance, 58*(1), 80-96.

Qureshi, S. (2019). Perspectives on development: why does studying information and communication technology for development (ICT4D) matter?. *Inf. Technol. Dev., 25*(3), 381-389. [http://dx.doi.org/10.1080/02681102.2019.1658478]

Rainer, R.K., Prince, B. (2021). Introduction to information systems. *John Wiley & Sons.*

Singh, J., Singh, S., Kumari, M. (2020). Role of ICT in supply chain management. *J. Interdisciplinary Cycle Res., 12*(10), 992-1007.

Song, S., Hwang, D.H.S., Park, S. (2025). What makes hospitality and tourism executives adopt information and communication technology: Antecedents and outcomes of ICT adoption. *Tour. Econ., 31*(2), 245-262. [http://dx.doi.org/10.1177/13548166241259519]

Srivastava, A.K. (2019). *Information technology and its application in Business.*Agra: Sahitya Bhawan Publications.

Szymczyk, A., El Emary, A. (2021). Advanced trends in ICT for innovative business management. *CRC Press.*
[http://dx.doi.org/10.1201/9781003028932]

Taleb, N., Mohamed, E.A. (2020). Cloud computing trends: A literature review. *Acad. J. Interdiscip. Stud., 9*(1), 91.
[http://dx.doi.org/10.36941/ajis-2020-0008]

Tapscott, D., Tapscott, A. (2016). *Blockchain revolution: How the technology behind bitcoin is changing money, business, and the world.* Penguin, New York.

Tuan, P.T., Le Thu, H.D., Hang, N.T.D. (2024). Influence of electronic word of mouth on consumers' intention to purchase airline e-tickets in the Mekong Delta. *Transp. Res. Procedia, 80*, 103-110. [http://dx.doi.org/10.1016/j.trpro.2024.09.014]

Talafha, T (2024). E-commerce payment systems: a review and comparison. *International Journal of Commerce and Management Research, 10*(5), 78-82. Available from: https://www.managejournal.com/assets/archives/2024/vol10issue5/10114.pdf

CHAPTER 6

From Theory to Practice: Integrating Sociology, Economics, and Political Science in Social Research

Tahir Hussain Ansari[1,*], **Aleena Alam Khan**[1] and **Bushra Saiyeda**[1]

[1] *Department of Humanities and Social Sciences, Integral University, Lucknow, Uttar Pradesh 226026, India*

Abstract: Many of the world's biggest problems—such as poverty, inequality, unemployment, climate change, and weak governance—are complex and deeply connected. These issues cannot be understood fully by using only one subject. This chapter explains why combining the strengths of sociology, economics, and political science through interdisciplinary research is important for understanding and solving real-world problems. Each of these subjects looks at society in a different way. Sociology helps us understand people, communities, culture, and social systems. Economics focuses on how resources are produced and shared, and how choices are made. Political science explains how power works, how decisions are made in governments, and how public policies are created. When we bring these fields together, we get a fuller and more realistic picture of the world. This chapter also discusses the challenges of doing interdisciplinary work—such as differences in methods, academic structures, and lack of training. It suggests practical solutions like encouraging team-based research, reforming education to support broad learning, and supporting flexible research systems. Interdisciplinary research helps connect theory with real-life issues and can lead to better ideas, stronger policies, and more useful knowledge for societies around the world.

Keywords: Economics, Interdisciplinarity, Political science, Social research, Sociology.

INTRODUCTION

In today's rapidly changing world, many of the challenges societies face are profoundly connected and difficult to solve. Problems such as poverty, social inequality, environmental crises, weak governance, and global health emergencies cannot be fully understood by focusing on a single field of study. These are not just economic problems or political problems—they are also social problems. They affect people's lives in different ways and are shaped by history, power,

* **Corresponding Author Tahir Hussain Ansari:** Department of Humanities and Social Sciences, Integral University, Lucknow, Uttar Pradesh 226026, India; E-mail: thansari@iul.ac.in

Tahmeena Khan, Manisha Singh, & Saman Raza (Eds.)

culture, and behaviour. As a result, they require knowledge from different academic disciplines (Repko & Szostak, 2020; Bolton, 2023).

Traditionally, the social sciences—especially sociology, economics, and political science—have developed separately. Each field has its way of thinking, its own tools and theories, and its own questions. This separation has helped scholars go deep into their subjects, but it has also created boundaries that can be limiting. Increasingly, researchers and policy-makers are realizing that many real-world problems happen at the intersection of society, economy, and politics. To understand and solve them, we need interdisciplinary research—that is, work that combines ideas and methods from multiple fields to create a more complete picture (Repko & Szostak, 2020; Nowotny *et al.*, 2001). As Gerring and Yesnowitz (2006) point out, the division of social science into separate disciplines has led to fragmented understandings of holistic problems. Similarly, Repko and Szostak define interdisciplinary research as the process of answering a question, solving a problem, or addressing a topic that is too broad or complex to be dealt with by one discipline alone. It involves drawing on different disciplines, integrating their insights, and creating a more unified understanding (Repko & Szostak, 2020; Gerring & Yesnowitz, 2006).

Each of the three main disciplines discussed in this chapter has unique strengths. Sociology focuses on how people interact, how communities are formed, and how social norms and institutions shape behaviour. It looks at inequality, culture, gender, identity, and the structure of society. Economics, on the other hand, studies how people and governments make choices about scarce resources. It uses models and data to understand production, consumption, markets, and policy outcomes. Political science looks at power, governance, institutions, public policy, and decision-making. It helps explain how governments function, how laws are made, and how political systems affect people's lives (Jacobs, 2014); (Shahidi Hamedani, *et al.*, 2024). For example, if we want to understand educational inequality, a sociologist might study how family background or cultural expectations shape students' success. An economist might analyze how income or costs affect school access or educational returns. A political scientist might examine how government policies, public funding, or election cycles affect educational services. Each of these views is valid—but when combined, they offer a richer and more complete explanation (Becher & Trowler, 2001). According to Frodeman, Klein, and Mitcham, interdisciplinarity is not a threat to academic traditions but a creative response to their limitations. It invites collaboration and dialogue and helps break down artificial boundaries that prevent deeper understanding (Frodeman *et al.*, 2010). The rise of global problems such as climate change, migration, and pandemics has further pushed the need for interdisciplinary work. The COVID-19 pandemic is a good example. It was not just a health crisis. It also affected employment, education, mental health, trust in government, and international relations. Understanding it required knowledge from many fields—public health, economics, sociology, psychology, and political science. As Greer *et al.* noted, countries that integrated social science perspectives into their public health strategies—especially by understanding community behaviour and political decision-making—were often more successful in managing the crisis (Greer *et al.*, 2020).

Despite the growing demand for interdisciplinary research, many challenges remain. Most universities and research institutions still organize knowledge along disciplinary lines. Students are trained to specialize in one field, and academic careers are often shaped by deep knowledge in a single area. Journals, funding agencies, and promotions still tend to value narrow expertise. As Jacobs observes, these structures make it hard for researchers—especially younger scholars—to cross disciplinary boundaries. This can reduce innovation and weaken the link between research and real-world problem-solving (Jacobs, 2014). Another challenge is methodological differences. Disciplines do not only ask different questions—they also use different methods. Economists often use quantitative models and statistical analysis. Sociologists might rely on interviews, observations, or textual analysis. Political scientists might use a mix of case studies, surveys, or institutional analysis. Combining these methods requires flexibility, trust, and a shared language. It also requires time, training, and resources, which are not always available. Tashakkori and Teddlie argue that mixed-methods research—where quantitative and qualitative tools are used together—can produce findings that are both broad and deep and are more useful for policy-makers (Tashakkori & Teddlie, 2010).

This chapter aims to explain how the integration of sociology, economics, and political science can help us better understand and address complex social problems. It does not try to replace the three disciplines or argue that one is better than the others. Instead, it shows how they can be combined in thoughtful and respectful ways. As Nicolescu suggests, real-world problems are often so complex that they require transdisciplinary thinking—an approach that goes beyond disciplines to connect knowledge, values, and action (Nicolescu, 2002).

The chapter begins by outlining the key contributions of each discipline. It then explains why and how integration improves our ability to understand the world. It also discusses the barriers to interdisciplinary work, such as institutional structures, teaching practices, and publication systems. While the focus is global, the lessons are relevant across many contexts—especially for researchers, students, and educators who want to make their work more meaningful and impactful. The chapter ends by offering practical suggestions for building stronger interdisciplinary practices in research and education. These include reforming university curricula, supporting collaborative research teams, encouraging new publication models, and designing policies that value diverse knowledge systems.

In short, this chapter makes the case that bringing together sociology, economics, and political science is not just an academic exercise—it is a necessary step toward producing knowledge that is useful, ethical, and grounded in real-world needs. As the world becomes more connected and uncertain, the ability to think across disciplines will be essential—not just for researchers, but for everyone who wants to understand and improve society.

DISCIPLINARY CONTRIBUTIONS TO SOCIAL RESEARCH

Understanding social issues from multiple angles begins with knowing what each discipline offers. Sociology, economics, and political science are three of the most widely used social sciences in the world. Each has developed unique tools, concepts, and theories to study society. Together, they help explain how individuals behave, how institutions function, and how policies affect people's lives (Repko & Szostak, 2020; Frodeman *et al.*, 2010).

Sociology: Society, Relationships, and Social Structures

Sociology is the study of human society, social behaviour, and the patterns that shape everyday life. Sociologists look at how people interact with each other and how their relationships are influenced by larger systems such as class, race, gender, religion, and institutions. Sociology helps us understand why social inequality exists, how cultures change, and how norms and values influence behaviour. Essential theories in sociology include Durkheim's concept of social solidarity, which explains how societies hold together, and Weber's idea of social action, which looks at how people give meaning to their behaviour (Qi *et al.*, 2024). Another influential thinker, Pierre Bourdieu, introduced the concepts of habitus, social capital, and cultural capital to explain how inequality is passed down through generations (Knott *et al.*, 2022). These ideas show that individual choices are often shaped by background, social environment, and power structures. Sociology also provides rich qualitative methods, such as ethnography, interviews, and case studies, which are useful for understanding people's experiences in depth. This kind of research helps us hear voices that are often ignored, especially those of marginalized groups (Manza, 2023).

Economics: Resources, Incentives, and Outcomes

Economics is the study of how individuals, firms, and governments use limited resources. It focuses on the production, consumption, and distribution of goods and services. Economics is often associated with numbers and models, but it also deals with important human questions: Who gets what? Why are some people rich and others poor? What happens when prices rise, or jobs are lost? Mainstream or neoclassical economics uses models to predict how people respond to incentives. However, critics argue that it assumes people are always rational and markets always work well (Benedikt *et al.*, 2025 & Tullock, 1962). In response, newer approaches like behavioural economics and institutional economics have become more popular.

Amartya Sen, for instance, moved beyond income to talk about human well-being in terms of capabilities—what people are actually able to do and be. He argued that development should not only be measured by GDP but by the freedom and dignity of individuals (Sen, 1999). Economics uses quantitative methods, including statistics, econometrics, and mathematical modelling. These methods help identify patterns and test policies, making economics useful for designing solutions to poverty, unemployment, and inflation (Mitchell, 2009).

Political Science: Power, Governance, and Policy

Political science explores how power is used, how decisions are made, and how institutions work. It studies governments, elections, policies, international relations, and political behaviour. Political scientists ask questions like: Who holds power? How do laws get made? How do citizens influence politics? Classic theories include Easton's systems theory, which sees politics as a process of input (demands), conversion (decisions), and output (policies) (Ansell *et al.*, 2023). Institutionalism looks at how formal and informal rules shape behaviour, while public choice theory applies economic ideas to understand how politicians and officials make decisions (Tashakkori & Teddlie, 2010).

Political science is especially useful for studying accountability, corruption, democracy, and public trust. Like the other disciplines, political science uses both qualitative and quantitative methods. It often includes case studies, surveys, and policy analysis. These tools help scholars and policymakers evaluate the success or failure of political systems and governance structures (Nicolescu, 2002).

Bringing the Disciplines Together

While these three disciplines are different, they often deal with overlapping problems. For example, a social protection program might be evaluated by:

a. Sociologists for its effects on community norms and social inclusion.
b. Economists for its efficiency and impact on income.

c. Political scientists study how it is governed and whether it improves trust in institutions.

Despite different theories and tools, these disciplines ask common questions about justice, inequality, power, and well-being. Scholars such as Mitchell argue that many real-world problems go beyond the "analytical grasp" of a single field, which is why interdisciplinary approaches are essential.

According to Choi and Pak, combining disciplines can improve both the quality and usefulness of research. They explain that each discipline brings a specific angle, and by combining these, researchers can better understand the complexity of social issues (Allen *et al.*, 2021). Similarly, Nowotny, Scott, and Gibbons describe interdisciplinary research as a way to produce "socially robust knowledge"—knowledge that is not only true but relevant and applicable to real life (Nowotny *et al.*, 2001).

The Case for Integration

In recent years, there has been a growing interest in interdisciplinary research, especially in the social sciences. Many researchers now recognize that single-discipline approaches are often not enough to understand complex social problems. While sociology, economics, and political science each offer important insights, combining their ideas and methods allows for a more complete and realistic view of the world (Repko & Szostak, 2020; Frodeman *et al.*, 2010).

Why Integration Matters

Social issues are rarely one-dimensional. For example, inequality is not just about income (a topic in economics), but also about access to education, social identity, political representation, and discrimination. Climate change is not only an environmental problem—it also affects jobs, migration, and political decisions. In such cases, using only one discipline limits our understanding. As Repko and Szostak explain, interdisciplinary research helps us see "the whole picture" by connecting ideas across fields (Repko & Szostak, 2020). It allows scholars to ask broader questions, challenge assumptions, and explore new connections between people, systems, and institutions.

Nicolescu refers to this as a "transdisciplinary mindset," which accepts that real life is messy and cannot be explained by one kind of knowledge alone (Nicolescu, 2002). This kind of integration is especially useful when studying "wicked problems"—issues that are complex, hard to define, and have no easy solutions (Gerring & Yesnowitz, 2006). Examples include urban poverty, gender inequality, public health crises, and migration. These problems are deeply connected to values, politics, and social norms. According to Rittel and Webber, such problems cannot be solved using technical tools alone (Rittel & Webber, 1973). Interdisciplinary approaches are better suited for such tasks because they include multiple perspectives and recognize the importance of context (Frodeman *et al.*, 2010).

What Integration Offers?

Combining sociology, economics, and political science has many advantages. Each discipline asks different questions, uses different methods, and focuses on different levels of analysis. When used together, they offer a more balanced and multi-layered view of a problem. Sociology helps us understand the lived experiences of people. It brings attention to culture, identity, and relationships. Economics provides models to study behaviour, costs, incentives, and outcomes. It focuses on efficiency, equity, and choice. Political science explains how rules, power, and decisions shape public life. It studies institutions, policy-making, and governance (Jacobs, 2014; Rittel & Webber, 1973; March & Olsen, 1984). These different viewpoints can be complementary rather than contradictory. As Frodeman, Klein, and Mitcham argue, interdisciplinarity does not weaken academic standards—it strengthens them by encouraging dialogue and innovation (Frodeman *et al.*, 2010). It allows researchers to build theories and models that are more realistic and useful for solving social problems. Integration also helps in policy-making. Governments and organizations often deal with issues that cut across sectors. A good policy needs to be socially acceptable, economically feasible, and politically possible. Using insights from all three disciplines helps design better programs and policies that match the real needs of people and institutions (Repko & Szostak, 2020; Choi & Pak, 2006).

Common Challenges to Integration

While integration has many benefits, it is not easy to achieve. One major challenge is the difference in language and methods between disciplines. For example, economists may rely on numbers and models, while sociologists use stories and interviews. Political scientists might prefer comparative studies or institutional analysis. Working together means learning to understand and respect other ways of thinking (Jacobs, 2014; Mitchell, 2009).

Another problem is disciplinary boundaries in universities and research institutions. Departments are often separated, and students are trained in just one field. Research projects, journals, and funding programs are usually organized by discipline. As Becher and Trowler describe, academic disciplines act like "tribes" with their own cultures, languages, and values (Becher & Trowler, 2001). This makes it hard for scholars to cross over or collaborate freely. Evaluation systems can also create problems. Scholars who try interdisciplinary work may find it harder to publish, get funding, or receive recognition. According to Jacobs, current academic reward systems favour specialization, which discourages younger researchers from taking risks (Jacobs, 2014).

Overcoming the Barriers

Despite these challenges, integration is possible and becoming more common. Scholars are finding ways to combine theories and methods by working in teams, attending cross-disciplinary workshops, or creating joint degree programs. Journals and funding agencies are slowly beginning to support interdisciplinary work, especially for research that deals with major global challenges (Tashakkori & Teddlie, 2010). As Choi and Pak argue, successful interdisciplinary work depends on building trust, understanding different worldviews, and focusing on common goals [8]. Interdisciplinary research is not about merging disciplines into one but about using their strengths together in thoughtful and creative ways. This kind of collaboration helps create knowledge that is not only more robust but also more socially relevant.

METHODOLOGICAL APPROACH

This chapter is based on a conceptual and exploratory approach, rather than field-based research. Its main goal is to show how sociology, economics, and political science can be brought together in social research. Instead of testing a specific theory or analyzing new data, it focuses on combining ideas from different disciplines to better understand complex social issues (Repko & Szostak, 2020; Frodeman *et al.*, 2010).

The discussion draws on secondary sources such as published academic work, policy documents, textbooks, and institutional reports. These sources help identify the key contributions of each discipline and highlight both the common ground and differences in their approaches. This method allows for critical reflection on how disciplinary knowledge is generated, communicated, and applied in diverse contexts (Frodeman *et al.*, 2010; Nicolescu, 2002).

A thematic analysis is used to organize insights around recurring themes—for example, disciplinary boundaries, methodological contrasts, and institutional constraints to collaboration. This approach is particularly useful for mapping intersections between fields and highlighting opportunities for integration (Lattuca, 2001).

Since this is a globally oriented chapter, the discussion does not focus on one country or region. Instead, it identifies general trends in interdisciplinary research and emphasizes how institutional structures and academic cultures influence collaboration across disciplines. These insights are relevant to educators, researchers, and policy-makers in different social and academic settings (Nicolescu, 2002).

The chapter adopts an interpretive framework that emphasizes meaning-making, context, and the blending of different epistemologies. Interpretive methods are especially useful in interdisciplinary research because they allow scholars to navigate the diverse assumptions, goals, and values embedded in different fields. As Tashakkori and Teddlie explain, combining qualitative and quantitative approaches through mixed-methods research enhances the relevance and depth of inquiry, especially when dealing with social complexity (Tashakkori & Teddlie, 2010).

Such a methodological stance requires flexibility, critical thinking, and openness to alternative ways of knowing. By acknowledging these epistemological differences, the chapter promotes a more inclusive and reflective approach to knowledge production—one that is sensitive to the challenges and potential of interdisciplinary research (Nicolescu, 2002; Sen, 1999).

Institutional and Pedagogical Challenges

While interdisciplinary research is becoming more popular, it still faces many barriers—especially within academic institutions. Most universities and research systems are built around separate subjects or departments. Students are trained in a single discipline, researchers are expected to specialize, and funding and publishing often favour narrow topics. These patterns make it hard for people to work across fields, even when they want to (Repko & Szostak, 2020 ; Frodeman *et al.*, 2010). This section explores the main institutional and educational challenges that limit interdisciplinary work in the social sciences.

- **Disciplinary Silos in Universities**

One of the biggest obstacles is the way universities are structured. Academic departments are usually organized by disciplines—such as sociology, economics, or political science—and have their own staff, budgets, and teaching programs. This system encourages depth, which is important, but it often discourages breadth or collaboration. As Becher and Trowler describe, disciplines are like academic "tribes." Each tribe has its own language, traditions, and rules (Nowotny *et al.*, 2001). For example, the kind of questions asked in economics might not make sense to a sociologist, and vice versa. This separation creates boundaries that are hard to cross, especially for students and early-career researchers. Because of this, many students finish their degrees without ever engaging with ideas from other fields. Even when they are curious, they may not have access to courses or mentors outside their department (Repko & Szostak, 2020 ; Lattuca, 2001).

- **Limited Interdisciplinary Training**

Most educational systems do not provide adequate interdisciplinary training. Students are usually taught the theories and methods of a single subject. They may not learn how to combine knowledge from different areas or how to work on complex problems that require multiple perspectives. Even interdisciplinary programs often fall short. In many cases, they simply offer a few courses from different departments, without helping students make connections between them (Jacobs, 2014).nAccording to Lattuca, true interdisciplinary learning requires integration—not just variety. Students must be taught how to compare methods, reflect on assumptions, and understand how knowledge is produced differently in each field (Lattuca, 2001). Teachers also face challenges. They may not have training in other disciplines or may worry that stepping outside their field will hurt their career. Without institutional support, it is difficult to design and deliver effective interdisciplinary courses (Nowotny *et al.*, 2001).

- **Conflicts in Research Methods and Approaches**

Different disciplines have different ways of doing research. Economists often rely on quantitative models and large datasets. Sociologists may use qualitative methods like interviews, ethnography, or textual analysis. Political scientists use a mix—some focus on institutions and laws, while others study public opinion or political behaviour. These differences are not just about tools—they reflect deeper ideas about what counts as knowledge. Economists may focus on patterns and averages, while sociologists look at meaning and experience. Political scientists often ask "who benefits" and "how decisions are made" (March & Olsen, 1984).

Bringing these approaches together takes careful planning and mutual respect. As Choi and Pak point out, successful interdisciplinary research requires researchers to be open-minded, flexible, and willing to learn from each other (Choi & Pak, 2006). Without this, projects can become unbalanced or unclear.

- **Funding and Publication Barriers**

Many funding systems are built for single-discipline research. Research grants are often reviewed by experts from one field, who may not understand or value interdisciplinary proposals. This makes it harder for scholars to get support for innovative work that crosses boundaries. Publishing is also difficult. Most high-ranking journals are discipline-specific and expect papers to follow certain norms. Interdisciplinary research may not fit these expectations or may be seen as less rigorous (Nowotny *et al.*, 2001; March & Olsen, 1984). As a result, scholars—especially early-career researchers—may avoid interdisciplinary work to protect their careers. Klein warned about these problems many years ago, and while things are slowly changing, many of the same issues still exist today (March & Olsen, 1984). Without clear pathways for success, interdisciplinary research can feel risky and unrewarded.

- **Evaluation and Career Progression**

Academic careers are shaped by evaluations—such as tenure, promotion, and hiring decisions. These often depend on discipline-specific achievements, like publishing in certain journals or getting grants in a specific field. Interdisciplinary researchers may not fit these expectations, which can hurt their chances of success. As Jacobs notes, institutions need to rethink how they measure academic work. Interdisciplinary contributions—like collaborative research, cross-disciplinary teaching, or public engagement—should be recognized and rewarded (Jacobs, 2014). Otherwise, scholars may feel forced to stay within their disciplinary "safe zones."

In summary, the institutional and pedagogical challenges facing interdisciplinary research are real but not impossible to overcome. Universities, funding bodies, and academic journals must update their systems to support collaboration and innovation. Students should be given opportunities to explore multiple fields, and teachers should be supported in creating new forms of learning. Interdisciplinary work takes more time, effort, and flexibility—but the rewards are greater. By breaking down barriers between disciplines, we can produce knowledge that is more realistic, more useful, and more connected to the real problems people face around the world.

Relevance and the Road Ahead

The integration of sociology, economics, and political science has gained increasing attention as researchers seek more comprehensive approaches to studying complex social issues. This section reflects on the current relevance of interdisciplinary thinking in social research and outlines the steps needed to advance this approach in academic and policy contexts.

In today's interconnected world, social problems are often multifaceted and deeply intertwined. For instance, inequality is not only an economic concern about income and wealth, but also a social issue involving access to education and healthcare, and a political issue related to representation and participation.

Similarly, climate change, health crises, and forced migration cut across multiple domains, requiring insights that span traditional disciplinary boundaries. As Nowotny, Scott, and Gibbons argue, there is a growing need for "socially robust knowledge"—research that is both academically sound and practically useful in real-world contexts (Nowotny *et al.*, 2001). Interdisciplinary approaches are uniquely positioned to produce such knowledge by combining analytical strengths across fields (Frodeman *et al.*, 2010). Interdisciplinary research enhances the relevance and impact of academic inquiry by allowing for broader problem framing, more inclusive analysis, and more practical outcomes. Repko and Szostak emphasize that interdisciplinary thinking is particularly suited to problems that involve uncertainty, competing values, or institutional complexity—conditions that characterize many of the world's most pressing challenges (Repko & Szostak, 2020). When researchers integrate theories, methods, and perspectives from sociology, economics, and political science, they can better understand both the causes and consequences of such challenges, leading to more effective and just solutions.

The value of integration is especially clear when examining how it changes the way research is conducted. First, it broadens the questions researchers ask, moving beyond narrow definitions of success or failure. Second, it encourages a mix of quantitative and qualitative data, capturing both patterns and lived experiences. Third, it supports analysis that connects individual behaviour to structural systems, and policy decisions to social outcomes (Tashakkori & Teddlie, 2010). This layered understanding is essential for designing interventions that are not only technically efficient but also socially acceptable and politically feasible.

A clear example of interdisciplinary thinking in action can be seen in urban development research. Economists might study housing markets and affordability; sociologists examine displacement, community identity, and inequality; while political scientists investigate zoning laws, governance, and participation. Studying these factors in isolation leads to partial understanding. In contrast, an integrated approach captures the social, economic, and political dynamics of urban life, enabling better planning and policymaking (Jacobs, 2014).

To make interdisciplinary thinking more common, it is important to start with education and training. Universities play a central role in preparing students to work across disciplines. This includes offering integrated curricula, supporting team teaching and collaborative assignments, and encouraging students to reflect on different worldviews. As Lattuca explains, interdisciplinary education should not merely present multiple subjects side by side but should actively help students connect ideas, challenge assumptions, and build new frameworks (Lattuca, 2001).

Graduates of such programs are better prepared to address complex problems in policy, business, civil society, and beyond.

At the same time, academic institutions must create a more supportive environment for interdisciplinary research. This means reforming funding structures to allow cross-departmental projects, recognizing collaborative work in academic evaluations, and promoting interdisciplinary journals and publication platforms. Jacobs notes that the current academic reward system often penalizes interdisciplinary researchers by favouring specialization and individual achievement (Jacobs, 2014). Changing this system is essential if scholars are to feel free to pursue integrative research without risking their careers.

Partnerships beyond the university also matter. Collaborations between researchers, governments, NGOs, and local communities can make research more grounded, inclusive, and relevant. For example, including community members in the design of research projects or inviting policy-makers to engage with academic findings can increase trust, improve data quality, and enhance the impact of the work (Choi & Pak, 2006). Interdisciplinary research is well-suited for such partnerships because it is inherently flexible, open, and responsive to diverse forms of knowledge (Nicolescu, 2002).

In conclusion, interdisciplinary research is more than a methodological preference—it is a necessary response to the challenges of the modern world. By combining the insights of sociology, economics, and political science, researchers can better understand and respond to complex issues that affect people's lives. To make this work more effective, changes are needed in how students are trained, how research is supported, and how academic contributions are evaluated. With stronger institutional support and a shared commitment to collaboration, the road ahead for interdisciplinary research is full of promise.

CONCLUDING REMARKS

This chapter has highlighted the growing importance of interdisciplinary research in understanding and responding to complex social problems. The integration of sociology, economics, and political science offers a more comprehensive framework for analyzing issues such as inequality, development, governance, and global crises. Each discipline brings unique tools and perspectives—sociology focuses on social behaviour and structure, economics on resource use and decision-making, and political science on power and institutions. When combined, these perspectives help researchers see the complete picture of how society functions and how different forces interact (Repko & Szostak, 2020; Mitchell, 2009; Frodeman *et al.*, 2010).

Interdisciplinary research is particularly valuable in today's world, where social, economic, and political challenges are increasingly interconnected. Problems like poverty, climate change, migration, and health inequities do not follow disciplinary boundaries. Addressing these issues effectively requires a broad and integrated approach. Interdisciplinary thinking allows researchers to connect micro-level experiences with macro-level structures, and to develop solutions that are not only technically sound but also socially and politically informed (Nowotny *et al.*, 2001; Rittel & Webber, 1973).

Despite its advantages, interdisciplinary research continues to face significant institutional and pedagogical challenges. Most universities are still organized around traditional disciplines, and students are often trained within a narrow academic framework. This structure limits exposure to alternative methods and perspectives. Furthermore, academic career paths, funding systems, and publication norms tend to reward specialization over collaboration, making it difficult for scholars—especially early-career researchers—to pursue interdisciplinary work (Jacobs, 2014; Becher & Trowler, 2001; Lattuca, 2001).

To address these challenges, several practical steps are needed. First, academic institutions should reform their curricula to promote interdisciplinary learning. This includes designing courses that cut across disciplinary boundaries, encouraging joint supervision of student research, and fostering critical thinking about different ways of knowing (Lattuca, 2001). Second, universities and research centres should provide structural support for interdisciplinary projects through collaborative spaces, flexible administrative arrangements, and shared research agendas (Becher & Trowler, 2001).

Third, evaluation systems must be updated to recognize the value of interdisciplinary contributions. Academic hiring, promotion, and funding decisions should take into account co-authored publications, interdisciplinary outputs, and research that engages with multiple audiences (Jacobs, 2014).

Funding agencies also have a crucial role to play. Dedicated grant programs for interdisciplinary research should be established, and review panels should include members from diverse backgrounds who understand the nature and value of integrated approaches. Similarly, academic journals and conferences should actively welcome interdisciplinary submissions and encourage dialogue across fields (March & Olsen, 1984). Open-access and digital platforms can also help increase the visibility and impact of interdisciplinary work (Choi & Pak, 2006).

In conclusion, the integration of sociology, economics, and political science is essential for producing research that is both academically rigorous and socially meaningful. While challenges remain, the benefits of interdisciplinary research, in terms of deeper understanding, better policy-making, and greater public relevance, are clear. By reforming educational practices, institutional structures, and evaluation systems, the social sciences can move toward a more connected and collaborative future.

REFERENCES

Allen, K. A., Kern, M. L., Rozek, C. S., McInerney, D. M., Slavich, G. M. (2021). Belonging: A review of conceptual issues, an integrative framework, and directions for future research. *Australian Journal of Psychology, 73*(1), 87-102.

Ansell, C., Sørensen, E., Torfing, J. (2023). Public administration and politics meet turbulence: The search for robust governance responses. *Public Administration, 101*(1), 3-22.

Benedikt, M., Zimmermann, F., Auchmann, B., Bartmann, W., Burnet, J. P., Carli, C., Belli, E. (2025). Future Circular Collider Feasibility Study Report: Volume 1 Physics, Experiments, Detectors. *The European Physical Journal C, 85*(12), 1468.

Bolton, D. (2023). A revitalized biopsychosocial model: core theory, research paradigms, and clinical implications. *Psychological Medicine, 53*(16), 7504-7511.

Gerring, J., Yesnowitz, J. (2006). A normative turn in political science?. *Polity, 38*(1), 101-133.
[http://dx.doi.org/10.1057/palgrave.polity.2300054]

Greer, S.L., King, E.J., da Fonseca, E.M., Peralta-Santos, A. (2020). The comparative politics of COVID-19: The need to understand government responses. *Glob. Public Health, 15*(9), 1413-1416.
[http://dx.doi.org/10.1080/17441692.2020.1783340] [PMID: 32564670]

Jacobs, A.J. (2014). *In defense of disciplines: Interdisciplinarity and specialization in the research university.*University of Chicago Press.
[http://dx.doi.org/10.7208/chicago/9780226069463.001.0001]

Klein, J.T. (1990). *Interdisciplinarity: History, theory, and practice*Wayne State University Press.

Knott, E., Rao, A. H., Summers, K., Teeger, C. (2022). Interviews in the social sciences. *Nature Reviews Methods Primers, 2*(1), 73.

Lattuca, L.R. (2001). *Creating interdisciplinarity: Interdisciplinary research and teaching among college and university faculty.*Vanderbilt University Press.
[http://dx.doi.org/10.2307/j.ctv167563f]

Manza, J. (2023). Sociology for Beginners. *Annual Review of Sociology, 49*(1), 241-261.

March, J.G., Olsen, J.P. (1983). The new institutionalism: Organizational factors in political life. *Am. Polit. Sci. Rev., 78*(3), 734-749.
[http://dx.doi.org/10.2307/1961840]

Mitchell, T. (2009). How neoliberalism makes its world: The urban property rights project in Peru. *The Urban Social Landscape: Historical and Comparative Perspectives.*
[http://dx.doi.org/10.4159/9780674054264-012]

Nicolescu, B. (2002). *Manifesto of transdisciplinarity.* SUNY Press.

Nowotny, H., Scott, P., Gibbons, M. (2001). *Re-thinking science: Knowledge and the public in an age of uncertainty.*Cambridge, UK: Polity Press.

Qi, J., Mazumdar, S., Vasconcelos, A. C. (2024). Understanding the relationship between urban public space and social cohesion: A systematic review. *International Journal of Community Well-Being, 7*(2), 155-212.

Repko, A.F., Szostak, R. (2020). *Interdisciplinary research: Process and theory.*Sage Publications.

Rittel, H.W.J., Webber, M.M. (1973). Dilemmas in a general theory of planning. *Policy Sci., 4*(2), 155-169.

[http://dx.doi.org/10.1007/BF01405730]

Sen, A. (1999). *Development as freedom.* Oxford University Press.

Shahidi Hamedani, S., Aslam, S., Mundher Oraibi, B. A., Wah, Y. B., Shahidi Hamedani, S. (2024). Transitioning towards tomorrow's workforce: Education 5.0 in the landscape of society 5.0: A systematic literature review. *Education Sciences, 14*(10), 1041.

Tashakkori, A., Teddlie, C. (2010). *SAGE handbook of mixed methods in social & behavioral research.* SAGE Publications.
[http://dx.doi.org/10.4135/9781506335193]

CHAPTER 7

Revisiting the Power of Language: An Interdisciplinary Inquiry into Foreign Language Anxiety, Cultural Sensitivity, and Identity in English Studies

Prashant Kumar Astalin[1], **Leena Rajak**[2] and **Sangeeta Chauhan**[2,*]

[1] *School of Education, Uttar Pradesh Rajarshi Tandon Open University, Prayagraj, Uttar Pradesh 211021, India*

[2] *Department of Education, Babasaheb Bhimrao Ambedkar University, Lucknow, Uttar Pradesh 226025, India*

Abstract: Language is a transformative power that not just facilitates communication but also constructs knowledge and identities, and mediates social interaction. English studies is a multidisciplinary discipline that investigates the complex role of language in human experience by drawing on ideas from linguistics, literature education, and psychology. But when we talk about an interdisciplinary framework, this chapter addresses the power of language, emphasizing its social, cognitive, and educational aspects. As stated by one of the renowned researchers Horwitz that 'Foreign language anxiety' (FLA) is one of the most serious hindrances to effective contemporary language instruction, an affective barrier which lessens the power of learners to communicate and cope with the host culture. Even under the umbrella of the paradigm of English as the dominant world language, FLA restricts learners' self-image, interest, and achievement. It speaks about interactions between language learning anxiety, cultural awareness, and identity construction in English-dominant environments from applied linguistics and educational psychology. Interdisciplinary inquiry positions addressing FLA as a multilateral approach targeting instruction methods, sociocultural impacts, and affective and cognitive factors. Thus, through syntheses of other fields' viewpoints, the current scenario calls for an open education paradigm that encourages intercultural awareness and language skills. The findings presumably emphasize the importance of providing students with secure learning spaces that eliminate fear and enable them to utilize English as a tool for culture and communication. Thereby, it adds to the general debate surrounding English studies *via* the contention that an interdisciplinary approach improves our understanding of language learning difficulties, especially among non-native speakers. It also adds in advancing learning development that is sensitive to psychological well-being and linguistic differences, ultimately revisiting the power of language in shaping global learning.

[*] **Corresponding author Sangeeta Chauhan:** Department of Education, Babasaheb Bhimrao Ambedkar University, Lucknow, Uttar Pradesh 226025, India; Email: c.sangeeta27@gmail.com

Tahmeena Khan, Manisha Singh, & Saman Raza (Eds.)

Keywords: Cognitive factors, Cultural awareness, Cultural sensitivity, English studies , Foreign language anxiety, Identity construction, Interdisciplinary, Language, Language proficiency, Linguistics, Monodisciplinary, Multilateral approach.

INTRODUCTION

Language as Transformative Power

Language is a mode of transfer through which an individual commutes their feelings, insights. It is a fundamental force that shapes thought, mediates human experience, and constructs individual and collective identities. It enables us to interact, but with this, it also allows us to make sense of our world and make a place in it. By birth, language becomes the very primary tool that we acquire from our surroundings, as we learn, grow, express, and negotiate our existence. Language serves as a foundation upon which societies stand firm and sustain a living system that carries with it the culture, values, and worldview of its speakers. The vital transformative power of a language always lies in its ability to influence not just the external communication, but also relies on the internal cognition and the realities of the society. As Vygotsky & Cole (1978) asserted in their sociocultural theory, a lingo serves as a semiotic tool through which a certain individual internalizes knowledge. Thought and language, in their framework, are critically intertwined; as cognitive development always occurs on the cognitive plane and later on moves to the individual plane through linguistic mediation. This view marks a departure from the complex traditional views that regarded language as a static code or mechanical structure. If we get into the historical context, structuralist linguistics, exemplified by Ferdinand de Saussure, approached language as a system that is closed off from signs governed by the internal rules (Saussure, 2011). In this paradigm, language was analyzed in terms of its formal properties- phonology, syntax, and morphology – without attending to meaning as it emerges in social context. Structuralism emphasized langue (the overarching structure) over parole (individual speech acts), and in doing so, reduced language to a system of oppositions and relations. While this approach offered a substantial insight into linguistic form, it largely overlooked the dynamic, situated, and identity- forming nature of language in use.

The limitations of structuralism gave rise to what is now termed the ‘sociocognitive’ turn in linguistic and language studies. The interdisciplinary shift began to integrate insights from different disciplines- cognitive psychology, sociology, anthropology, and discourse studies. Language should be seen as a function, as a system of meaning-making shaped by the social purposes (Soylu *et al.*, 2023). Systemic Functional Linguistics (SFL) pressures that language serves the three key functions: ideational (representing experience), interpersonal (enacting social relations), and finally, textual (organizing discourse). This marked the vital reorientation toward understanding language in context and as a resource for construing reality. Parallelly, the field of critical applied linguistics, which was led by eminent scholars like Pennycook (2001), questioned the neutrality of language and emphasized its embeddedness in its power, ideology, and identity politics. Language was not only seen as a mirror to reality but as a tool for constructing one. From this certain perspective, language view itself became a space of negotiation- where learners struggle with dominant discourses, identity reformation, and intercultural awareness. Moving along, sociolinguists highlighted how language practices are socially regulated and ideologically charged. Linguistic norms are not neutral; they reflect societal power dynamics, gender ideologies, and cultural scripts. In classrooms, for instance, the dominance of 'native speaker norms' reinforces linguistic hierarchies and marginalizes diverse linguistic identities. In the recent past decades, poststructuralist and sociocultural theorists have focused increasingly on language's role in identity formation. Norton, (2013) introduced the very concept of 'investment' in language learning, arguing that learners position themselves in relation to others, negotiate social roles, and resist or reproduce cultural norms. Thus, the movement from structuralist to socio-cognitive perspectives of language represents a profound thinking of its purpose and power. Language is not merely a container for thoughts but an active agent in shaping them. It helps us mediate our interactions with others; it also successfully influences how we conceptualize our world and molds how we see ourselves. This conceptual shift has enormous implications for English studies, particularly in contexts of second language acquisition, intercultural communication, and educational policy.

Hence, to understand the language in its social, cognitive, and affective dimensions is crucial, especially in a globalized world where English play both as a bridge and a barrier, as (Kramsch, 1993) asserted, the teaching of language must go beyond just teaching and learning the grammar and vocab to engage learners in their cultural as well as the emotional nuances of discourse. It is through this lens that this particular chapter seeks to explore the very complex interactions between language anxiety, cultural sensitivity, and identity, ultimately advocating for pedagogies that are responsive to the psychological and sociocultural realities of learners.

The Rationale for an Interdisciplinary Framework

Language has been the bedrock of human civilization- not only as a tool for simple to complex communication, but it serves as a mechanism that also shapes knowledge, mediates relationships, and constructs personal and cultural identities.

In recent decades, scholarly inquiry into language has transcended disciplinary silos, recognizing the rich intersections among linguistics, psychology, education, and literary studies. The very subject of English studies has developed to accommodate an interdisciplinary ethos responsive to the complexity of modern language learning and usage, making this particular convergence especially clear. The evolving dynamics of transnational identities, multilingualism, and global communication underpin the need for an interdisciplinary framework. English serves as a site of power, ideology, and identity negotiation in addition to being a language for communication (Pennycook, 2001). It is now frequently referred to as a worldwide lingua France. A framework that goes beyond the structuralist emphasis of traditional language instruction to include the emotive, cognitive, and social aspects of learning is necessary in such a setting in order to comprehend the specific difficulties encountered by non- native speakers, particularly in educational settings.

Linguistics and Beyond: Language as a Multidimensional Construct

The ground of linguistics has contributed significantly to our understanding of language structures, acquisition patterns, and usage. However, when one is isolated from other disciplines, its ability to fully account for language's role in shaping identity and experience is limited. As Widdowson (2003) contends, language cannot be studied merely as a system of signs or rules; it must be seen as a socially situated practice embedded in context. Applied linguistics, in particular, has emerged as a bridge between theoretical linguistics and practical concerns, incorporating insights from education, psychology, and anthropology to understand how language functions in real-life situations (Elshahwy *et al.*, 2023).

This disciplinary expansion is not merely academic but essential. Foreign Language Anxiety (FLA), for example, is a psychological phenomenon that cannot be understood through linguistic analysis alone. Introduced by Horwitz *et al.*, (1986), FLA is described as 'a distinct complex of self- perceptions, beliefs, feelings, and behaviours related to classroom language learning arising from the uniqueness of the language learning arising from the uniqueness of language learning process' (Horwitz *et al.*, 1986). It sheds light on how language learning is deeply personal and emotionally charged, often entangled with learners' self-esteem, identity, and social positioning.

The Role of Psychology and Educational Theory

Educational psychology, particularly in its focus on motivation, emotion, and also on cognitive development, provides crucial insights into why some learners thrive while others struggle in language classrooms. There are theories, such as Krashen's (1982) Affective Filter Hypothesis (AFH), which proposes that emotional variables, like anxiety, motivation, and confidence, can either facilitate or hinder language acquisition. In Krashen's model, a high affective filter (resulting from high anxiety or low motivation) acts as a barrier to input processing, thus preventing successful language acquisition. These particular psychological factors cannot be addressed through linguistic curricula alone. They demand pedagogical approaches that only consider the whole learner, their backgrounds, aspirations, insecurities, and also the cultural orientations. At the same time, integrating these vital elements into the practice in a classroom. It also requires collaboration across several disciplines.

Education and the Ethics of Teaching Language

The educational implications of interdisciplinary research are very intense. Teaching language, especially while teaching English as a foreign or second language, is not a neutral act. It is rather a political and ethical endeavour shaped by several issues of identity, power, and representation (Saiyara, 2022). Educators who discover the cultural and emotional risks involved in language acquisition can foster more inclusive and effective learning environments. The dialogic pedagogy, for example, as a concept introduced by Freire (2018), puts emphasis on the co-construction of knowledge through dialogue, respect, and shared inquiry. This approach challenges the banking model of education and instead positions learners as active agents in their linguistic and cultural development.

Literature and Identity: English Studies in a Globalized World

English studies, with its roots deep down in literary criticism, has rapidly grasped the interdisciplinary methodologies that engage with identity politics, postcolonial theory, and cultural studies. Literature offers us a very unique window into the lived experiences of language learners and the sociopolitical contexts of English use. When literature is intermingled into language teaching, it humanizes the process and offers narratives through which learners can explore complex emotional and cultural terrains (Kramsch, 1993), aptly emphasizing the significance of context and culture in language teaching, arguing that language cannot be separated from the social context in which it is embedded. This work urges the teachers to incorporate authentic texts, cultural artefacts, and dialogic methods to build intercultural competence.

The Need for Integration

The ever-evolving demands of globalization, migration, and multicultural communication have created a pressing need for educational frameworks that are responsive, flexible, and inclusive. A narrowly defined discipline can no longer suffice in addressing the complexities of English language learning in the 21St

century. What is needed is an integrated approach – one that synthesizes insights from linguistics, psychology, education, and also literature to develop a holistic understanding and environment for the language learners and their world. Overall, an interdisciplinary approach should not only be seen as an academic preference but as a necessary epistemology for addressing the emotional, social, and cognitive realities of English language learners. It calls for a pedagogical shift towards greater empathy, intercultural sensitivity, and learner-centeredness, which is essential in evolving both effective language acquisition and honouring the identities and experiences of diverse student populations.

Theoretical Foundations

Language has transcended from being far more than a system of sounds, symbols, and rules. It is a medium through which individuals communicate, construct knowledge, negotiate social identities, and mediate intercultural experiences. English studies, emerging at the intersection of linguistics, literature, education, and psychology, offers a multidisciplinary lens for us to explore the very transformative power of language in shaping human experiences. In recent academic discourse, researchers have rapidly discovered the significance of addressing language learning not just as an isolated cognitive exercise but as a deeply social, emotional, and cultural process (Kramsch, 1993). Particularly, the phenomenon of Foreign Language Anxiety (FLA) (Horwitz *et al.*, 1986) has drawn immense attention towards the affective barriers that can significantly impair learners' abilities to communicate and integrate into target language communities, despite global English's dominance. We desire to synthesize the theoretical foundations from linguistics, educational psychology, and intercultural communication studies, while arguing for an interdisciplinary approach to understanding the complex challenges of language acquisition. By drawing from the diverse yet connected fields, the discussion highlights how cognitive, emotional, and sociocultural dimensions intertwine in the language acquisition process. Through this framework, we desire to advocate for pedagogical models that are not only linguistically enriching but also emotionally affirming and culturally responsive.

Linguistic Turn in English Studies

The very linguistic turn in the field of English studies makes a vital paradigm shift, moving scholarly attention from the study of language as a static system (philology) to a dynamic force that helps construct reality and meaning. Traditionally, philology emphasises the historical development and structure of languages, focusing heavily on etymology, textual criticism, and grammar (Turner, 2015). However, with the advances of modern linguistics in the 20th century, particularly influenced by scholars such as Saussure (2011), language began to be perceived as a system of signs that is grounded within the social power structures (Foucault, 1972). The concept of discourse heavily emphasizes how language not only reflects but also produces knowledge, framing what can be said, thought, and understood within a given historical moment. Moving on, Ballard (1980), through his theory of systemic functional linguistics, further contributed by proposing that language parallelly enacts the ideational, interpersonal, and the textual functions, thus emphasizing its centric role in social life. These perspectives revolutionized English studies, grounding an intense understanding of language's formative influence on human cognition, identity, and social structures.

Educational Psychology and Language Learning

The field of educational psychology has immensely contributed to the field of language acquisition from understanding the internal processes that affect language learning, particularly through cognitive-affective frameworks. Learning a second or foreign language is not purely a mechanical process; it engages emotional, psychological, and social domains that can either facilitate or hinder acquisition (Dornyei, 2013). One of the pivotal contributions in this area is Horwitz *et al.*'s (1986) articulation of Foreign Language Anxiety (FLA), conceptualised as a distinct set of self-perceptions, beliefs, feelings, and behaviours related to classroom language learning arising from the uniqueness of the language learning acquisition. Their Foreign Language Classroom Anxiety Scale (FLCAS) provided a validated instrument to measure learners' anxiety and also revealed how FLA could negatively impact performance, participation, and long-term motivation. Importantly, FLA has been recognized as a significant affective filter that can inhibit input processing, limit output, and reinforce negative self-conceptions among learners

Intercultural Communication and Identity Construction

As language learning increasingly occurs within globalized, multicultural environments, intercultural communication studies have gained prominence in understanding how learners navigate complex cultural terrains. Byram's (2020) model of Intercultural Communicative Competence (ICC) stands as a seminal contribution, proposing that language education should not only revolve around linguistic fluency but also around fostering attitudes of openness, skills of interpreting and relating, critical cultural awareness, and knowledge about other cultures. He also emphasized that true language mastery involves the ability to mediate between cultures, challenging ethnocentric perspectives and developing empathy towards cultures.

The Problem of Foreign Language Anxiety (FLA)

Foreign Language Anxiety is widely known as one of the most pervasive affective barriers impeding second language acquisition. It is not merely a passing feeling of nervousness but a deep-seated emotional response that can significantly alter learners cognitive processing, academic achievement and social integration that alters cognitive processing, academic achievement, and social integration (Horwitz *et al.*, 1986). As the process of language acquisition is getting embedded within the multicultural and globalized contexts, while also addressing FLA has never been more urgent. Understanding its psychological foundations, cultural intersections, and its profound implications for learner identity and motivation is crucial to developing inclusive and effective pedagogies.

Defining and Diagnosing FLA

FLA is always characterized by psychological symptoms such as apprehension, self-doubt, an inferiority complex, and fear of negative evaluation, which often translate into tangible academic consequences, including lower performance and withdrawal from language learning tasks. Learners affected by FLA may exhibit avoidance behaviours, reduced classroom participation, and impaired communicative competence (Krashen, 1982). Affective Filter Hypotheses, as also mentioned before, provides us with a theoretical explanation for these phenomena; it suggests that emotional variables like anxiety act as a filter that limits or, say, blocks the intake of comprehensible input necessary for language acquisition. A heightened affective filter not only diminishes learning efficiency but also becomes a barrier and compounds learners' feelings of incompetence, creating a self-perpetuating cycle of anxiety and academic underachievement.

The Role of FLA in English-Dominant Contexts

In environments where English is dominant, it functions not merely as a means of communication but as a marker of cultural capital and prestige. FLA can be exacerbated by feelings of cultural intimidation and linguistic imperialism (Phillipson, 1992). Learners who hail from a non-English background may internalize certain feelings of inferiority, resulting in a lack of self-confidence, a diminished sense of self-perception, and weakened motivation. The omnipresence of English as a lingua franca often marginalises other languages and cultural expressions while creating an implicit hierarchy that alienates learners and reinforces anxiety. Consequently, students may perceive their identities as deficient or incomplete, particularly when their linguistic backgrounds are devalued within educational or social settings.

Intersections with Cultural Sensitivity

The nexus between the FLA and cultural sensitivity is complex but often underexplored. Anxiety related to language is often rooted in cultural dissonance- an internal conflict arising when learners' native cultural frameworks clash with those dominant in the target language community (Maclntyre, 2007). A lack of perceived belonging can intensify feelings of alienation and helplessness, further aggravating anxiety. Socio-cultural competence, defined as the ability to navigate through and negotiate different cultural norms and values, is thus not just an ancillary skill but a central factor in mitigating FLA. Enabling the learners to feel appreciated, accepted, and understood in the multicultural settings while fostering the sense of belonging that is essential for lowering the affective filter and promoting the very authentic engagement with the language.

Identity, Language, and Learning in Global English Contexts

This is an interconnected world, and here English is seen as a passport to global opportunity- a language that opens doors to education, employment, and social mobility. Yet, beneath this optimistic narrative lies a more complicated reality. English as a global language carries not only opportunities but also deep-seated inequalities that have been carried on for centuries. For many language learners, acquiring English is not just about mastering vocabulary or grammar; it is about navigating a world where power, identity, and belonging are constantly negotiated. We explore how language learning, especially in English-dominant contexts, shapes and reshapes learners' identities, sometimes offering empowerment and sometimes reinforcing marginalization. However, learners have come to terms with the English studies.

English as a Global Language: Opportunity or Oppression?

The spread of English has been celebrated for connecting people across cultures. However, critical scholars have drawn attention to the phenomenon of native speakerism- idealization of native English speakers as the ultimate language models, often at the expense of multilingual and non-native speakers (Hopkyns, S. (2022)). This particular ideology perpetuates a hierarchy where certain accents, backgrounds, and cultural expressions are very privileged while others are seen as deficient.

When we closely link, the very issue of linguistic hegemony arises, where English dominance can suppress local languages and cultural identities. Phillipson, (1992) famously also argued that the global promotion of English often functions as a form of linguistic imperialism, creating environments where non-native speakers feel pressured to abandon their linguistic roots to gain acceptance or advancement. In these contexts, language learning is not just a neutral act, as it carries on it becomes a site of identity struggle, often leading learners to internalize feelings of inadequacy or alienation.

Identity Construction in the Language Classroom

Within the walls of the language classroom, these broader global forces come into instant focus. Post-structuralist theorists, as mentioned (Norton, 2013), have argued that identity is not fixed but constantly negotiated through language use. For learners, every interaction in the target language is a negotiation of who they are and how they wish to be perceived. Language classrooms become microcosms of the world's cultural politics, where issues of race, gender, class, and nationality intersect and influence learning experiences. Norton introduced the idea of investment to describe how learners might resist or reshape imposed identities, finding new voices in their second languages. These perspectives invite educators and facilitators to see language acquisition as deeply intertwined with learners' hopes and struggles and an evolving sense of self.

Toward Interdisciplinary Interventions

Addressing FLA through Instructional Design

As the complexities of language learning unfold slowly in the rapidly evolving world and a culturally diverse setting, interdisciplinary interventions emerge as critical strategies to address Foreign Language Anxiety (FLA) (Fig. **1**), promote intercultural sensitivity, and foster the development of identity. Traditional models, which often emphasized the process of rote learning and monolingual standards, are inadequate in cultivating emotionally secure, culturally responsive, and linguistically empowered learners (Horwitz, 2001). Thus, contemporary language education must pivot toward humanizing pedagogies, culturally embedded instructional design, and an open education paradigm that recognizes linguistic and cultural plurality as assets rather than to be seen as barriers.

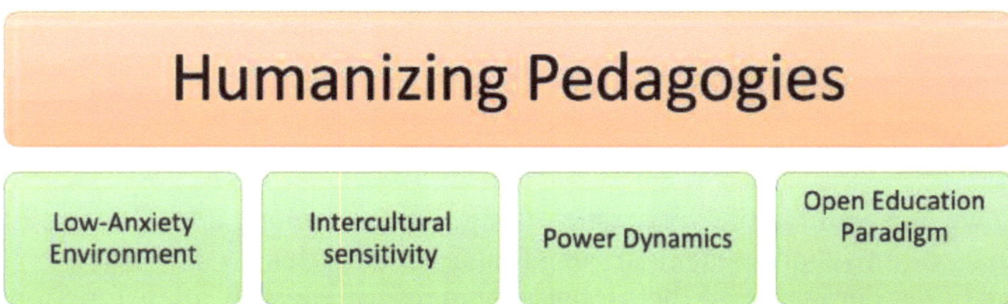

Fig. (1). Foreign language anxiety: Instructional design.

Humanizing pedagogies, as shown in Fig. (**2**), as posited by Freire (2018), advocate for classroom environments where students' emotional well-being is prioritized along with cognitive development. In language instruction, this means creating low-anxiety environments that mitigate the fear of making mistakes and encourage linguistic risk-taking. Language learners must feel safe to experiment and reframe their identities without the oppressive burden of native speaker norms. Research shows that reducing anxiety not only improves learner engagement but also enhances retention and creative use of language (Peng, 2022).

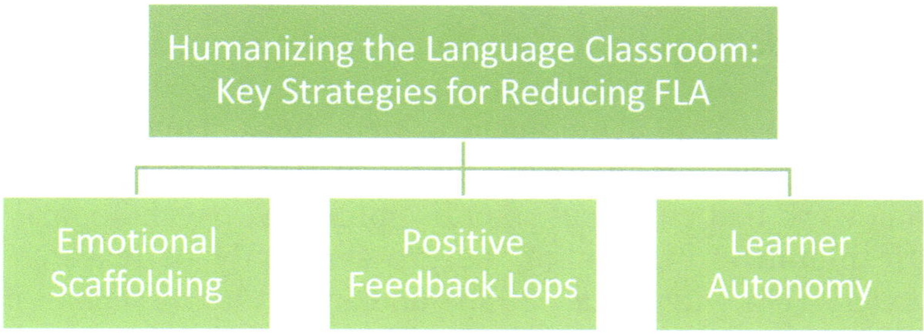

Fig. (2). Humanizing the language classroom: Key strategies for reducing FLA" — A flowchart showing techniques like emotional scaffolding, positive feedback loops, and learner autonomy.

Parallel to this is the integration of intercultural sensitivity models into instructional design. Traditional language textbooks often present sanitized or monolithic versions of culture. Instead, literature, authentic texts, and cultural artefacts serve as a dynamic entry point into the lived realities of target language communities (Kramsch, 1993). Engaging with the real-world materials nurtures critical consciousness, making students not just absorbers but also active interpreters of the cultural narratives. Dialogic teaching practices (Bakhtin, 1981) encourage open-ended conversation, where cultural misunderstandings become opportunities for deeper reflection rather than failures to be corrected. Moving forward, critical pedagogy urges teachers to deconstruct power dynamics inherent in language hierarchies. By situating learners as co-creators of knowledge, facilitators can empower them to navigate and challenge linguistic imperialism and ethnocentrism (Freire, 2018). As students recognize the political dimensions of language, they develop stronger intercultural competencies essential for global citizenship.

Finally, an open education paradigm must underpin these interventions. Equity inclusion, and multilingual affordances are non-negotiable in 21st-century classrooms. Open educational resources, community-based projects, and plurilingual strategies validate learners' diverse linguistic backgrounds rather than marginalize them (Weller, 2020). Students must be encouraged to see themselves not as passive recipients of English but as cultural agents, capable of shaping a discourse within their local contexts. Summing up and moving toward interdisciplinary interventions is not merely a pedagogical choice but a moral imperative. By synthesizing insights from applied linguistics, educational psychology, critical theory, and cultural studies, we can reimagine language learning spaces that are emotionally safe, culturally rich, and socially transformative.

Implications and Future Directions

English studies can no longer afford to stick to static or monodisciplinary models of instruction (Fig. **3**). Traditionally, the field has privileged literary analysis and linguistics norms rooted in Anglo-centric paradigms. However, contemporary scholarship urges a radical rethinking: English studies must evolve into a dialogic, dynamic framework that embraces interdisciplinarity, psychological well-being, and linguistic plurality. In light of the growing body of research on foreign language anxiety (Horwitz *et al.*, 1986) and intercultural sensitivity (Chen & Starosta, 2000), this chapter explores the transformative potential of an interdisciplinary English studies model that integrates affective, cognitive, and sociocultural dimensions.

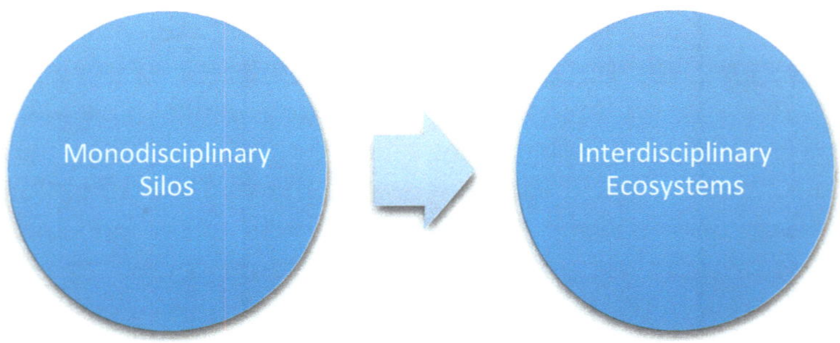

Fig. (3). Evolving english studies: From mono-disciplinary silos to interdisciplinary ecosystems.

Rethinking English Studies

The imperative for reimaging English studies stems from its current inadequacies in trying to address the complexities of globalized communication and identity negotiation. Also, moving forward, dialogic concept maps means acknowledging the view of Bakhtin, its meaning is duly constructed in conversation and interaction. This relational stance demands today's facilitators to not only teach English studies, language and literature but also foster dialogic competencies, encouraging students to navigate and negotiate across linguistic and cultural borders. Adding to the psychological well-being must be woven into the pedagogical fabric, so that we can also shift our focus from the rote learning of English studies and genuinely cultivate the willingness to learn from within. Thus, English studies must create spaces that support linguistic risk-taking, validate emotional experiences and discover the multilingual identities as assets rather than deficiencies (Garcia & Wei, 2014).

Policy and Curriculum Recommendations

Shifts in institutions are vital to implement this vision. Policies as framed by the policymakers must prioritise the intercultural competence, emotional resilience, and sociocultural awareness as an outcome of English language education, through redesigning assessment frameworks to value the reflective, experimental, and dialogic learning over the technique of rote memorization and prescriptive correctness.

The several programs for teacher education must also undergo a very significant transformation. Educators need training that equips them to address not just the linguistic skill but also the affective and sociocultural dimensions of language learning. Programs should incorporate modules on mitigating foreign language anxiety, intercultural communication, and inclusive pedagogies (Kramsch, 1993; Freire, 2018). Teachers must prepare to recognize and respond to the emotional landscapes of their learners, creating classrooms that are humanizing, empowering, and critically conscious. Hence, the future of English studies lies in the interdisciplinarity, centring emotional well-being, and valorising linguistic diversity. This particular paradigm shift demands intentional, systemic changes in curricula, policies, and pedagogical practices. Therefore, English studies can truly harness a great deal of the power of language as a tool for personal evolution, social justice, and global citizenship.

Suggestions for Future Research

As the landscape of English language learning continues to evolve, future research must adopt more nuanced, long-term and intersectional approaches to fully understand the complexity of foreign language anxiety (FLA) and its implications for identity construction. One of the most compelling avenues for future inquiry lies in longitudinal studies that examine how learners' language-related identities develop over time in relation to anxiety. Although cross-sectional data has been valuable in identifying correlational patterns, it cannot capture the shifting nature of language learner identity across stages of proficiency, institutional transitions or socio-political changes. Such longitudinal designs could unpack the temporal dimensions of FLA, revealing how coping mechanisms, motivation, and cultural integration evolve, thereby informing more sustainable pedagogical strategies. In tandem with temporal depth, there is an urgent need for comparative research across diverse classroom ecologies. While FLA has been extensively studied in Western and East Asian contexts, less attention has been given to underrepresented and multilingual regions in the Global South. Comparative studies involving classrooms from different geopolitical, socio-linguistic, and educational settings can illuminate the role of local pedagogical cultures, teacher beliefs, and language policies in either exacerbating or mitigating anxiety (Widdowson, 2003). By juxtaposing experiences of FLA in, for example, Scandinavian cooperative learning environments versus exam-oriented systems in South Asia, researchers can more critically examine how institutional structures intersect with emotional and psychological learner variables.

Moreover, it is imperative to explore further the intersectionality of gender, race, and language anxiety, particularly in transnational or migratory educational settings. Most existing studies tend to fear language learners as a homogeneous group, overlooking how identity markers such as race, gender, and class mediate their emotional experiences. For instance, female students or students from marginalized racial backgrounds might experience heightened vulnerability in English-dominant classrooms due to linguistic stereotyping or discriminatory attitudes (Ballard, 1980; Norton, 2013). Addressing this gap requires interdisciplinary collaboration between applied linguistics, gender studies, and critical pedagogy to foreground learners' diverse subjectivities in research and practice.

To summarize, future research must move beyond surface-level diagnostics of FLA and embrace multi-scalar frameworks that account for time, place, and identity. This holistic perspective is essential not only for advancing the theoretical terrain of language anxiety but also for building inclusive, culturally sustaining educational practices that recognize learners' diverse needs and backgrounds.

CONCLUDING REMARKS

Hence, even though English studies today is a chance to global opportunities, it is also a doorway to several barriers as mentioned above in the chapter but with the evolving pedagogical design and the constant efforts of the facilitators and stakeholders English can shift from being just a study of vocab and grammar and harness a great opportunity to mingle with the different disciplines, this indulgence will open a number of pathways for the learners to excel, and also understand the junctions that they face while learning a certain language. It is only through the interdisciplinary approach that English studies can be truly humanized. Thus, revisiting the power of language means fostering environments where learners become conscious cultural agents rather than passive recipients. Humanizing language education is fundamental to shaping not only better communicators but also empathetic citizens.

REFERENCES

Byram, M. (2020). *Teaching and assessing intercultural communicative competence: Revisited*Multilingual Matters.

Chen, G. M, Starosta, W. J (2000). The development and validation of the intercultural sensitivity scale. *Human Communication, 3*, 1-15.

Dornyei, Z. (2013). *The psychology of second language acquisition*Oxford University Press. Available form: https://global.oup.com/academic/product/the-psychology-of-second-language-acquis-tion-9780194421973

Elshahawy, K. E. M. (2023). Language Reconciliation: Is It Theoretical, Applied or L2 (ESL/EFL) Curriculum & Instruction?. *Journal of English Language Teaching and Applied Linguistics, 5*(4), 147-162.

Freire, P. (2018). Pedagogy of the Oppressed. 50th Anniversary Edition. *Bloomsbury Academic.*

Garcia, O., Wei, L. (2014). *Translanguaging: Language, bilingualism and education.*Palgrave Macmillan. [http://dx.doi.org/10.1057/9781137385765]

Hopkyns, S. (2022). A global conversation on native-speakerism: Toward promoting diversity in English language teaching. *World Englishes, Global Classrooms: The Future of English Literary and Linguistic Studies,* Singapore: Springer Nature Singapore.17-33.

Horwitz, E. (2001). Language anxiety and achievement. *Ann. Rev. Appl. Linguist., 21*, 112-126. [http://dx.doi.org/10.1017/S0267190501000071]

Horwitz, E.K., Horwitz, M.B., Cope, J. (1986). Foreign language classroom anxiety. *Mod. Lang. J., 70*(2), 125-132.
[http://dx.doi.org/10.1111/j.1540-4781.1986.tb05256.x]

Kramsch, C. (1993). *Context and Culture in Language Teaching.* Oxford University Press.

Krashen, S.D. (1982). *Principles and Practice in Second Language Acquisition.*Pergamon Press. Available form: http://www.sdkrashen.com/content/books/principles_and_practice.pdf

Norton, B. (2013). *Identity and Language Learning: Extending the Conversation.*Multilingual Matters & Channel View Publications. Available form: https://www.jstor.org/stable/jj.18799909

Pavlenko, A. (2001). Language learning memoirs as a gendered genre. *Appl. Linguist., 22*(2), 213-240. [http://dx.doi.org/10.1093/applin/22.2.213]

Peng, J. E. (2022). Willingness to communicate. *The Routledge Handbook of Second Language Acquisition and Individual Differences,* 159-172.

Phillipson, R. (1992). *Linguistic imperialism.*Oxford University Press.

Saiyara, N. (2022). The pedagogical procedures and indicators of post-methods for English language teaching: A theoretical perspective. *J Lang Linguistic Stud, 18*(1), 454-65.

Saussure, F. de (2011). Course in general linguistics. *Columbia University Press.*

Soylu, M., Soylu, A., Das, R. (2023). A new approach to recognizing the use of attitude markers by authors of academic journal articles. *Expert Systems with Applications, 230*, 120538. [http://dx.doi.org/10.1016/j.eswa.2023.120538]

Turner, J. (2015). *Philology: The Forgotten Origins of the Modern Humanities.*Princeton University Press. Available form: https://www.jstor.org/stable/j.ctt5hhrxf

Vygotsky, L.S., Cole, M. (1978). *Mind in society: Development of higher psychological processes.*Harvard University Press.

Weller, M. (2020). 25 Years of Ed Tech. *Athabasca University Press.* Available form: https://www.aupress.ca/books/120290-25-years-of-ed-tech/

Widdowson, H. (2003). *Defining issues in English language teaching.*Oxford University Press.

CHAPTER 8

Biology as an Interdisciplinary Subject: Bridging Disciplines for a Deeper Understanding of Life

Shifa Suhail[1], **Kashifa Khatoon**[1], **Arpita Tripathi**[2], **Pallavi Awasthi**[1], **Reena Vishvakarma**[1] and **Swati Sharma**[1,*]

[1] *Department of Biosciences, Integral University, Lucknow, Uttar Pradesh 226026, India*

[2] *Colvin Talukedar College, Lucknow, Uttar Pradesh 226007, India*

Abstract: Biological science has always been portrayed as an interdisciplinary subject. Essentially, botany, zoology, and microbiology deal with all the life forms, while biochemistry relates to all the chemical reactions in living beings. Biophysics and biostatistics have always been a part of biological assays and investigations. From school education to higher studies, currently, biology is a part of the STEM (Science, technology, engineering, and mathematics) education system and curriculum. Although they are distinct, these are related technical disciplines. The term is typically used in the context of education policy or curriculum choices in schools and colleges. The STEM curriculum with experiential learning and appropriate content surely increases students' competency levels. Nowadays, the relation of biological sciences with social, behavioural or neurosciences has also been well established and studied. With the advent of technology, biology was redefined, and new areas of computational biology emerged. Moreover, AI has further elevated the levels and created a lot more sub-branches. Though all these areas of interdisciplinary research were growing in specific research fields, the interdisciplinary nature of the subject is yet to be established in the education system. With the advent of new educational policies such as NEP 2020 in India, these gaps might be filled, and this can lead to a change in students' overall learning and research experience for lifelong learning.

Keywords: AI, Biological science, Interdisciplinary, Lifelong learning, STEM, Teaching-learning.

INTRODUCTION

Biology, defined as 'A branch of science that deals with living organisms and their vital processes' (Webster Dictionary), incorporates the study of plants, animals, and microbes, *i.e.*, Botany, Zoology, Microbiology, along with biochemistry, physiology, genetics, and ecology. It includes the structure and

* **Corresponding authors Swati Sharma:** Department of Biosciences, Integral University, Lucknow, Uttar Pradesh 226026, India; E-mail: sw_sh@rediffmail.com

function as well as their growth and distribution, and also the origin and evolution of life itself. These domains contain many specializations within themselves. The nature of these specialized areas is interdisciplinary in itself because of the complexity of biological systems. Developing the connection between biology and other scientific branches is one of the most important paradigm shifts in education.

Biology is now a part of the STEM education system and curriculum. Although they are distinct, these are related technical disciplines. The term is typically used in the context of education policy or curriculum choices in schools and colleges. These areas are grouped as STEM-based teaching and learning offers an interdisciplinary approach (Fig. **1**) that removes traditional barriers among four disciplines: science, technology, engineering, and mathematics, and makes learning more experiential, integrating them into real-life experiences (McDonald, 2016).

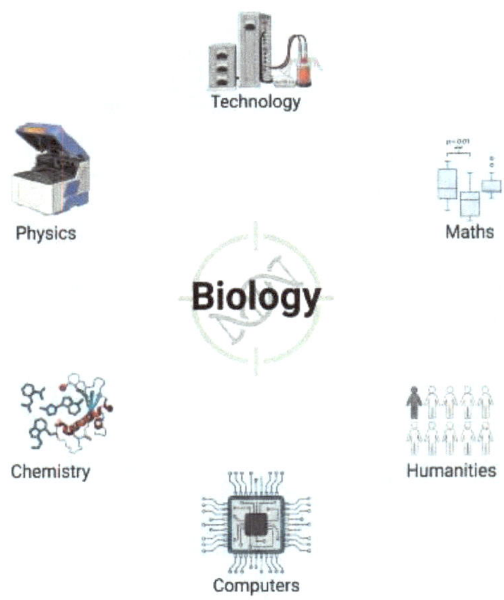

Fig. (1). Interdisciplinary nature of biology.

Meanwhile, the teaching-learning process has also advanced with the use of computers and internet-automated systems. The students nowadays have grown up in a world of technology with online social networking and electronic-based resources as a way of life. These are being used to enhance student understanding and interest in biology as well as create newer methods of research and analysis, such as sequencing-based meta-analysis or bioinformatics-based studies.

Present-era biologists or new biologists will be those who know about all disciplines, and have deep knowledge and understanding of biology while working in interdisciplinary related fields (Vermeulen *et al.*, 2013). This requires interdisciplinary approaches across different disciplines apart from physics, chemistry, and mathematics, such as engineering and computer science, to deal with complex higher-level problems related to food, health, energy, and the environment. They need to go far beyond their biology content knowledge and acquire new methods, approaches, new techniques, and new tools (Kleinman, 2020; Myser, 2011). Henceforth, they are required to understand the connections between the disciplines. Also, they should learn how to make these connections. Thus, preparing future biologists will need exposure as well as experience with other fields, such as engineering and technology, without which they will fail to comprehend and might not survive in the ever-competitive world.

This chapter aims to explore how integrated interdisciplinary biology as a subject can impact the overall learning experience of students and result in a deeper understanding of life.

Biology in STEM

Biological sciences are an intersection of biology, chemistry, physics, and mathematics referred to as biochemistry, biomaths/biostats, biophysics, and biotechnology (Fig. **2**). It includes the study of biochemistry, inorganic as well as organic chemistry, physical chemistry, physics, and mathematics-statistics for instrumentation, data-handling, statistical-modelling, while computational studies for visualization, and computational techniques. Technological advances further established the field of study and research.

Chemistry and Biology

The importance of chemistry in biology can be easily judged by the presence of a chapter, or at least a brief introduction to chemistry, among the introductory topics in biology textbooks. It is a common saying that life is a set of chemical reactions. It is said to have originated from inorganic chemicals, which formed organic ones via a series of complex chemical reactions. Proteins, carbohydrates, lipids, and nucleic acids (DNA and RNA) (Fig. **3**) are all organic molecules that go through complex chemical reactions in a life form (Adams, 2012).

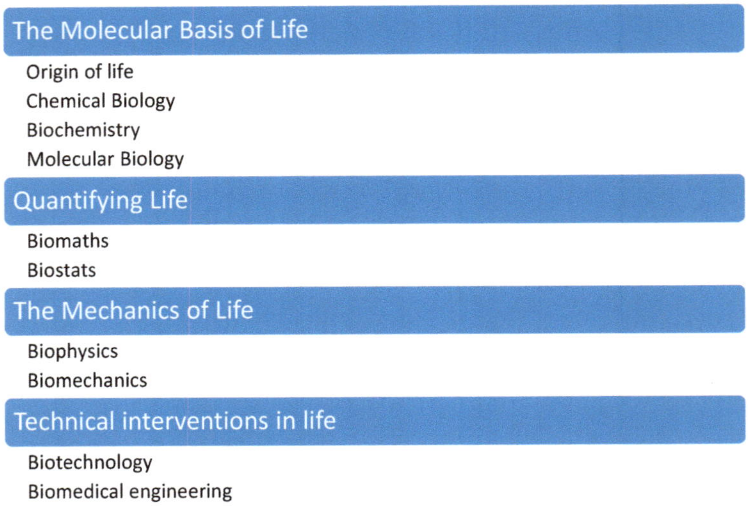

Fig. (2). STEM and disciplines of biology: the study of life.

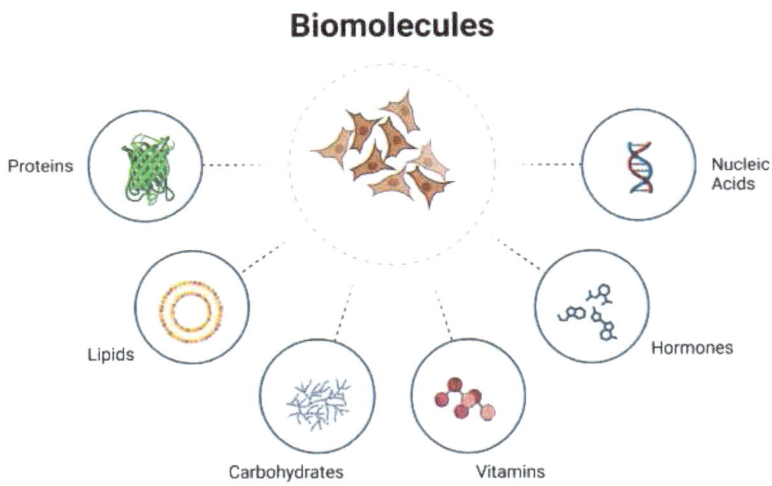

Fig. (3). Biomolecules: molecular basis of life.

Homeostasis involves the regulation of body fluids, pH, hormones, and gas exchanges. On looking closely, organs, tissues, or cells are made up of these biomolecules only, while hormones, neurotransmitters, and signalling factors are also chemicals (Srivastava *et al.*, 2020). Some of these reactions include:

- Building of bones, muscles, and tissues
- Respiration
- Photosynthesis

- Digestion
- Germination
- fruit ripening
- Cell cycle and cell growth
- Senescence
- Inheritance and genetics
- Action of medications, supplements, and toxins

The relationship between biology and chemistry has been studied widely. The umbrella expanded to give rise to sub-areas, such as:

- Chemical Biology
- Biochemistry
- Molecular biology

Chemical Biology

Chemical Biology is termed as the application of chemical techniques and analysis to biological systems, and the study and manipulation of biological systems by novel synthetically produced small molecules.

Biochemistry

Biochemistry is the sub-discipline of both chemistry and biology, and it involves the study of chemical processes in living organisms. It may be subdivided into processes explaining life in terms of chemistry and classified as structural biology, enzymology, and metabolism.

Molecular biology

Molecular biology is defined as 'the discipline describing and investigating the mechanisms whereby biomolecules, upon their assembly, contribute to the execution and/or regulation of biological processes essential for the homeostasis of cells, tissues/organs and living organisms, and crucial for their response to internal or external perturbations' (Lio *et al.*, 2025). It includes the structural, functional, and expressional study of macromolecules such as DNA, RNA, or proteins. Molecular biology further combines physics and chemistry with biochemistry and genetics, for the lowest resolvable level of explanation of biochemical reactions.

Biology and Physics

Physics is inseparable from biology, as biological systems work on the laws of physics. Moreover, all the analysis techniques involve physics. Biophysics deals with the methods of physics driving biological systems such as cell movement and functions of the brain, circulatory, and immune system. Biophysicists have interdisciplinary knowledge of math, physics, and chemistry, and they can understand areas ranging from cellular mechanisms to how plant cells change light into energy to how the DNA of healthy cells transforms into cancer cells (Noble, 2010).

Biology and Mathematics

All the experiments and analyses in Biology require basic knowledge of maths. Jungck *et al.* have reviewed the role of mathematical biologists as leaders in programmatic research over the past 60 years, and the subsequent reforms in both the mathematics and biology fields, and their education (Steen, 2005). Tetaj and Viirman (2023) have recently reviewed the difference between mathematics and biology narratives in assignments. Biomaths and Biostats are an integral part of the biology curriculum. Statistical models and analysis are essential in most experiments.

Biology and Environmental Science

Ecology and environmental science almost mirror each other. With the advent of remote sensing technologies, this area has grown exponentially (Purkis *et al.*, 2011).

Biology and Engineering: Biotechnology and Biomedical Engineering

Biotechnology and biomedical engineering are vibrant, dynamic, interdisciplinary fields at the intersection of biology, technology, and engineering, aimed at improving human health, agriculture, and industry. From the age-old practices of fermentation to the cutting–edge innovations in synthetic biology and computational biology, these domains show how science and technology can work together to solve problems in the real world (Fatima *et al.*, 2024). Whether enhancing or improving food production through agricultural biology technology or identifying and treating diseases with molecular and computational tools, these fields demonstrate how science and technology converge to address real-world challenges. As global concerns like food security, environmental sustainability, and healthcare demands intensify, biotechnology and biomedical engineering continue to play an important role in shaping our future. Biotechnology is a field of science that uses living cells, or living systems, to develop new technologies that enhance our lives. It is a blend of biology and technology (Gupta *et al.*, 2025).

Fermentation technology

Fermentation is a biological process in which microorganisms like yeast and bacteria convert or break down organic substances and carbohydrates into alcohol (Fig. **4**), lactic acid, or carbon dioxide (Soto, 2019). Traditionally, it has been used to increase the microbiological features, health benefits, and shelf life of foods, leading to the development of many dietary staple foods around the world, such as yoghurt, cheese, kimchi, tempeh, and kombucha. These fermented foods are characterised by their unique sensory attributes and physiological benefits, including improved intestinal health and immune-modulating effects that are increasingly identified in research (Niyigaba *et al.*, 2025).

Glucose ($C_6H_{12}O_6$)

Glycolysis (enzyme action)

Pyruvate ($C_3H_4O_3$)

Anaerobic conditions

Ethanol (C_2H_5OH) + CO_2

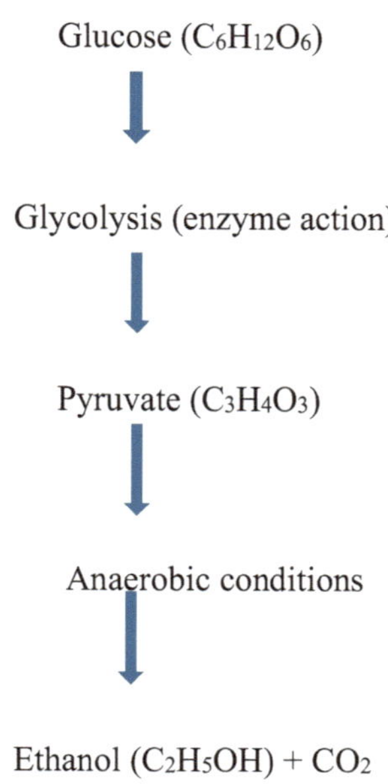

Fig. (4). Ethanol fermentation pathway.

Synthetic Biology

In 2000, Eric Kool and other speakers gave the term 'synthetic biology' at the American Chemical Society in San Francisco. It has mainly two subfields, which are growing disciplines. One uses artificial molecules to produce emergent behaviours from native biology, with the aim of generating unnatural life. The other seeks interchangeable parts from natural biology to assemble into systems that act unnaturally. It has used biochemical pathway design and hereditary components to develop organisms that can synthesize important chemicals and precursors for antibiotics and polymers (Samanta *et al.*, 2024).

Agricultural Biotechnology

Biotechnology plays an important role in promoting agricultural productivity and plant protection by integrating cutting-edge genetic techniques, including genetic engineering and marker-assisted selection. CRISPR-Cas9 is a powerful gene editing method that precisely cuts the DNA sequences. Using this technology has led to improved genetically modified plants, disease, abiotic stress tolerance, *etc.* (Kumar *et al.*, 2023). In addition, bioinformatics and simulation modelling have innovative ideas, including residual harvesting, without damaging environmental impacts like emissions or generating harmful greenhouse gases or soil degradation. Biotechnology applications, such as precision breeding and other biotechnology uses, have changed agriculture by bringing traits or features that include resistance to drought and heat, nutrient efficiency, and enhanced yields. Successful case studies of this technology, such as Bt (*Bacillus thuringiensis*) cotton, blight-resistant potatoes, golden rice, *etc.*, are illustrative of the potential biotechnology has to solve challenges in the worldwide protection of food (Yadav *et al.*, 2024).

Biomedical Engineering

A combination of the fields, such as engineering, biology, and medicine, is used to develop technologies to improve human welfare. Biomedical engineers (BG) work on a wide range of innovations, including medical devices such as pacemakers, artificial organs, imaging devices, MRI and CT scanners (Fig. **5**), and innovative prosthetics. Tissue engineering and regenerative medicine are the most transformative areas within biomedical engineering, which create artificial organs like skin and tissues that can replace damaged ones (Dzobo *et al.*, 2018) BG are at the forefront of robotic surgery and medical automation, robotic systems are designed that assist surgeons in performing complex procedures with higher precision and minimal invasiveness, due to faster recovery times and fewer complications. Another subfield, neural engineering, is about understanding and interfacing with the nervous system to develop treatments for conditions like paralysis, epilepsy, and Parkinson's disease. It is a field that offers complex solutions to medical problems (Ereifej *et al.*, 2019).

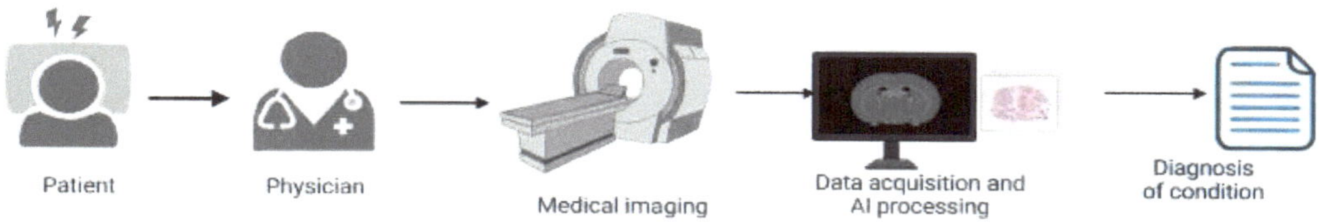

Fig. (5). AI methods as predictive AI tools in biomedical Imaging.

Molecular medicine

Molecular medicine is a branch of medicine that focuses on understanding diseases at the molecular and genetic levels. It involves studying how molecules like DNA, RNA, proteins, and other cellular components interact in the body and how changes in these molecules can lead to diseases (Kurreck *et al.*, 2025). The aim of molecular medicine is to enable personalized or precision medicine. For example, in cancer treatment, tumours can reveal specific mutations that drive cancer growth. Molecular diagnostic techniques, such as PCR and DNA microarrays, are widely used to detect diseases. Molecular diagnostics like PCR and DNA microarray are widely used to detect diseases. These technologies have been important in controlling diseases like COVID-19 (Afzal, 2020).

BIOLOGY AND COMPUTER SCIENCE: COMPUTATIONAL BIOLOGY

Computational biology has been defined by Hallam Stevens, in 'Life Out of Sequence', as "biology adapted itself to the computer, not the computer to biology." He also stated that "Computers do not just raise the ancient biology, they also introduce fully fresh instruments and questions, like simulations, statistics, and how data is managed, that completely re-model the way biological research is being completed".

In computational biology, complex datasets such as DNA/RNA/Protein sequences, complex molecular structures, as well as gene expression analysis, can be analyzed and interpreted in real-time, which would otherwise be exhaustive, complex, and expensive (Way *et al.*, 2021)

One of the important examples of how computers remould biological research is the use of databases and strongest matches. Biological knowledge today is defined, well-organised, and easily accessible through computation. Another way computers have remoulded biology is by introducing statistics and data analysis methods. A very good example is understanding how mutational processes shape genomes. The larger patterns of Mutational processes, like cigarette smoke, sunlight, or defects in homologous recombination, are not visible in individual mutations but only in their global patterns.

Large-scale attempts to collect the genetic constitution and physical composition of model organisms and people are essential to the success of computational biology since these studies need vast data sets (Markowetz, 2017). Genetic variations between cells in the same tumour can help to make cancer resistant to treatment. These can be compared statistically or using AI with clinical information on treatment resistance.

Interdisciplinary Research for Life Science

Interdisciplinary research in life science refers to the integration of knowledge, tools, techniques, and ideas from different fields (Wagner *et al.*, 2011). Collaboration between interdisciplinary fields such as physics, chemistry, biology, computer science, engineering, and mathematics has grown crucial as life science problems arise. If bioinformatics merges with computer science, it analyses large-scale genomic data; similarly, a combination of biomedical engineering with genetic and medical sciences develops innovative ideas or technologies, including artificial organs and therapeutic gene therapy. So, these merging interdisciplinary fields in life science approach not only foster scientific innovation but also provide more comprehensive solutions to worldwide issues such as disease prevention, climate change, and sustainable agriculture. It also enhances our ability to solve real-world problems through these merging combinations. It provides the complexity of biological systems that often require knowledge and methodologies from different fields. When scientists from different fields collaborate to solve complex biological problems. Instead of just using biology, researchers combine knowledge and techniques from multiple disciplines, like physics, computer science, engineering, and mathematics, with life sciences. Some interdisciplinary areas in life science research are shown in Fig. (**6**).

Computer-aided drug designing and Health informatics

Health informatics merges with epidemiology, computer science and social science to improve healthcare delivery through the analysis of large datasets such as the COVID-19 pandemic, where rapid analysis of data can inform and guide containment strategies (Thiébaut *et al.*, 2017). Computer-aided drug designing has helped screen thousands of candidate molecules *in silico,* saving a lot of money and time (Fig. **7**).

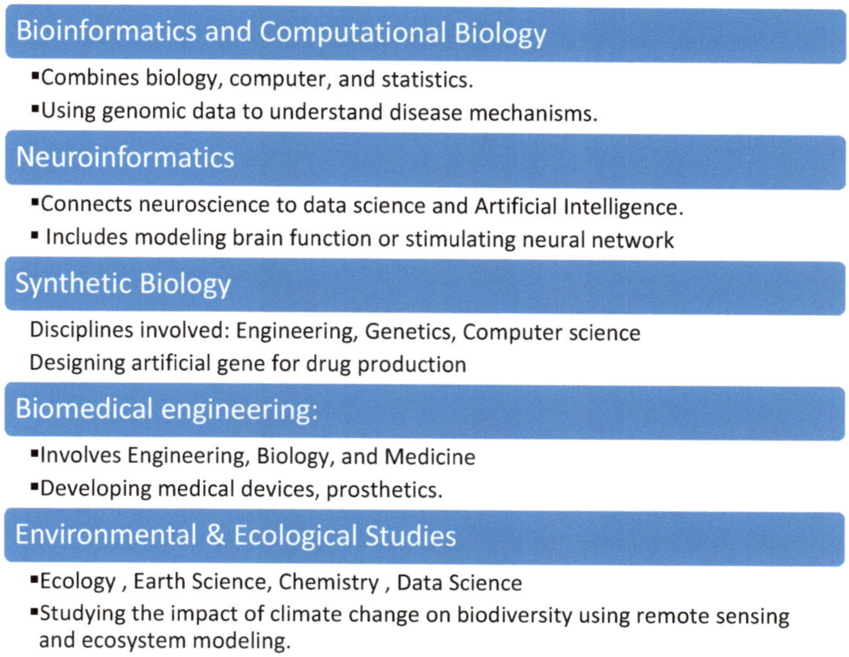

Fig. (6). Some interdisciplinary areas in life science research.

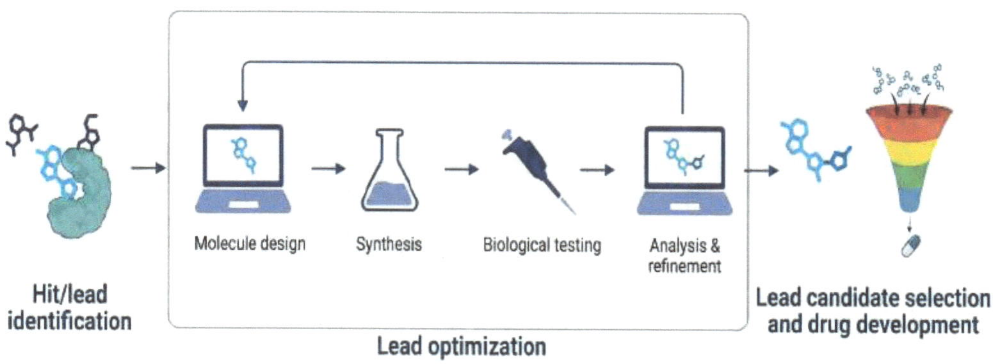

Fig. (7). Computer-aided drug designing: screening of lead drug molecules.

Biology and Social Sciences: The Human Connection

The changing relationship between biology and the human sciences is examined in this chapter, with a focus on how multidisciplinary research deepens our comprehension of what it means to be human (Rose, 2013). The chapter examines how moral reasoning, cultural background, and biological knowledge interact to shape answers to current issues in genetics, neurology, healthcare, and public health, with bioethics as its main focus. The chapter concludes by advocating for an integrative framework that places human dignity, agency, and justice at the forefront of biological scientific application.

Bridging Biology and the Human Experience

From DNA to ecosystems, biology has made incredible progress in understanding the structures and workings of life. However, the human sciences come into play when we pose queries like "What is a good life?" "When does life begin or end?" and "How should we treat one another in light of biological differences?" Philosophy, anthropology, sociology, history, and psychology are among the fields that investigate not just how life functions but also why it is important.

There is more than just multidisciplinary collaboration at the nexus of biology and human sciences; there is mutual dependence. The ethical compass, cultural nuance, and interpretive skills required to apply biological information responsibly are provided by human sciences. In an era of genetic engineering, artificial intelligence, reproductive technologies, and biopolitical decision-making, this relationship is becoming more and more significant.

Abuse of biological and medical authority was a major factor in the development of the area of bioethics in the middle of the 20th century (Czech, *et al.*, 2023). The Tuskegee Syphilis Study in the United States and the atrocities of Nazi medical experiments sparked international discussion about the moral limits of research.

The Arts and Literature

To humanize medicine and investigate suffering, empathy, and healing, the medical humanities employ storytelling, art, and performance. The human sciences and biology must jointly develop a framework that recognizes the diversity of the world by eschewing ethnocentric frameworks in favour of cultural humility, tackles injustice by empowering just systems through biology rather than by reinforcing hierarchy, and embraces complexity by acknowledging that human concerns cannot be adequately addressed by a single viewpoint (Patel, 2023).

Wisdom must equal the strength of biology as it continues to influence our futures in the lab, the clinic, the genome, and society. The human sciences, which remind us that every data point has a life, a narrative, or a decision behind it, are the source of that wisdom (Jeste, 2020). Being human means having more than just a biological makeup. It means living with purpose, autonomy, vulnerability, and accountability. We can only fully comprehend and respect what it means to be human when biology and human sciences work together.

CONTEMPORARY ETHICAL CHALLENGES IN BIOLOGY

Different perspectives on death, disease, and care are revealed by anthropological research, demonstrating the need for culturally sensitive ethical judgments (Kleinman, 1978). The study of biocultural anthropology examines how human development is jointly shaped by biology and culture Baum, *et al.*, (2024). For instance, food availability, which is frequently governed by societal structures, and genetic predispositions both influence nutritional status. The study of epigenetics shows how societal contexts can impact gene expression. Discrimination, poverty, stress, and trauma can leave molecular traces that affect health for generations. This reinforces the insight that biology is not destiny, but rather a responsive system shaped by human experience. Biology gives us tools; human sciences help us decide how to use them. Several disciplines play a vital role in ethical reflection. Cultural differences in beliefs, methods of healing, familial structures, and personal definitions are exposed by anthropology. Sociology studies the role of technology in society, the medicalization of life, and power dynamics in healthcare facilities. Psychology studies how people deal with disease, ethical quandaries, and identity shifts brought on by biological interventions. Currently, the interdisciplinary aspect of biology can be categorised into several sub-areas as indicated in Table **1**. All of these fields have a bright future.

Table 1. The A to Z of interdisciplinary branches and sub-branches in biology.

S. No.	Branch	Interdisciplinary Areas	Subbranches
1.	Agricultural Science	Biology + Agronomy	• Plant Breeding, • Soil Microbiology, • Crop Protection
2.	Astrobiology	Biology + Astronomy	• Origin of Life, • Extremophile study, • Planetary/Space Biology
3.	Biochemical Engineering	Biology + Chemical Engineering	• Bioreactor Design, • Downstream Processing, • Metabolic Engineering
4.	Bioinformatics	Biology + Computer Science	• Genomics, • Proteomics, • Computational Biology, • Systems Biology
5.	Biomedical Science	Biology + Medicine	• Pharmacology, • Molecular Medicine, • Clinical Diagnostics

(Table 1) cont.....

S. No.	Branch	Interdisciplinary Areas	Subbranches
6.	Biostatistics	Biology + Statistics	• Epidemiology, • Population Studies, • Clinical Trial Design
7.	Biotechnology	Biology + Engineering	• Fermentation Technology, • Synthetic Biology, • Agricultural Biotechnology
8.	Environmental Science	Biology + Earth Sciences	• Climate Change Biology, • Environmental Toxicology, • Restoration Ecology • Remote sensing data analysis
9.	Forensic Science	Biology + Criminology	• DNA Fingerprinting, • Forensic Entomology, • Toxicology
10.	Genomic Medicine	Biology + Genomics	• Precision Medicine, • Cancer Genomics, • Gene Therapy
11.	Molecular Medicine	Biology + Chemistry	• Drug Development, • Drug designing, • Biomarker Discovery, • Molecular Diagnostics • Toxicology
12.	Neuroscience	Biology + Psychology/Physics	• Cognitive Neuroscience, • Neurogenetics, • Neuropharmacology
13.	Pharmacogenomics	Biology + Pharmacology	• Personalized Medicine, • Drug-Gene Interactions
14.	Synthetic Biology	Biology + Design Engineering	• Genetic Circuits, • Minimal Genomes, • Artificial Cells
15.	Systems Biology	Biology + Mathematics	Δ Network Biology, Δ Dynamic Modelling, Δ Omics Integration

CONCLUDING REMARKS

This chapter has provided insight into the interdisciplinary nature of biological sciences. Learners nowadays need to solve complex, interdisciplinary problems that will require them to go far beyond the core biology syllabus and understand the essence of many other disciplines, including AI and ML. The future biologist should be able to use engineering and technology to survive in a highly competitive environment. This change in pedagogy and research will lead to bridging the gap between different streams, will result in a change in students' aptitude and interest in the field, and will help in the development of lifelong learning.

REFERENCES

Afzal, A. (2020). Molecular diagnostic technologies for COVID-19: Limitations and challenges. *J. Adv. Res., 26*, 149-159.
[http://dx.doi.org/10.1016/j.jare.2020.08.002] [PMID: 32837738]

Gupta, V., Sengupta, M., Prakash, J. P., & Tripathi, B. D. (Eds.). (2017). *Basic and applied aspects of biotechnology*Springer Nature. Available from: https://link.springer.com/book/10.1007/978-981-10-0875-7

Baum, C. M., Christiansen, C. H., & Bass, J. D. (2024). The person-environment-occupation-performance (PEOP) model. In *Occupational therapy* (pp. 47-55). Routledge.

Czech, H., Hildebrandt, S., Reis, S. P., Chelouche, T., Fox, M., González-López, E., & Roelcke, V. (2023). The Lancet Commission on medicine, Nazism, and the Holocaust: historical evidence, implications for today, teaching for tomorrow. *The Lancet, 402*(10415), 1867-1940.

Dzobo, K., Thomford, N. E., Senthebane, D. A., Shipanga, H., Rowe, A., Dandara, C., Pillay, M., Motaung, K. S. C. M. (2018). Advances in regenerative medicine and tissue engineering: Innovation and transformation of medicine. *Stem Cells International, 2018*(1)
[http://dx.doi.org/10.1155/2018/2495848]

Ereifej, E.S., Shell, C.E., Schofield, J.S., Charkhkar, H., Cuberovic, I., Dorval, A.D., Graczyk, E.L., Kozai, T.D.Y., Otto, K.J., Tyler, D.J., Welle, C.G., Widge, A.S., Zariffa, J., Moritz, C.T., Bourbeau, D.J., Marasco, P.D. (2019). Neural engineering: the process, applications, and its role in the future of medicine. *J. Neural Eng., 16*(6), 063002.
[http://dx.doi.org/10.1088/1741-2552/ab4869] [PMID: 31557730]

Fatima, D.G., Magomedova, D.A., Parvez, D.S. (2024). *Biotechnology and Sustainable Development.*. Shineeks Publishers.

Gupta, R.K., Matharani, J. (2025). Scrutinizing India's regulatory framework on GM crops: Aligning sustainable environmental practices with food security demands. *J. Neonatal Surg., 14*(6), 320-327.
[http://dx.doi.org/10.63682/jns.v14i6.3321]

Jeste, D.V. (2020). Human wisdom and progress: Integrating biology, psychology, and social sciences. *Am. J. Geriatr. Psychiatry, 28*(1), 1-2.

Kleinman, A. (2020). *The Illness Narratives: Suffering.*. Hachette, UK: Healing, And The Human Condition.

Kumar, M., Prusty, M.R., Pandey, M.K., Singh, P.K., Bohra, A., Guo, B., Varshney, R.K. (2023). Application of CRISPR/Cas9-mediated gene editing for abiotic stress management in crop plants. *Front. Plant Sci., 14*, 1157678.
[http://dx.doi.org/10.3389/fpls.2023.1157678] [PMID: 37143874]

Kurreck, J., Stein, C.A. (2025). *Molecular Medicine: An Introduction.*. John Wiley & Sons.

Liao, C., Doilom, M., Jeewon, R., Hyde, K. D., Manawasinghe, I. S., Chethana, K. T., & Dong, W. (2025). Challenges and update on fungal endophytes: classification, definition, diversity, ecology, evolution and functions. *Fungal Diversity, 131*(1), 301-367.

Markowetz, F. (2017). All biology is computational biology. *PLoS Biol., 15*(3), e2002050.
[http://dx.doi.org/10.1371/journal.pbio.2002050] [PMID: 28278152]

McDonald, C. (2016). STEM Education: A review of the contribution of the disciplines of science, technology, engineering and mathematics. *Science Education International,, 27*(4), 530-569.

Jeste, D. V., Lee, E. E., Palmer, B. W., Treichler, E. B. H. (2020). Moving from Humanities to Sciences: A New Model of Wisdom Fortified by Sciences of Neurobiology, Medicine, and Evolution. *Psychological Inquiry, 31*(2), 134-143.

[http://dx.doi.org/10.1080/1047840X.2020.1757984]

Myser, C. (2011). *Bioethics Around the Globe.*. Oxford University Press.
[http://dx.doi.org/10.1093/acprof:osobl/9780195386097.001.0001]

Niyigaba, T., Küçükgöz, K., Kołożyn-Krajewska, D., Królikowski, T., Trząskowska, M. (2025). Advances in Fermentation Technology: A Focus on Health and Safety. *Appl. Sci. (Basel), 15*(6), 3001.
[http://dx.doi.org/10.3390/app15063001]

Noble, D. (2010). Biophysics and systems biology. *Philosophical Transactions of the Royal Society A: Mathematical, Physical and Engineering Sciences, 368*(1914), 1125-1139.
[http://dx.doi.org/10.1098/rsta.2009.0245]

Patel, P. (2023). *The ethical contribution of cultural diversity as a social determinant of health.*Electronic Theses and Dissertations..https://dsc.duq.edu/etd/2248

Purkis, S.J., Klemas, V.V. (2011). *Remote Sensing and Global Environmental Change.*. John Wiley & Sons.
[http://dx.doi.org/10.1002/9781118687659]

Soto, B. (2019). *Fermentation processes.* UK: ED-Tech Press.

Samanta, A., Baranda Pellejero, L., Masukawa, M., & Walther, A. (2024). DNA-empowered synthetic cells as minimalistic life forms. *Nature Reviews Chemistry, 8*(6), 454-470.

Srivastava, A., Kumari, M., Gond, D. P. Subhashini. Basic overview of human physiology. In S. Paul & D. Bhatia (Eds.), *Smart Healthcare for Disease Diagnosis and Prevention* (pp. 193–212).
[http://dx.doi.org/10.1016/B978-0-12-817913-0.00019-5]

Steen, L. A (2005). *Math and bio 2010: linking undergraduate disciplines*Mathematical Association of America. Available from:
https://search.library.stonybrook.edu/permalink/01SUNY_STB/1s7p8na/alma990011378250204856

Thiébaut, R., Thiessard, F., Informatics, S.E. (2017). Public Health and Epidemiology Informatics. *Yearb. Med. Inform., 26*(1), 248-251.
[http://dx.doi.org/10.15265/IY-2017-036] [PMID: 29063574]

Vermeulen, N., Parker, J.N., Penders, B. (2013). Understanding life together: A brief history of collaboration in biology. *Endeavour, 37*(3), 162-171.
[http://dx.doi.org/10.1016/j.endeavour.2013.03.001] [PMID: 23578694]

Wagner, C.S., Roessner, J.D., Bobb, K., Klein, J.T., Boyack, K.W., Keyton, J., Rafols, I., Börner, K. (2011). Approaches to understanding and measuring interdisciplinary scientific research (IDR): A review of the literature. *J. Informetrics, 5*(1), 14-26.
[http://dx.doi.org/10.1016/j.joi.2010.06.004]

Way, G.P., Greene, C.S., Carninci, P., Carvalho, B.S., de Hoon, M., Finley, S.D., Gosline, S.J.C., Lê Cao, K.A., Lee, J.S.H., Marchionni, L., Robine, N., Sindi, S.S., Theis, F.J., Yang, J.Y.H., Carpenter, A.E., Fertig, E.J. (2021). A field guide to cultivating computational biology. *PLoS Biol., 19*(10), e3001419.
[http://dx.doi.org/10.1371/journal.pbio.3001419] [PMID: 34618807]

Yadav, V., Pal, D., Poonia, A.K. (2024). A Study on Genetically Engineered Foods: Need, Benefits, Risk, and Current Knowledge. *Cell Biochem. Biophys., 82*(3), 1931-1946.
[http://dx.doi.org/10.1007/s12013-024-01390-x] [PMID: 39020085]

CHAPTER 9

Bridging Disciplines with a Particular Emphasis on Chemistry

Kulsum Hashmi[1], **Tahmeena Khan**[2,*], **Manisha Singh**[3], **Saman Raza**[1], **Seema Joshi**[1] and **Abdul Rahman Khan**[2]

[1] *Department of Chemistry, Isabella Thoburn College, Lucknow, Uttar Pradesh 226007, India*

[2] *Department of Chemistry, Integral University, Lucknow, Uttar Pradesh 226006, India*

[3] *Department of Education, Integral University, Lucknow, Uttar Pradesh 226006, India*

Abstract: Interdisciplinary education has gained considerable attention in higher education institutions. The focus of interdisciplinary education is to integrate diverse perspectives and methods to solve complex real-life challenges. By integrating insights, methods, and knowledge from multiple academic fields, interdisciplinary teaching provides students with a comprehensive and integrated learning experience. Interdisciplinary learning is a highly complex area in modern education that, despite its importance, has not been sufficiently theorized or researched. The conventional teaching model limits students' development of comprehensive qualities and fails to fully engage students' interests. Several challenges are also presented by interdisciplinary education, despite its potential benefits. Integrating multiple disciplinary perspectives into a cohesive curriculum is one of the main obstacles. Faculty members are required to collaborate across departments and create innovative pedagogical approaches to bridge the gaps between disciplines. Integrated interdisciplinary teaching models have gradually emerged as effective strategies to address these challenges as a result of the continuous evolution of modern educational concepts. This paper investigates the aids and methods of interdisciplinary teaching, including project-based learning, problem-based learning, competency-based learning, and thematic learning. This study aims to explore the effects of interdisciplinary science learning on sustainability-focused education, with a particular emphasis on chemistry's inherent relationship to other disciplines. The integration of sustainable development concerns in teaching presents a productive opportunity as well as a challenge for chemistry education to strengthen its societal dimensions.

Keywords: Biology, Cheminformatics, Chemistry, Informatics, Interdisciplinary education, Methodologies, Sustainable development.

[*] **Corresponding Author Tahmeena Khan:** Department of Chemistry, Integral University, Lucknow, Uttar Pradesh 226006, India; E-mail: tahmeenakhan30@yahoo.com

Tahmeena Khan, Manisha Singh, & Saman Raza (Eds.)

INTRODUCTION: POTENTIAL OF BRIDGING DISCIPLINES

Individuals can enhance their knowledge, intellectual abilities, and critical thinking skills through higher education, which also plays an essential role in promoting personal growth (Manoj & Hussain, 2023). It fosters curiosity and self-exploration and facilitates a better understanding of oneself and the world, with a focus on sustainability (Serafini *et al.*, 2022). In the dynamic realm of higher education, the conventional academic discipline silos are gradually being replaced by a more collaborative and interconnected method termed interdisciplinary education (Kanmaz, 2022). The paradigm shift aims to create a comprehensive learning environment that encourages students to explore various perspectives, approaches, and knowledge domains beyond their disciplines (Mansilla & Duraising, 2007). The advantages and obstacles of interdisciplinary education come to the forefront of pedagogical discussions as universities try to equip their graduates with the flexibility and adaptability required by an increasingly complex world. The motivation for interdisciplinary education comes from the awareness that real-world challenges are intricate and require a wide range of solutions from different disciplines (Repko, 2008).

Recently, interdisciplinary education has gained more attention in higher education institutions. The focus of interdisciplinary education is to integrate diverse perspectives and methods to solve complex real-life challenges (Brassler & Dettmers, 2017). This learning approach is becoming more popular as policymakers and educators acknowledge the need for graduates who can think critically, solve complex problems, and work with others in a variety of disciplines (Guo *et al.*, 2020). By combining insights across various disciplines, students not only develop a better understanding of complex problems but also enhance their critical thinking abilities needed to negotiate the difficulties of a global society that is interconnected (Newell *et al.*, 2001). This perspective aligns with the views of John Dewey, the educational philosopher, who emphasized a curriculum that reflects the interconnectedness of knowledge and prepares students for a multifaceted, dynamic world (Dewey, 1938). There are numerous benefits of interdisciplinary education. First of all, it encourages innovative thinking and creativity by providing exposure to a wide variety of approaches and thought processes (Mansilla & Duraising, 2007). It fosters a comprehensive understanding of the intricate problems, enabling students to approach problem-solving from a variety of perspectives.

Interdisciplinary collaboration reflects the real-world dynamics of professions, in which individuals must collaborate across disciplines to solve complex problems. Therefore, graduates with interdisciplinary skills are in a better position to succeed in a workforce that is becoming more interdisciplinary. The transition to interdisciplinary education is a complex process that presents numerous challenges. When trying to lead institutional structures that usually prioritize discipline-specific silos, faculty members may face challenges.

By integrating insights, methods, and knowledge from multiple academic fields, interdisciplinary teaching provides students with a comprehensive and integrated learning experience. In the field of language learning, interdisciplinary techniques provide opportunities to improve language skills while developing critical thinking, creativity, and problem-solving abilities. Interdisciplinary education addresses complex problems or real-world challenges by integrating knowledge and skills from various subject areas. For example, studying Martin Luther King Jr.'s speech "I Have a Dream," which combines English and History, not only helped students grasp rhetorical skills but also promoted historical and cultural awareness (Repko & Szostak, 2020). Interdisciplinary education allows students to integrate knowledge from various disciplines and apply language proficiency in a variety of contexts in today's globalized world, where problems often span multiple fields. The importance of fostering language skills through interdisciplinary approaches is as follows:

- Improving Vocabulary and Content Knowledge: By integrating science into language lessons, technical terms can be introduced while contextualising learning. For instance, in English, students may encounter terms like renewable resources, carbon footprint, and sustainability while discussing renewable energy. A documentary like *Before the Flood* could be shown by the teacher, followed by student discussions or written reflections about its content (Kunwar & Acharya, 2025).
- Developing Critical Thinking and Analytical Skills: Learners are required to analyse and understand complex information when engaging with interdisciplinary content. For example, global economics could be explored through English by analysing inflation-related articles and writing essays with proposed solutions. This activity simultaneously enhanced the writing, reading, and analytical skills (Kerr *et al.*, 2023).
- Fostering Communication and Collaboration: Collaborative projects that combine various subjects foster teamwork and communication skills. For instance, a project like Create Your Own Business that integrates Economics and English could include brainstorming product concepts, designing presentations, and pitching ideas to classmates acting as mock investors. Authentic use of persuasive language is promoted, and soft skills are developed through this setup (Tomlinson, 1995).
- Enhancing Cultural Awareness: By integrating history and literature, students can analyse cultural nuances by exploring global folklore. Chinua Achebe's

Things Fall Apart not only enhances reading skills but also promotes discussions on cultural conflicts, colonization, and the role of language in identity (Farrell & Jacobs, 2020).

Additionally, students can be better prepared through interdisciplinary education to meet the needs of the modern workforce, where the ability to collaborate across various disciplines is highly valued. However, several challenges are also presented by interdisciplinary education, despite its potential benefits. Integrating multiple disciplinary perspectives into a cohesive curriculum is one of the main obstacles. Faculty members are required to collaborate across departments and create innovative pedagogical approaches to bridge the gaps between disciplines. Additionally, it can be difficult to assess and evaluate interdisciplinary learning outcomes because traditional metrics may not adequately reflect the diverse knowledge and skills that are developed through interdisciplinary education (Hutchings 2023). The theoretical framework of interdisciplinarity is represented in Fig. (**1**).

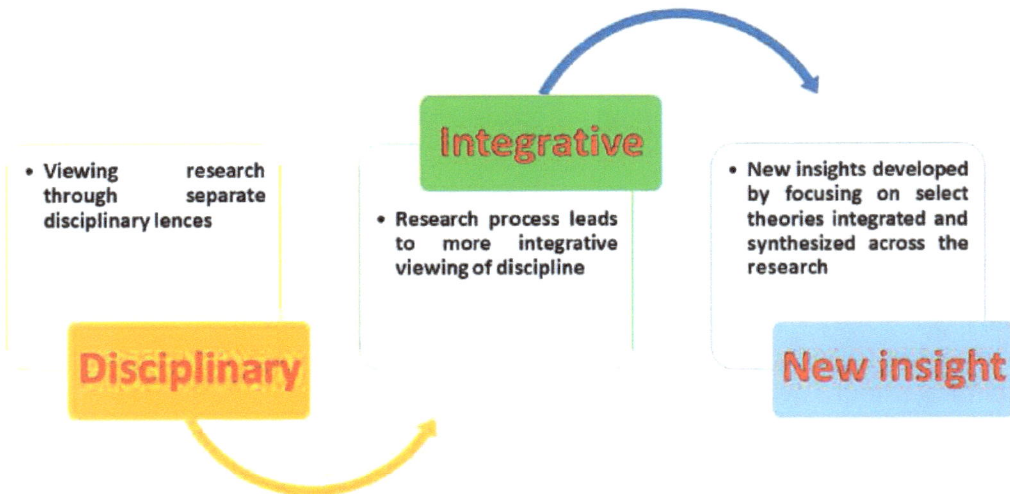

Fig. (1). Steps towards an interdisciplinary theoretical framework.

Teaching Science as an Interdisciplinary Discipline: Aids and Methodologies

In science teaching, experiments are essential, and demonstration experiments are a vital part of the instructional approach. The use of effective demonstration aids can improve the quality of teaching and facilitate a better understanding of the subject matter by students (Kazachkova 2023; Liu & Yang, 2022). Teaching through textbooks and videos can be challenging due to the lack of clear communication and understanding among students. Teaching aids may be effectively used in such cases to reinforce instruction and support students in acquiring new information (Tsai *et al.*, 2020; Yusof & Hasnan 2023; Mao, 2024). Sometimes, teaching aids are designed to stimulate experimental effects that cannot be practically demonstrated or to visualize microscopic phenomena. Interdisciplinary fusion can also be applied to create experimental teaching aids that help to make experimental phenomena more concrete and visual (Nghitoolwa *et al.*, 2024; Akhtar *et al.*, 2017; Jin & Yang, 2022).

STEAM education is a comprehensive approach that focuses on fostering students' creativity, problem-solving abilities, and skills. The quality of students' learning is influenced by learning burnout, which is an important factor. Thus, how STEAM teaching aids can be designed to mitigate learning burnout and enhance students' interest and learning quality has become a topic widely discussed among various researchers (Yee *et al.*, 2021; Kuvondikovna and Hakima, 2024; Xu *et al.*, 2023). Many scholars are currently researching teaching aids on the physiological and psychological aspects of students. The reasons that affect students' experience and learning needs are more complicated because of the variety of ways in which students interact with teaching aids (Mairizwan *et al.*, 2022; Kamaruddin & Hussain, 2024). The challenge of maintaining objectivity with the general weighting methods leads to teaching aid designs that are not closely aligned with users' real needs. In contrast, Structural Equation Modelling (SEM) is effective in evaluating variables like attitude, motivation, and psychological factors that cannot be easily measured in practice. For instance, indirect indicators can be used to measure attitude, motivation, psychology, and other factors, while SEM can be used to examine the intricate relationships among various dependent and independent variables (Yu *et al.*, 2024; Wu *et al.*, 2019).

Analysis of Interdisciplinary Teaching Aid Design

- **Analysis of design objectives**
  - Integration of subject knowledge: Teaching aids should be designed to cover the core knowledge of various disciplines and integrate the concepts and principles of different fields, to foster an interdisciplinary knowledge system among students. For instance, when a teaching aid on environmental protection is designed, it is essential to combine knowledge of biology, chemistry, engineering, and other disciplines. This approach helps students grasp the concept of ecosystems, the chemistry behind pollution control, and the engineering of environmental protection systems.
  - Objectives of fostering students' abilities: The emphasis is on enhancing students' problem-solving, teamwork, innovation, and critical thinking skills. With the help of teaching aids, students are motivated to think independently, explore innovative solutions, and collaborate with peers during project-based learning, thereby enhancing their comprehensive skills

(Singh *et al.*, 2024).

 - Objective of cost control: To maintain the quality and educational impact of teaching aids, the design, production, and use costs of these teaching aids are minimized to enhance their popularity and generalizability. These include post-maintenance costs, labour costs, raw material costs, and other factors.

- **Constraints**
  - Curriculum and syllabus standards and constraints: Teaching aids should be designed according to the appropriate syllabus and course standards to ensure that the covered knowledge and skills are related to the teaching content and can efficiently support the development of teaching activities.
  - Safety constraints: To ensure the safety of students, teaching aids must be free from sharp corners and edges, harmful substances, and other safety risks. Structural design and material selection should be carefully tested and evaluated for safety.
  - Technical feasibility constraints: Based on the present technological and production capacity, teaching aids should be designed for practical implementation using available tools and resources. Teaching aids should be designed in a way that avoids overly intricate designs and the reliance on immature technologies, ensuring they can be produced and implemented on time with high quality.

Researchers proposed various methods for interdisciplinary education. These methods are focused on boosting students' active participation and relating their academic knowledge to practical use. The two most popular learning methods were project and problem-based learning. Career and technical education (CTE), competency-based learning, and thematic learning were also mentioned. Interviewed practitioners reported experience with thematic learning, problem and project-based learning, and CTE.

Methods for Interdisciplinary Education

Project-Based Learning

Project-based learning is a student-centred teaching method. It allows students to experience the process of problem formulation, design, implementation, and evaluation of a project. While doing this, they take the initiative to acquire knowledge, improve their skills, and strengthen their attitudes. Collaboration, inquiry, comprehensiveness, and authenticity are qualities of project-based learning that can successfully encourage students' deep learning and critical thinking abilities (Gao *et al.*, 2024).

Problem or project-based learning is regarded as an effective technique for interdisciplinary education, as it starts with a pressing issue that needs to be addressed. Researchers suggested that project-based learning reflects the belief that interdisciplinary education is the most effective way to teach children (Sutton & Knuth, 2017). This is our best answer to that question: What is interdisciplinary learning?" Veronica Boix Mansilla emphasized the effectiveness of project-based learning because it begins with the problem. It starts with the problem and proceeds by identifying the disciplines that can aid in addressing it (Mansilla, 2006). It has been observed by some researchers that project- or problem-based learning is not inherently interdisciplinary and must be properly designed to effectively support student learning. Leslie Eaves recommends that teachers start with project-based learning in one academic discipline to gain experience before integrating it across multiple subjects. Ron Berger suggested "that a rigorous standard is beneficial for project-based learning. The PBLWorks Gold Standard Project-Based Learning model can be used in place of other ones, as it is considered similar to those of Big Picture, High Tech High, EL, and New Tech." (Berger *et al.*, 2016). PBL may be among the best methods for interdisciplinary teaching, but its effectiveness ultimately relies on how well it is implemented, which depends on the training provided. Does the instructional strategy still need to focus on what students are trying to solve and what instruments they need to do it?" (Warkentien *et al.*, 2022).

Other methods of Interdisciplinarity

- Thematic learning or thematic integration, a well-established interdisciplinary method, is typically implemented in elementary classrooms where one teacher provides instruction across subjects throughout the day. Science, Math, and literacy experiences are integrated by teachers within thematic units. A district leader stated that "selecting a theme becomes the intersectionality for all disciplines." For instance, students explore rainforests by examining the history, products, and economy of the rainforest, and ecosystems with diverse fauna and flora. As explained by the district leader, thematic learning differs from project-based learning in that it may not involve producing a notable product or project for audiences beyond the classroom.
- Researchers have cited competency-based education as an approach that promotes and facilitates interdisciplinary education. Well-developed competencies often integrate multiple subject areas. Many standards are transferable, cross-competency, and wide in the sense that they can be used in any field after being taught in another. According to Eliot Levine of the Aurora Institute, project work that includes performance-based assessments plays a central role in competency-based education. For instance, in New York, a competency-based school categorized competencies into ten broad groups that

are fundamentally cross-disciplinary. Teachers in every subject implemented Shared Outcomes that embed critical "21st-century skills" needed for success in college, careers, and beyond, such as argument, precision, collaboration, communication, conclusion, creativity, discernment, innovation, investigation, and planning (Levine & Patrick, 2019).

- Unnamed interdisciplinary learning: Ron Berger (Berger *et al.*, 2016) referred to "unnamed" interdisciplinary learning that is already present in regular school classes and activities. For example, learners are researching and writing a science class report using their English skills. Drama, orchestra, band, debate club, student government, school newspaper, Model United Nations, and many more extracurricular activities are interdisciplinary. The "unnamed" interdisciplinary experiences are frequently the most impactful experiences of high school because they offer a realistic glimpse into life beyond school. Engaging in activities like writing papers, performing, acting, or debating means you are creating something meaningful, just as you would outside the classroom. These experiences are not interdisciplinary, but they are interdisciplinary learning and are beneficial for students.

CHEMISTRY FOR SUSTAINABLE DEVELOPMENT

Owing to the technical and practical utility of the subject, chemistry has acquired a crucial role in a sustainable future. The meeting of Sustainable Development Goals (SDGs) to solve various environmental problems largely relies on the solutions offered by chemistry and interdisciplinary approaches to mitigate climate change and biochemical flows. The realization of SDGs needs interdisciplinary collaborations as they cannot be fulfilled by a single discipline, practically or theoretically, in the context of teaching and learning. Geographical perspectives must also be added for the strengthening of education for sustainable development (ESD). Geography as a subject is devoted to the analysis of the human-environment interaction and also takes into account the social perspectives, as a part of natural sciences. This multidisciplinary method can enhance the educational value and quality of science learning, particularly chemistry education.

Chemistry is an important subject for a sustainable future and accomplishing SDGs (Matlin *et al.*, 2022). However, as the subject is related to the study and emission of chemicals, it is often perceived in a contradictory direction. Nevertheless, chemistry is making an impact in the technological field with a drastic change in conventional chemical practices for sustainability. Green Chemistry or Green and Sustainable Chemistry (GSC) (UNEP 2019) are some safe and benign chemistry practices. Despite these advancements, the concept of sustainability is not fully covered and developed in chemistry curricula and books

(Herranen *et al.*, 2021) in secondary as well as in higher chemistry education, and the gap remains there to implement the ESD in chemistry education (Millar *et al.*, 2025). There are connections established between chemistry education and the idea of the Anthropocene era (Lowa *et al.*, 2014). The integration of sustainable development concerns in teaching offers a productive opportunity as well as a challenge for chemistry teaching to strengthen societal dimensions. The potential role of chemistry in meeting SDGs is presented in Table **1** (Pan *et al.*, 2019; Water, 2018; Stuckenberg & Contento, 2018; Vaughan, 2012; Linder, 2017; Kümmerer *et al.*, 2020; Slootweg, 2020; Iaquaniello *et al.*, 2018).

Table 1. Chemistry and the SDGs.

SDG	Role of Chemistry
1 NO POVERTY	Employment generation in chemical industries and plants.
2 ZERO HUNGER	Generation of sustainable and novel solutions for agricultural practices and high crop yields.
3 GOOD HEALTH AND WELL-BEING	Drug discovery and the affordability of medicines.
4 QUALITY EDUCATION	Lifelong learning of basic chemistry principles and their implementation in daily life.
5 GENDER EQUALITY	Providing equal opportunities for employment, particularly in small and medium-scale industries.

(Table 1) cont.....

SDG	Role of Chemistry
6 CLEAN WATER AND SANITATION	Development of novel water purification methodologies and materials used for water purification on a large scale.
7 AFFORDABLE AND CLEAN ENERGY	Creation of new materials and substances as clean energy sources.
8 DECENT WORK AND ECONOMIC GROWTH	More job opportunities for chemistry graduates in different sectors.
9 INDUSTRY, INNOVATION AND INFRASTRUCTURE	Development of industries and promotion of entrepreneurship, invention.
10 REDUCED INEQUALITIES	Reduction of disparity between developed and developing nations by promoting more job opportunities for chemistry students working on the interface with other allied streams.
11 SUSTAINABLE CITIES AND COMMUNITIES	Promotion of sustainable sources of energy and resources.
12 RESPONSIBLE CONSUMPTION AND PRODUCTION	Conservation of non-renewable resources by inventing new approaches and materials.

(Table 1) cont.....

SDG	Role of Chemistry
13 CLIMATE ACTION	Combating climate change by promoting green chemistry principles and sustainable environmental protection.
14 LIFE BELOW WATER	Curb water pollution to save aquatic flora and fauna.
15 LIFE ON LAND	Conservation and restoration of ecosystems and land animals through environment-friendly practices involving chemistry and other allied fields.
16 PEACE, JUSTICE AND STRONG INSTITUTIONS	Conception of inclusive societies with special measurable goals based on principles of chemistry.
17 PARTNERSHIPS FOR THE GOALS	Establishment of a balance between industry, academia, and communities.

Chemistry as an Interdisciplinary Subject

Chemistry, biology, mathematics, physics, environmental science, *etc*., are all different branches of science that are related to each other. The reaction of photosynthesis is a chemical process and a classic example of the conjugation of biology with chemistry. Animals synthesize proteins and amino acids, produce glucose, *etc*. The animal bodies are a storehouse of chemicals, relating biology with chemistry. Chemistry is related to physics as well; Physical chemistry is a branch of chemistry, and the underlying theories of physical chemistry and formulas of physics are interrelated. Like physical chemistry, mathematical formulas and theories are the basis of the formulas and measurements used in chemistry.

Chemistry as a subject has acquired a multidisciplinary status as it is subsumed into many different subjects like molecular biology (Bustin, 2015), biochemistry (Lang and Bodner, 2020), and pharmacy (Gupta *et al.*, 2017). The practical implementation of chemistry in biosciences has integrated it with bioengineering and biomedicine curricula. Chemistry holds significant potential in realizing the SDGs adopted by the United Nations (UN) member states to end poverty and protect the planet (Axon & James, 2018; Anastas & Zimmerman, 2018). Chemistry, through its integration with biology, engineering, mathematics, and physics, plays a pivotal role in economic sectors and industries with a multibillion-dollar turnover. Maslow's hierarchy of needs can also explain the real-world problems involving chemistry Neubauer & Martskvishvili, 2018). The industries catering to the population with sustainable resources work on the interface of chemistry with biology, engineering, mathematics, and physics. The safety and security needs also involve multidisciplinary research on, for example, health (medicines, personal protective equipment). The advancements in computer hardware and software have offered opportunities in curricula worldwide. Computational aspects of chemistry have offered implementations in online teaching.

Cheminformatics is a field that combines chemistry and computer science, focusing on the indexing, classification, and retrieval of information about various chemical compounds (Harvey, 2018; Willett, 2020; Voicu *et al.*, 2020).

It is used to deal with large data sets through artificial intelligence (AI), and machine learning (ML), particularly in drug discovery and development, and in advanced genetics and molecular biology studies in search of new drug targets and medicines, to increase productivity (Hathout & Metwally, 2016). The correlations obtained from the ML/AI approaches in deep learning are used to facilitate the identification of biological processes, the discovery of possible drug molecules, to know the biological consequences of the compounds, and to help in clinical trials.

Computational chemistry provides students with a virtual platform to learn important concepts such as thermodynamics, kinetics, stereochemistry, and molecular descriptors—including physicochemical, electronic, and constitutional parameters—while also enabling the exploration of molecular interactions with biological carriers or receptors for drug delivery using minimal resources and limited chemical exposure (Hathout *et al.*, 2020). Computational chemistry can be taught with the integration of mathematics and computational programming. The complexity must be explained, taking into account the background of the student, to facilitate the delivery of the concepts of computational chemistry for real-life multidisciplinary teamwork in different areas.

Coordination Chemistry is another interdisciplinary domain of chemistry through which the complexation of inorganic metal ions with organic ligands leads to the formation of new molecules with different properties from precursors. The practical applicability of these new molecules can be used through artificial modelling in biological systems, biotechnologies for the targeted synthesis of enzymes, and in the pharmaceutical and medicine industries, synthesis of materials with essential properties like semiconduction, photoluminescence, *etc.* The prime goal of contemporary methodologies is to identify the link between education and research, to create a space to implement the research findings in teaching methodologies, and to motivate students towards research. Through quantum calculations, the energy of the chemical systems can be predicted. For studying interactions between different molecules, several programs are used, of which GAMESS is an important program containing different aspects of molecular dynamics and mechanics.

Chemistry, Geography, and Ecology

Developing awareness in the students of environmental culture requires teaching geography and ecology with respect to biology, chemistry, and mathematics. The mitigation of environmental problems needs complex measures and must be dealt with from different aspects, which requires knowledge obtained from different disciplines. For instance, biology focuses on the interactions between living organisms and their environment, including the biosphere and ecosystems, whereas chemistry is a subject dealing with the release of chemicals in the atmosphere and their impacts, along with the study of mineral resources, fertilisers, *etc.* Physics provides an overview of the radioactive substances and how they affect living organisms. The integration of chemistry with ecology and geography contributes to the development of flexible thinking and provides practical solutions to the theoretical concepts, and to organise the content in an improved way, resulting in the acquisition of knowledge and blending disciplines. The interdisciplinarity of the problem and its address depend on the nature of the transfer among different disciplines. To promote interdisciplinarity among the students, the teachers can implement compulsory correlations stipulated in the curricula based on logic. Lesson plans can also be developed through teamwork and group work (Petrova & Fadeeva, 2018).

Chemistry and Biology

Scientists exploring new frontiers in highly qualified, specialized fields are responsible for dividing our knowledge into discrete faculties. Traditional scientific disciplines are insufficient in addressing real-world problems like climate change, economic development, and overcoming systemic diseases due to their rigid boundaries. Scientists are encouraged to engage in multidisciplinary, interdisciplinary, and transdisciplinary collaboration to tackle today's most complex problems by combining their knowledge and expertise (Frodeman *et al.*, 2017). Multidisciplinarity is the phenomenon where people from different disciplines work together, each using their own knowledge. Interdisciplinarity refers to the integration of knowledge and techniques from various disciplines through a real synthesis of methods. A phenomenon that combines intellectual structures outside of the disciplinary perspective is called trans disciplinarity (Menken & Keestra, 2016).

One well-known example of transcending traditional domains of expertise and producing innovative, very significant research findings is the study of the interface between chemistry and biology. Over the past few decades, significant progress has been made in understanding mechanistic biological chemistry, and the conventional perspective on the roles of metals and their complexes in biological systems has steadily expanded—reflected in their growing presence at conferences such as Metals in Biology, Metallomics, *etc*. The primary target of multidisciplinary research is to investigate the interactions between inorganic complexes and biomolecules. In biological systems, the study of polyoxometalates (POMs) has led to unprecedented and unpredictable discoveries, resulting from interdisciplinary efforts. The examples demonstrate that off-type chemistry can only be found at intersections across various disciplines, such as POMs-biomolecules research, which can be expanded and transferred to other research areas.

The use of POMs in multiple biological systems has been a rapidly expanding field of research since 1970, when the inhibitory effect of the silicotungstic acid $H_4[Si^{IV}W^{VI}{}_{12}O_{40}]$ on sarcoma viruses and murine leukaemia was first reported (Chermann *et al.*, 1970). In the past decade, papers on POMs published in the field of biological applications have tripled. There were 115 publications in 2009 to 443 in 2019, according to a Scopus database search in August 2020 for the term "polyoxometalate AND application". These publications were later limited to the disciplines of "Biochemistry, Genetics and Molecular Biology", "Pharmacology, Toxicology and Pharmacy", and "Medicine". The antiviral (Zhang *et al.*, 2023), insulin-mimicking (Dinčić *et al.*, 2020), anticancer (Bijelic *et al.*, 2019), and antibiotic (Bijelic *et al.*, 2018) activities of different POMs have been extensively reported in the literature. Various biological processes, including β-amyloid aggregation (Zhao *et al.*, 2019) and P-type ATPase activity (Gumerova *et al.*, 2018), are inhibited by POMs. They can also function as artificial enzymes (Van Rompuy & Parac-Vogt, 2019). Due to their flexibility and reactivity, similar to other inorganic coordination compounds, the identity and stability of POMs depend on reaction conditions such as pH, buffer type, and interactions with other components in the environment, making an interdisciplinary approach essential to better understand their role in biological systems (Gumerova & Rompel, 2020). A comprehensive understanding of inorganic chemistry, biochemistry, and biology is necessary for the integration and assessment of the biological activity of POMs.

In interdisciplinary teaching, students gain both traditional knowledge and a deeper, multidimensional perspective on complex issues by integrating insights from various fields. This model of education prompts learners to look beyond individual subjects and develop an integrated cognitive framework. Students can improve their ability to solve real-world problems by integrating their knowledge from chemistry and biology in more complex situations (Fig. **2**). The education sector is shifting its focus from content delivery to skill development, focusing on critical analysis, innovative thinking, and effective problem-solving skills. By combining knowledge from multiple disciplines, learning outcomes are improved, and students gain the necessary abilities to solve modern societal problems, providing a strong basis for their future development (Ding, 2024).

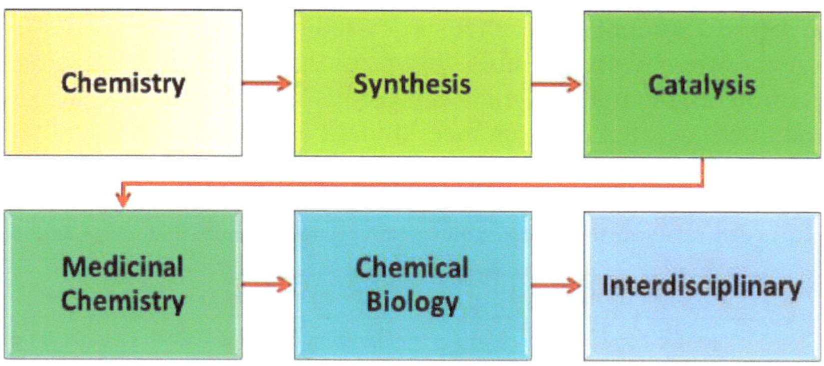

Fig. (2). Interdisciplinarity between chemistry and biology.

Biology, an essential branch of natural science, is intertwined with other disciplines like chemistry. Biology embraces various chemical concepts, including the reactions involved in cellular respiration and photosynthesis, as well as the structure and function of biological molecules like proteins and DNA. The integration of biology and chemistry helps students comprehend the mechanisms of chemical reactions in biological processes, as well as fundamental concepts such as metabolic pathways and molecular interactions.

The implementation of integrated chemistry-biology courses at some schools presents valuable learning experiences, but it also requires dedicated faculty support and flexible curricular planning (Wolfson, 1998; Mahaffy *et al.*, 2019). Schools often have strict curriculum standards for general chemistry, which are not conducive to course redesign. For instance, interdisciplinary units are typically excluded from high school courses with state or AP syllabi and from college courses that complete general chemistry within one semester. Students are ideally motivated and prepared through an introductory chemistry course to understand chemical concepts across disciplines and address interdisciplinary challenges, without the course content being diminished.

Table **2** (Taylor *et al.*, 2009) provides a summary of course lecture topics and related interdisciplinary examples. Examples frequently illustrate the relevance or practical implications of a topic. For instance, when atomic size and periodic trends are introduced, the importance of a fraction of a nm in proper neuron signalling is emphasized by neurons' ion channel selectivity, which can distinguish between sodium and potassium ions. This helps answer the question, 'Why should we care about periodic trends?' Other examples are given to the students to introduce the types of problems that chemistry might one day be used to solve. When the Nernst equation for redox chemistry was first presented in class, the exercise was replaced with a problem on how vitamin B_{12} is reduced inside the body. Students use their knowledge of polar covalent bonds to differentiate between water-soluble vitamins that are easily excreted and fat-soluble vitamins that can accumulate dangerously, in another lecture. Classroom response devices, or clickers, enhance student engagement in problem-solving examples in the classroom.

Table 2. Topics in general chemistry lectures, along with related examples in biology and medicine.

S.No.	Chemistry lecture topics	Biology-related examples
1.	Periodic trends	The atomic size of Na^+ ion channels in neurons
2.	Lewis structures, Covalent bonds	Thionyl chloride is used for the production of Novocaine. Cyanide ion is found in cassava plants and cigarettes..
3.	Exceptions to Lewis structure rules	Free radicals in biology: role in DNA damage and essential for life
4.	Polar covalent bonds, ionic bonds	The study compares folic acid and vitamin A, focusing on water-soluble vitamins and fat-soluble vitamins.
5.	VSEPR theory	Molecular shape in enzyme substrate complexes
6.	Thermodynamics	Oxidation of glucose to derive energy from plants
7.	Balancing redox equations, electrochemical cells	Oxidative metabolism of drugs
8.	Oxidation-reduction reactions	Vitamin B_{12} is reduced within the body.
9.	Rate laws	Kinetics of glucose oxidation in the body.
10.	Biochemistry	Case study on methionine synthase.

Students are expected to apply their chemistry problem-solving abilities in biology-based assignments, where interdisciplinary thinking is reinforced as a core aspect of the course. For instance, by using Lewis structure rules for free radicals, learners can predict which metabolism by-products are highly reactive radical species. The interconnection between biology and medicine provides a framework for chemistry problems, such as identifying hydrogen bond donors or acceptor atoms in cancer drugs.

Chemistry has been integrated into a wide range of biological subjects due to its success as a method of understanding the natural world. Biochemistry (Mehler, 1983; Lang & Bodner, 2020), Molecular biology (Nguyen *et al.*, 2025), and pharmacology and pharmacy (Mozafari *et al.*, 2021) are all areas that are influenced by chemistry, but have long since grown into separate, independent fields. The focus on observing biological processes as systems makes it impractical to teach every aspect from a detailed chemical perspective. For instance, the majority of bacterial gene expression depends on the binding interactions between DNA and macromolecules, but teaching and comprehending gene expression at this level would take too long. Chemistry-related bioscience subjects have been significantly integrated into other areas of biology, bioengineering, and biomedicine due to their practical value.

Informatics and its Association with Chemistry and Biology

Informatics is the field that studies the structure, behaviour, and interaction of natural and synthetic systems. Chemistry, biology, and biomedical sciences need data obtained by different methodologies, such as experimental observations or computer-aided simulations (Gasteiger, 2020). High-throughput synthesis and proteomics, glycomics, *etc.*, are different aspects of informatics. With the advent of "big data," the field of informatics is also growing to strengthen, store, and process models and their analysis to predict information (Jabeen *et al.*, 2023). For instance, between 2006 and 2018, the DrugBank database experienced a significant increase in drug-related data, with a 300% in investigated drugs and 600% in drug-drug interaction data (Wishart *et al.*, 2018). Within informatics, specific research disciplines working on the interface have been covered. In today's times, it has become unimaginable to develop chemistry and biology without the help of computers and specialized software, making informatics an indispensable tool to store, process and analyse heterogeneous data. Based on different algorithms, predictive models are derived, complementing the experimental information.

The studies of molecular systems, including their chemical and biological properties, and their simulation, are integral to applied research, like designing new schemes of synthesis in chemistry and simulation of physiological processes and their dynamics, yielding new hypotheses in biological sciences. Both require computational visualization to represent chemical and biological spaces of increased complexity (López-López *et al.*, 2020).

Cheminformatics

Cheminformatics uses computers and formulates computational techniques in chemistry for better decision making (Azad *et al.*, 2021) and designing, analysing, and disseminating chemical and allied information. It is an interdisciplinary domain, explored and taught in different universities globally. A huge amount of data is produced in chemical science, which is too complex to comprehend the structure of compounds and their biological activity, and requires processing and analysis through theoretical calculation. For better processing and visualization of results, Cheminformatics has been built. The history of cheminformatics dates to the 1950s, when it was established to address problems related to the representation, elucidation of structures, and designing of new synthesis schemes, the collection of data, and its analysis with a particular emphasis on drug discovery. The field of cheminformatics mixes this information to convert data into information and knowledge for better decision-making and drug identification. Cheminformatics works on the principles of computational and quantum chemistry and force field-based molecular modelling. It has been debated that cheminformatics and bioinformatics are parts of biomedical informatics (Romano & Tatonetti, 2019). However, cheminformatics focuses on drug discovery. Food informatics is a new branch of cheminformatics introduced in 2014 and encompassing the informatics of food chemicals (Rojas *et al.*, 2021). Polymer informatics and materials informatics are also emerging as the new subdisciplines (Rickman *et al.*, 2019), expanding cheminformatics into different areas of chemistry (Audus & Pablo, 2017). The increasing usefulness and popularity of the field has created a need to write textbooks, new courses, and degree programmes to aid formal training. With the high demand for researchers working on cheminformatics the chemistry curricula must be modified accordingly to introduce cheminformatics, either as a part of training or at the master's level.

The main point to be addressed through different informatics is the data to be processed and examined, such as chemical structures, reactions, peptides, and gene sequencing. Different data requires special representation to handle, process, and analyse domain-specific data. The complexity of the data poses challenges to organizing, archiving, processing, and analysing data for knowledge extraction. Bioinformatics deals with a much larger amount of information than cheminformatics (Pence & Williams, 2016). The potential applications of cheminformatics (Fig. **3**) are:

- In analytical chemistry, it is used to analyse data about the quality and other parameters related to the origin of the objects (Bajorath, 2024).
- In drug discovery, cheminformatics is used to identify new lead molecules and optimize their structures, and screening. (Khan *et al.*, 2019).
- It is required to store and retrieve chemical structures and sort large data sets computationally. This can also be used as an e-learning tool and as a platform where knowledge and data sharing are feasible between researchers and learners.

Cheminformatics also holds the potential to analyse and simplify large data sets, and promotes interdisciplinary research involving chemistry, information science, informatics, *etc*. The field also endorses molecular modelling, simulation of structure, and reaction presentation.

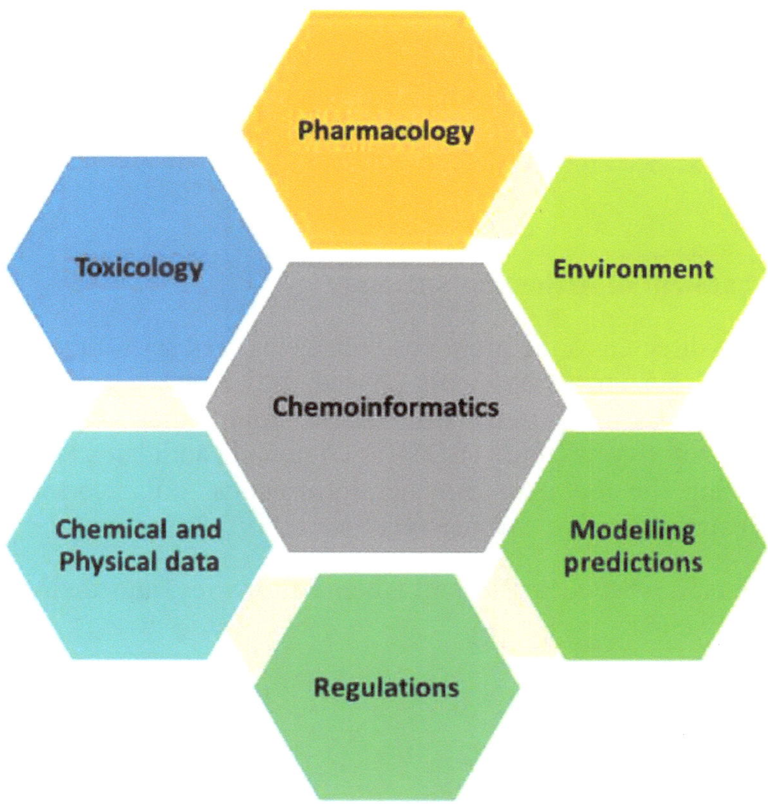

Fig. (3). Key applications of cheminformatics.

- **Bio and Cheminformatics in the Identification of Novel Pyrazole and Benzimidazole-Based Derivatives as Penicillin-Binding Protein 2a Inhibitors**

Synthesis of ten compounds based on pyrazole and benzimidazole was performed to determine their antibacterial activity against two strains of *S. aureus,* viz. MSSA ATTC6538 and MRSAUSA300. The compounds were docked with PBP2a and bound similarly as lead quinazolinone PBP2a inhibitors, exhibiting the same mechanistic approach (Shalaby *et al.*, 2019). Below are some examples of drugs and therapies developed on the interface of bio and cheminformatics.

- **Optimization of Drug Delivery in Targeting Meningitis**

Bacterial meningitis is caused by Group B *Streptococcus agalactiae, Listeria monocytogenes,* and *Escherichia coli*, and by Haemophilus influenzae. Acute cases are associated with morbidity and mortality (Oordt-Speets *et al.*, 2018). Bio and cheminformatics have been used to analyse antimicrobial drugs for an effective nasal to brain delivery formulation targeting meningitis, with cephalosporin antibiotics like cefotaxime and ceftriaxone working on biopharmaceutical levels. Molecular dynamics simulations were performed through GROMACS v4.6.5, showing high affinity of ceftriaxone (Hathout *et al.*, 2020).

- **For the Treatment of Alzheimer's Disease**

As medications directed to the brain pose obstacles, a nose-to-brain targeting has been developed. A research work based on the concepts of bio and cheminformatics has evaluated two natural compounds, curcumin and its congener bisdemethoxycurcumin (BDMC), as potent candidates for nose-to-brain therapy, showing the usage of bio/cheminformatics tools working in a cost, resource, and time-effective manner (Hathout *et al.*, 2018).

The intercorrelation between Bio and Cheminformatics, and their outcomes, is represented in Fig. (**4**)

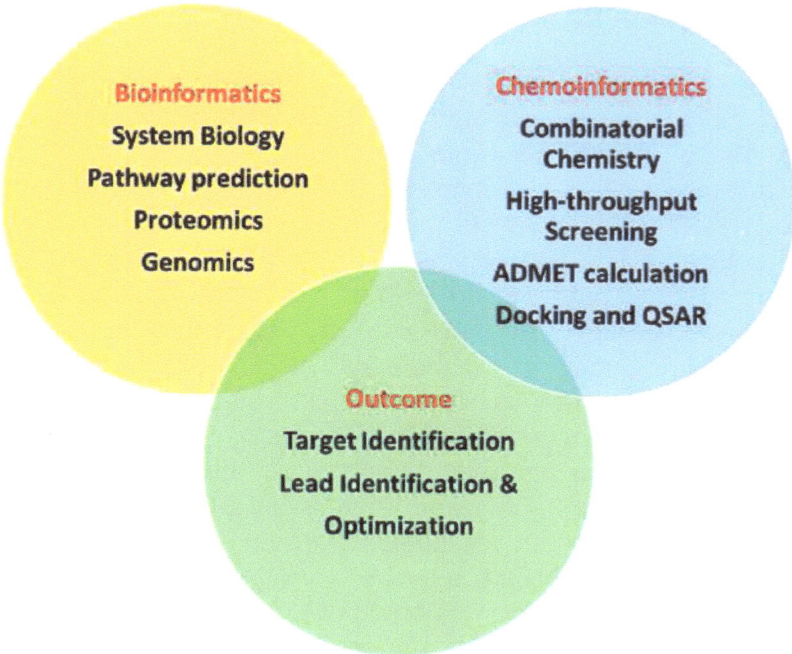

Fig. (4). The intercorrelation between Bio and cheminformatics.

Amalgamation of Cheminformatics and Bioinformatics Integration with Artificial Intelligence (AI)

Clinical trials suffer primarily due to the issues emerging from the unfavourable reactions in the body and the availability of the drugs. Since *in vitro* and *in vivo* experiments are costly and time-consuming, as a result, computational models anticipating the absorption, distribution, metabolism, excretion, and toxicity (ADME-Tox) properties have been established to streamline the process of drug development and risk assessment of new compounds (Raza *et al.*, 2024). AI models have significantly enhanced the accuracy of the prediction of drug efficacy and safety based on ADME-Tox data sets, working on the principles of machine learning (ML)/deep learning (DL) (Singh *et al.*, 2024).

Therapeutic Data Commons (TDC) is an open Python library designed for systematic access and evaluation of ML models (Huang *et al.*, 2021). Different models have been developed to determine properties that require limited experimental findings. The efficacy of multitask learning based on graph neural networks (MT-GNN) as compared to other ML techniques relying on *in vitro* data has been established (Hamzic *et al.*, 2022). ML and DL can also be integrated in disease genomics, systems biology, protein structure analysis, and molecular evolution.

CONCLUDING REMARKS

This chapter provided an overview of interdisciplinary education with a particular emphasis on chemistry. Interdisciplinary teaching provides students with a comprehensive and integrated learning experience. It is an effective method for enhancing language skills. The integration of various disciplines in learning activities fosters enhanced linguistic proficiency, critical thinking, and cultural awareness among students. Interdisciplinary teaching is a teaching approach that helps students develop their versatility and communication skills. This study investigates the theoretical framework and implementation practices of interdisciplinarity, focusing on techniques such as project-based learning, problem-based learning, competency-based learning, and thematic learning to achieve cohesive connections with other subjects. The study suggests that interdisciplinary integration significantly enhanced students' innovative thinking, cognitive abilities, and comprehensive application skills, leading to improved practical outcomes in chemistry teaching. Future investigations should focus on improving collaboration between teachers of various disciplines, enhancing their capacity to carry out interdisciplinary teaching, revising assessment approaches for integration, and utilizing technology in education to encourage the broad application of interdisciplinary teaching across various disciplines. Interdisciplinary integration, which takes place through ongoing exploration and innovation, will provide more opportunities for middle school students and promote the overall advancement of educational quality.

REFERENCES

Akhtar, S., Warburton, S., Xu, W. (2017). The use of an online learning and teaching system for monitoring computer aided design student participation and predicting student success. *Int. J. Technol. Des. Educ., 27*(2), 251-270.
[http://dx.doi.org/10.1007/s10798-015-9346-8]

Anastas, P.T., Zimmerman, J.B. (2018). The United Nations sustainability goals: How can sustainable chemistry contribute?. *Curr. Opin. Green Sustain. Chem., 13*, 150-153.
[http://dx.doi.org/10.1016/j.cogsc.2018.04.017]

Audus, D.J., de Pablo, J.J. (2017). Polymer informatics: opportunities and challenges. *ACS Macro Lett., 6*(10), 1078-1082.
[http://dx.doi.org/10.1021/acsmacrolett.7b00228] [PMID: 29201535]

Axon, S., James, D. (2018). The UN Sustainable Development Goals: How can sustainable chemistry contribute? A view from the chemical industry. *Curr. Opin. Green Sustain. Chem., 13*, 140-145.
[http://dx.doi.org/10.1016/j.cogsc.2018.04.010]

Azad, I., Khan, T., Maurya, A.K., Irfan Azad, M., Mishra, N., Alanazi, A.M. (2021). Identification of Severe Acute Respiratory Syndrome Coronavirus-2 inhibitors through in silico structure-based virtual screening and molecular interaction studies. *J. Mol. Recognit., 34*(10), e2918.
[http://dx.doi.org/10.1002/jmr.2918] [PMID: 34132436]

Bajorath, J. (2024). Milestones in chemoinformatics: global view of the field. *J. Cheminform., 16*(1), 124.
[http://dx.doi.org/10.1186/s13321-024-00922-0]

Berger, R., Woodfin, L., Vilen, A. (2016). Learning that lasts: Challenging, engaging, and empowering students with deeper instruction. John Wiley & Sons.

Bijelic, A., Aureliano, M., Rompel, A. (2018). The antibacterial activity of polyoxometalates: structures, antibiotic effects and future perspectives. *Chem. Commun. (Camb.), 54*(10), 1153-1169.
[http://dx.doi.org/10.1039/C7CC07549A] [PMID: 29355262]

Bijelic, A., Aureliano, M., Rompel, A. (2019). Polyoxometalates as potential next-generation metallodrugs in the combat against cancer. *Angew. Chem. Int. Ed., 58*(10), 2980-2999.
[http://dx.doi.org/10.1002/anie.201803868] [PMID: 29893459]

Brassler, M., Dettmers, J. (2017). How to enhance interdisciplinary competence—interdisciplinary problem-based learning versus interdisciplinary project-based learning. *Interdiscip. J. Probl. Based Learn., 11*(2)
[http://dx.doi.org/10.7771/1541-5015.1686]

Bustin, S.A. (2015). Molecular Biology of the Cell, Sixth Edition; ISBN: 9780815344643; and Molecular Biology of the Cell, Sixth Edition, The Problems Book; ISBN 9780815344537. *International Journal of Molecular Sciences, 16*, 28123-28125.

Chermann, J.C., Raynaud, M., Jasmin, C., Mathé, G. (1970). Powerful new inhibitor of murine leukaemia and sarcoma viruses. *Nature, 227*(5254), 173-174.
[http://dx.doi.org/10.1038/227173a0] [PMID: 4316954]

Dewey, J. (1938). *Experience and education.* Kappa Delta Pi.

Dinčić, M., Čolović, M.B., Matutinović, M.S., Ćetković, M., Stevović, T.K., Mougharbel, A.S., Todorović, J., Ignjatović, S., Radosavljević, B., Milisavljević, M., Kortz, U. (2020). In vivo toxicity evaluation of two polyoxotungstates with potential antidiabetic activity using Wistar rats as a model system. *RSC Adv., 10*(5), 2846-2855.
[http://dx.doi.org/10.1039/C9RA09790B]

Ding, Q. (2024). Interdisciplinary integration theory and practice paths in middle school biology education. *International Journal of Educational Teaching and Research, 1*, (2).
[http://dx.doi.org/10.70767/ijetr.v1i2.206]

Farrell, T.S., Jacobs, G.M. (2020). Essentials for successful English language teaching. Bloomsbury Publishing.

Frodeman, R., Klein, J.T., Pacheco, R.C.D.S. (2017). The Oxford handbook of interdisciplinarity. Oxford University Press.
[http://dx.doi.org/10.1093/oxfordhb/9780198733522.001.0001]

Gao, H., Hashim, H., Md Yunus, M. (2024). The effect of project-based learning (PjBL) Class Model on Chinese undergraduates' argumentative writing performance. *Cogent Education, 11*(1), 2396715.
[http://dx.doi.org/10.1080/2331186X.2024.2396715]

Gasteiger, J. (2020). Chemistry in times of artificial intelligence. *ChemPhysChem, 21*(20), 2233-2242.
[http://dx.doi.org/10.1002/cphc.202000518] [PMID: 32808729]

Gumerova, N., Krivosudský, L., Fraqueza, G., Breibeck, J., Al-Sayed, E., Tanuhadi, E., Bijelic, A., Fuentes, J., Aureliano, M., Rompel, A. (2018). The P-type ATPase inhibiting potential of polyoxotungstates. *Metallomics, 10*(2), 287-295.
[http://dx.doi.org/10.1039/C7MT00279C] [PMID: 29313547]

Gumerova, N.I., Rompel, A. (2020). Polyoxometalates in solution: speciation under spotlight. *Chem. Soc. Rev., 49*(21), 7568-7601.
[http://dx.doi.org/10.1039/D0CS00392A] [PMID: 32990698]

Guo, P., Saab, N., Post, L.S., Admiraal, W. (2020). A review of project-based learning in higher education: Student outcomes and measures. *Int. J. Educ. Res., 102*, 101586.
[http://dx.doi.org/10.1016/j.ijer.2020.101586]

Gupta, A., Low, W.L., Britland, S.T., Radecka, I., Martin, C. (2018). Physicochemical characterisation of biosynthetic bacterial cellulose as a potential wound dressing material. *British Journal of Pharmacy, 2*(2) [http://dx.doi.org/10.5920/bjpharm.2017.27]

Hamzic, S., Lewis, R., Desrayaud, S., Soylu, C., Fortunato, M., Gerebtzoff, G., Rodríguez-Pérez, R. (2022). Predicting in vivo compound brain penetration using multi-task graph neural networks. *J. Chem. Inf. Model., 62*(13), 3180-3190. [http://dx.doi.org/10.1021/acs.jcim.2c00412] [PMID: 35738004]

Mao, Haobo (2024). Research on the design of teaching aid resource system based on unsupervised representation learning. *Journal of Electrical Systems, 20*(2), 211-225. [http://dx.doi.org/10.52783/jes.1161]

Harvey, J. (2018). Computational chemistry. Oxford University Press. [http://dx.doi.org/10.1093/hesc/9780198755500.001.0001]

Hathout, R.M., Abdelhamid, S.G., El-Housseiny, G.S., Metwally, A.A. (2020). Comparing cefotaxime and ceftriaxone in combating meningitis through nose-to-brain delivery using bio/chemoinformatics tools. *Sci. Rep., 10*(1), 21250. [http://dx.doi.org/10.1038/s41598-020-78327-w] [PMID: 33277611]

Hathout, R.M., El-Ahmady, S.H., Metwally, A.A. (2018). Curcumin or bisdemethoxycurcumin for nose-to-brain treatment of Alzheimer disease? A bio/chemo-informatics case study. *Nat. Prod. Res., 32*(24), 2873-2881. [http://dx.doi.org/10.1080/14786419.2017.1385017] [PMID: 29022380]

Hathout, R.M., Metwally, A.A. (2016). Towards better modelling of drug-loading in solid lipid nanoparticles: Molecular dynamics, docking experiments and Gaussian Processes machine learning. *Eur. J. Pharm. Biopharm., 108*, 262-268. [http://dx.doi.org/10.1016/j.ejpb.2016.07.019] [PMID: 27449631]

Hathout, R.M., Metwally, A.A., Woodman, T.J., Hardy, J.G. (2020). Prediction of drug loading in the gelatin matrix using computational methods. *ACS Omega, 5*(3), 1549-1556. [http://dx.doi.org/10.1021/acsomega.9b03487] [PMID: 32010828]

Herranen, J., Yavuzkaya, M., Sjöström, J. (2021). Embedding chemistry education into environmental and sustainability education: Development of a didaktik model based on an eco-reflexive approach. *Sustainability, 13*(4), 1746. [http://dx.doi.org/10.3390/su13041746]

Huang, K., Fu, T., Gao, W., Zhao, Y., Roohani, Y., Leskovec, J., Coley, C. W., Xiao, C., Sun, J., Zitnik, M. (2021). Therapeutics data commons: Machine learning datasets and tasks for drug discovery and development. *arXiv.*

Hutchings, P. (2023). *The course portfolio: How faculty can examine their teaching to advance practice and improve student learning..* Taylor & Francis.

Iaquaniello, G., Centi, G., Salladini, A., Palo, E., Perathoner, S. (2018). Waste to chemicals for a circular economy. *Chemistry, 24*(46), 11831-11839. [http://dx.doi.org/10.1002/chem.201802903] [PMID: 29939452]

Petrova, N. N., Fadeeva, A. A., Bazanov, A. S. (2018). Implementation Of The Scientific And Methodological Approach In Natural Science Teaching Methods. In: Lo, S. K., (Ed.), *Education Environment for the Information Age. European Proceedings of Social and Behavioural Sciences,* Future Academy.587-595. [http://dx.doi.org/10.15405/epsbs.2018.09.02.68]

Jabeen, S., Siddiqui, V.U., Bala, S., Khan, A.R., Raza, S., Khan, T. (2023). Computational Nanotoxicology and its Applications. In: *Computational Toxicology for Drug Safety and a Sustainable Environment..* Bentham Science Publishers. [http://dx.doi.org/10.2174/9789815196986123010012]

Jin, H., Yang, J. (2021). Using computer-aided design software in teaching environmental art design.

Comput. Aided Des. Appl., 19(S1), 173-183.

Kamaruddin, M.H.H., Hussain, M.A.M. (2024). The Development of Teaching Aid for Design and Technology Learning. *Global Journal of Educational Research and Management, 4*(1), 34-43.

Kanmaz, A. (2022). A study on interdisciplinary teaching practices: Primary and secondary education curricula. *Afr. Educ. Res. J., 10*(2), 200-210.

Kazachkova, O. A. (2023). Design of Specialized Interactive Teaching Aid: the Main Aspects. 2023 3rd International Conference on Technology Enhanced Learning in Higher Education (TELE). Lipetsk, Russian Federation. 297-300.

Kerr, K.G., Edwards, K.E., Tweedy, J.F., Lichterman, H., Knerr, A.R. (2023). *The curricular approach to student affairs: A revolutionary shift for learning beyond the classroom.*. Routledge.

Khan, T., Lawrence, A.J., Azad, I., Raza, S., Joshi, S., Khan, A.R. (2019). Computational drug designing and prediction of important parameters using in silico methods-A review. *Curr. Computeraided Drug Des., 15*(5), 384-397.
[http://dx.doi.org/10.2174/1573399815666190326120006] [PMID: 30914032]

Kümmerer, K., Clark, J.H., Zuin, V.G. (2020). Rethinking chemistry for a circular economy. *Science, 367*(6476), 369-370.
[http://dx.doi.org/10.1126/science.aba4979] [PMID: 31974237]

Kunwar, R., Acharya, N.H. (2025). Analyzing integrated curriculum grade (1-3): Concepts, development and practices in nepal. *Teacher Half-Yearly Journal, 17*(1), 173-190.
[http://dx.doi.org/10.3312/thj.v17i1.77885]

Kuvondikovna, K.S., Hakima, B. (2023). Designing Visual Aid. *European Journal of Higher Education and Academic Advancement, 1*(2), 165-167.
[http://dx.doi.org/10.61796/ejheaa.v1i2.131]

Lang, F.K., Bodner, G.M. (2020). A review of biochemistry education research. *J. Chem. Educ., 97*(8), 2091-2103.
[http://dx.doi.org/10.1021/acs.jchemed.9b01175]

Levine, E., Patrick, S. (2019). What Is Competency-Based Education? An Updated Definition. Aurora Institute.

Linder, M. (2017). Ripe for disruption: reimagining the role of green chemistry in a circular economy. *Green Chem. Lett. Rev., 10*(4), 428-435.
[http://dx.doi.org/10.1080/17518253.2017.1392618]

Liu, F., Yang, K. (2021). Exploration on the Teaching Mode of Contemporary Art Computer Aided Design Centered on Creativity. *Comput. Aided Des. Appl., 19*(S1), 105-116.

López-López, E., Bajorath, J., Medina-Franco, J.L. (2021). Informatics for chemistry, biology, and biomedical sciences. *J. Chem. Inf. Model., 61*(1), 26-35.
[http://dx.doi.org/10.1021/acs.jcim.0c01301] [PMID: 33382611]

Lowa, W.L., Leea, C., Wilkesa, M., Robertsb, C., Hillb, D.J. (2014). Development of a rapid, effective method for seeding biofiltration systems using alginate bead-immobilized cells. *International Journal of Chemical and Environmental Engineering, 5*, 3-14.

Mahaffy, P.G., Ho, F. M., Haack, J.A., Brush, E.J. (2019). Can chemistry be a central science without systems thinking?. *J. Chem. Educ., 96*(12), 2679-2681.
[http://dx.doi.org/10.1021/acs.jchemed.9b00991]

Mairizwan, M., Hidayati, H., Dewi, W.S., Afrizon, R., Jarlis, R. (2022). Increasing the competence of physics teachers in designing PjBL-based teaching aids for the implementation of the Merdeka curriculum. *Journal Penelitian Pendidikan IPA, 8*(6), 2948-2953.

Manoj, K., Hussain, S. (2023). The Interdisciplinary Education in Higher Institutions: Trends, Challenges, And Future Directions. *International Research Journal Modernization in Engineering Technology and*

Science, 05, 87-90.

Mansilla, V.B. (2006). Interdisciplinary Work at the Frontier: An empirical examination of expert. *Science, 294*, 5547-1651.

Mansilla, V.B., Duraising, E.D. (2007). Targeted assessment of students' interdisciplinary work: An empirically grounded framework proposed. *J. Higher Educ., 78*(2), 215-237. [http://dx.doi.org/10.1080/00221546.2007.11780874]

Matlin, S.A., Cornell, S.E., Krief, A., Hopf, H., Mehta, G. (2022). Chemistry must respond to the crisis of transgression of planetary boundaries. *Chem. Sci., 13*(40), 11710-11720. [http://dx.doi.org/10.1039/D2SC03603G]

Mehler, A.H. (1983). Strategies of biochemical education. *Biochem. Educ., 11*(3), 95-118. [http://dx.doi.org/10.1016/0307-4412(83)90133-4]

Menken, S., Keestra, M. (2016). An introduction to interdisciplinary research: theory and practice. Amsterdam University Press.

Millar, V., Park, W., Dillon, J. (2025). The science curriculum: issues, tensions and future prospects. *Int. J. Sci. Edu., 47*(15-16), 1965-1971. [http://dx.doi.org/10.1080/09500693.2025.2504644]

Mozafari, M.R., Torkaman, S., Karamouzian, F.M., Rasti, B., Baral, B. (2021). Antimicrobial applications of nanoliposome encapsulated silver nanoparticles: A potential strategy to overcome bacterial resistance. *Current Nanoscience, 17*(1), 26-40. [http://dx.doi.org/10.2174/1573413716999200712184148]

Neubauer, A.C., Martskvishvili, K. (2018). Creativity and intelligence: A link to different levels of human needs hierarchy?. *Heliyon, 4*(5), e00623. [http://dx.doi.org/10.1016/j.heliyon.2018.e00623] [PMID: 29872757]

Newell, W. H. (2001). A theory of interdisciplinary studies. *Issues in Integrative Studies, 19*, 1-25.

Nghitoolwa, H.T.S., Chirimbana, M., Medusalem, M. (2024). The Impact of Teaching Aids on The Academic Performance of Learners in Natural Science. *Journal of Education for Sustainable Innovation, 2*(1), 74-83.

Nguyen, T.T., Jiang, Z., Nguyen, V.N., Le, N.Q.K., Chua, M.C.H. (2025). Integrating ESM-2 and graph neural networks with AlphaFold-2 structures for enhanced protein function prediction. *ACS Omega, 10*(33), 38103-38111. [http://dx.doi.org/10.1021/acsomega.5c05484]

Oordt-Speets, A.M., Bolijn, R., van Hoorn, R.C., Bhavsar, A., Kyaw, M.H. (2018). Global etiology of bacterial meningitis: A systematic review and meta-analysis. *PLoS One, 13*(6), e0198772. [http://dx.doi.org/10.1371/journal.pone.0198772] [PMID: 29889859]

Pan, A, Roy, S. G, Haldar, U, Mahapatra, R. D, Harper, G. R, Low, W. L, De, P, Hardy, J. G (2019). Uptake and release of species from carbohydrate containing organogels and hydrogels. *Gels, 5*(4), 43. [http://dx.doi.org/10.3390/gels5040043]

Pence, H.E., Williams, A.J. (2016). Big data and chemical education. *J. Chem. Educ., 93*(3), 504-508. [http://dx.doi.org/10.1021/acs.jchemed.5b00524]

Petrova, N. N., Fadeeva, A. A., Bazanov, A. S. (2018). Implementation of the scientific and methodological approach in natural science teaching methods. In: Lo, S. K., (Ed.), *Education Environment for the Information Age. European Proceedings of Social and Behavioural Sciences,* Future Academy.587-595.

Raza, S., Khan, T., Singh, M. (2024). AI Tools for Teaching-Learning Chemistry. In: *Artificial Intelligence: A Multidisciplinary Approach towards Teaching and Learning..* Bentham Science Publishers.

Repko, A.F. (2008). Assessing interdisciplinary learning outcomes. *Acad. Exch. Q., 12*(3), 171-178.

Repko, A.F., Szostak, R. (2020). *Interdisciplinary research: Process and theory.* Sage publications.

Rickman, J.M., Lookman, T., Kalinin, S.V. (2019). Materials informatics: From the atomic-level to the continuum. *Acta Materialia, 168*, 473-510.
[http://dx.doi.org/10.1016/j.actamat.2019.01.051]

Rojas, C., Aranda, J.F., Pacheco Jaramillo, E., Losilla, I., Tripaldi, P., Duchowicz, P.R., Castro, E.A. (2021). Foodinformatic prediction of the retention time of pesticide residues detected in fruits and vegetables using UHPLC/ESI Q-Orbitrap. *Food Chem., 342*, 128354.
[http://dx.doi.org/10.1016/j.foodchem.2020.128354] [PMID: 33268165]

Romano, J.D., Tatonetti, N.P. (2019). Informatics and computational methods in natural product drug discovery: a review and perspectives. *Front. Genet., 10*, 368.
[http://dx.doi.org/10.3389/fgene.2019.00368] [PMID: 31114606]

Serafini, P.G., Moura, J.M., Almeida, M.R., Rezende, J.F.D. (2022). Sustainable development goals in higher education institutions: a systematic literature review. *J. Clean. Prod., 370*, 133473.
[http://dx.doi.org/10.1016/j.jclepro.2022.133473]

Shalaby, M.A., Dokla, E., Serya, R., Abouzid, K. (2019). Identification of novel pyrazole and benzimidazole based derivatives as PBP2a inhibitors: Design, synthesis, and biological evaluation. *Archives of Pharmaceutical Sciences Ain Shams University, 3*(2), 228-245.
[http://dx.doi.org/10.21608/aps.2019.16625.1010]

Singh, M., Azad, I., Qayyoom, M.A., Khan, T. (2024). A study on perceptions and practices of STEAM-based education with university students. *Social Sciences & Humanities Open, 10*, 101162.
[http://dx.doi.org/10.1016/j.ssaho.2024.101162]

Singh, M., Jha, A.K., Khan, T., Raza, S. (2024). The evolution of artificial intelligence from philosophy to new frontier. In: *Artificial Intelligence: A Multidisciplinary Approach towards Teaching and Learning.*. Bentham Science Publishers.

Slootweg, J.C. (2020). Using waste as resource to realize a circular economy: Circular use of C, N and P. *Curr. Opin. Green Sustain. Chem., 23*, 61-66.
[http://dx.doi.org/10.1016/j.cogsc.2020.02.007]

Stuckenberg, D., Contento, A. (2018). Water scarcity: The most understated global security risk. *Harvard National Security Journal.* Available from: https://harvardnsj.org/2018/05/18/water-scarcity-the-m-st-understated-global-security-risk/

Sutton, P.S., Knuth, R. (2017). A schoolwide investment in problem-based learning. *Phi Delta Kappan, 99*(2), 65-70.
[http://dx.doi.org/10.1177/0031721717734193]

Tomlinson, C.A. (1995). *Differentiating instruction for advanced learners in the mixed-ability middle school classroom.*. Reston, VA: ERIC Clearinghouse on Disabilities and Gifted Education.

Tsai, C.Y., Liu, T.Y., Lu, Y.H., Nisar, H. (2020). A novel interactive assembly teaching aid using multi-template augmented reality. *Multimedia Tools Appl., 79*(43-44), 31981-32009.
[http://dx.doi.org/10.1007/s11042-020-09584-0]

United Nations Environment Programme. (2019). Global Chemicals Outlook II-From Legacies to Innovative Solutions: Implementing the 2030 Agenda for Sustainable Development. UNEP. Available from: https://www.unep.org/resources/report/global-chemicals-outlook-ii-legacies-innovative-solutions

Uptake and release of species from carbohydrate containing organogels and hydrogels. *Gels, 5*(4), 43.
[http://dx.doi.org/10.3390/gels5040043] [PMID: 31575001]

Van Rompuy, L.S., Parac-Vogt, T.N. (2019). Interactions between polyoxometalates and biological systems: from drug design to artificial enzymes. *Curr. Opin. Biotechnol., 58*, 92-99.
[http://dx.doi.org/10.1016/j.copbio.2018.11.013] [PMID: 30529815]

Vogel Taylor, E.M., Mitchell, R., Drennan, C.L. (2009). Creating an interdisciplinary introductory chemistry course without time-intensive curriculum changes. *ACS Chem. Biol., 4*(12), 979-982.

[http://dx.doi.org/10.1021/cb9002927] [PMID: 20017574]

Voicu, A., Duteanu, N., Voicu, M., Vlad, D., Dumitrascu, V. (2020). The rcdk and cluster R packages applied to drug candidate selection. *J. Cheminform., 12*(1), 3.
[http://dx.doi.org/10.1186/s13321-019-0405-0] [PMID: 33430987]

Warkentien, S., Goeking, J., Dilig, R., Knapp, L., Stanley, R. (2022). Interdisciplinary Education: literature review and landscape analysis. Washington: RTI International.

Water, U.N.. (2018). *Sustainable Development Goal 6 synthesis report on water and sanitation.*. New York: Published by the United Nations New York.

Willett, P. (2020). The literature of chemoinformatics: 1978–2018. *Int. J. Mol. Sci., 21*(15), 5576.
[http://dx.doi.org/10.3390/ijms21155576] [PMID: 32759729]

Wishart, D.S., Feunang, Y.D., Guo, A.C., Lo, E.J., Marcu, A., Grant, J.R., Sajed, T., Johnson, D., Li, C., Sayeeda, Z., Assempour, N., Iynkkaran, I., Liu, Y., Maciejewski, A., Gale, N., Wilson, A., Chin, L., Cummings, R., Le, D., Pon, A., Knox, C., Wilson, M. (2018). DrugBank 5.0: a major update to the DrugBank database for 2018. *Nucleic Acids Res., 46*(D1), D1074-D1082.
[http://dx.doi.org/10.1093/nar/gkx1037] [PMID: 29126136]

Wolfson, A.J., Hall, M.L., Allen, M.M. (1998). Introductory chemistry and biology taught as an interdisciplinary mini-cluster. *J. Chem. Educ., 75*(6), 737.
[http://dx.doi.org/10.1021/ed075p737]

Wu, C.K., Lee, Y.L., Lin, M.H., Wang, C.P., Wang, K.T. (2019). A Needs Analysis for Education and Learning in Young Child Teaching Aid Design. *Proceedings of the 3rd International Conference on Education and Multimedia Technology* New York, NY, USA.
[http://dx.doi.org/10.1145/3345120.3345182]

Xu, S., Zhang, Y., Yang, T., Wang, X. (2023). Research on the interaction design of teaching aid products based on the concept of modularity. *HCI International 2023 – Late Breaking Papers. HCII 2023. Lecture Notes in Computer Science,* Cham: Springer.*14060*
[http://dx.doi.org/10.1007/978-3-031-48060-7_27]

Yee, M. H., Ismail, A., Shah, S. Z. M. (2021). Super Structural Teaching Kits as Teaching Aid Materials for Construction Technology Students. *Research and Innovation in Technical and Vocational Education and Training, 1*(2), 165-173. Available from: https://publisher.uthm.edu.my/periodicals/index.php/ritvet/article/view/508

Yu, Y., Gong, Z., Zhou, Y., Wang, S., Jiao, Z. (2024). Design of an interactive multifunctional teaching aid. In: *Fourth International Conference on Mechanical, Electronics, and Electrical and Automation Control (METMS 2024), 13163*, 150-155.
[http://dx.doi.org/10.1117/12.3030329]

Yusof, M.S., Hasnan, K.A. (2023). Development of teaching aid kit for mechanical components function in mechanical design for form 2 design technology subject. *AIP Conf. Proc., 2750*, 050006.
[http://dx.doi.org/10.1063/5.0161030]

Zhang, C., Liu, R., Kong, X., Li, H., Yu, D., Fang, X., Wu, L., Wu, Y. (2023). Adaptive responses of a Peroxidase-like Polyoxometalate-based Tri-assembly to bacterial microenvironment (BME) significantly improved the anti-bacterial effects. *International Journal of Molecular Sciences, 24*(10), 8858.
[http://dx.doi.org/10.3390/ijms24108858]

Zhao, J., Li, K., Wan, K., Sun, T., Zheng, N., Zhu, F., Ma, J., Jiao, J., Li, T., Ni, J., Shi, X., Wang, H., Peng, Q., Ai, J., Xu, W., Liu, S. (2019). Organoplatinum-Substituted Polyoxometalate Inhibits β-amyloid Aggregation for Alzheimer's Therapy. *Angew. Chem. Int. Ed., 58*(50), 18032-18039.
[http://dx.doi.org/10.1002/anie.201910521] [PMID: 31591753]

CHAPTER 10

Mathematics Meets Art: Unlocking the Power of Analytical Creativity

Sangeeta Chauhan[1,*], **Michael Tarance Suraj**[1] and **Sonu Bara**[1]

[1] *Department of Education, Babasaheb Bhimrao Ambedkar University, Lucknow, Uttar Pradesh 226025, India*

Abstract: The integration of mathematics and art enables analytical methods to meet creative practices that have developed human understanding throughout many historical generations. This paper explores how mathematics and art merged from Renaissance perspective techniques to modern implementations of fractals and digital creative systems in art production. Artistic creations in visual arts, music, architecture, and digital media benefit from mathematical principles through the work of Leonardo da Vinci, M.C. Escher, and Benoit Mandelbrot, and through the application of the golden ratio, symmetry, and algorithms. The research supports interdisciplinary education through STEAM projects and project-based learning, which develops meaningful artistic products from mathematical principles to improve students' interest and boost critical intelligence. New technologies help with educational integration. Educators must overcome resistance and assessment challenges alongside budget constraints. Synthesis eliminates disciplinary barriers to produce adaptable thinkers who can solve complex problems with analytical and creative capabilities. This interdisciplinary relationship between mathematics and art sets new educational guidelines by creating comprehensive thinking methods that merge analytical strengths with creative faculties to promote innovative discoveries across the fields of logic and art.

Keywords: Algorithms, Analytical, Art, Challenges, Creativity, Critical thinking, Digital, Education, Geometry, Golden ratio, Integration, Interdisciplinary, Learning, Mathematics, Project-based learning, Renaissance, Steam, Symmetry, Teaching.

INTRODUCTION

Mathematics and Art have been perceived as divergent disciplines occupying distinct domains in human knowledge endeavours (Meng *et al.*, 2024).

* **Corresponding author Sangeeta Chauhan:** Department of Education, Babasaheb Bhimrao Ambedkar University, Lucknow, Uttar Pradesh 226025, India; E-mail: c.sangeeta27@gmail.com

Tahmeena Khan, Manisha Singh, & Saman Raza (Eds.)

Mathematics concepts are based on logic, abstraction, the science of patterns, structure, and relationships, which are governed by axioms and theorems. The aim of mathematics as a discipline is to seek universal truth with the help of systematic reasoning, quantifying phenomena, and solving problems with precision (Posy & Ben-Menahem, 2023). Art on hand uses subjective expression, channelising emotion, experiences, and cultural narratives, which are transmuted through music, painting, and scripture (Ahmed, 2023). The major focus is on aesthetics, interpretation, and drawing different meanings from a given context in a flexible framework.

The concept of analytical creativity integrates these two domains. It provides a creative and logical interdependent relationship between these two disciplines (Grey, 2017). This is based on practical integration, where mathematical practices use the arts. Over the years, many paintings and monuments have been prime examples of the use of geometrical shapes. Left-brain rationality and right-brain artistry have been challenged by analytical creativity, highlighting the interplay between structural thinking and imaginative exploration in both disciplines (Gardner *et al.*, 2008).

Both disciplines have collaborated, contributing to and developing each other. During the Renaissance, artists used concepts of linear perspective, which were grounded in geometry, to revolutionise realism (Barbin & Menghini, 2019). A prime example of this is the mathematician and artist Leonardo da Vinci, who drew inspiration from anatomical aesthetics. In the current era, innovations such as computer graphics, data visualisation, and generative art have been made possible by the interplay of both disciplines (Pleasants, 2023). This synthesis fosters holistic cognition, blending problem-solving with expressive innovation and reframing educational paradigms to nurture versatile thinkers.

The fusion of mathematics and art is a universal phenomenon, evident in diverse cultural practices that mix analytical exactness with creative expression. The intricate stone carvings of the Maya, of Mesopotamian culture found in Chichen Itza's Temple of Kukulcan, display the complex use of astronomical alignments and geometric proportions, echoing the mathematical calculations that harmonize with aesthetic and spiritual purposes (Šprajc *et al.*, 2023). Islamic mosaic designs of the Middle East, like adorning the Dome of the Rock in Jerusalem, showcase the complex use of the polygonal patterns and symmetry to symbolize divine order based on the principles of tessellation and group theory. The Aboriginal Australian dot paintings made by Warlpiri people use spatial patterns that translate environmental and cultural themes through proportional arrangements that align with mathematical concepts of balance and rhythm. All these traditions around the world showcase the shared human desire to use concepts of mathematics with artistic value to resonate with intellectuality and emotions. By using various examples in the curriculum, teachers can create a culturally responsive classroom where students are connected to the universal principles of analytical creativity.

They indicate that humans around the globe are attracted to using math in art to express deep thoughts and stir emotion. Making use of variety in texts in class helps teachers develop classrooms that nurture learning by bridging a student's heritage to basic principles of creativity. This style of teaching expands the connection between math and art, while also boosting each individual's sense of belonging in a world of invention and imagination.

Many researchers have proved involvement in activities that connect math with art to sharpen both the mind and the spirit. Designing a geometric kaleidoscope requires logical analysis and spatial skills, as does preparing a step-by-step dance that uses tempo changes (Buckley *et al.*, 2018). Since these activities stimulate logical-mathematical, spatial, and bodily-kinaesthetic skills, they become a productive way for students to learn and understand the subject. When designing a symmetrical kaleidoscope, students must precisely calculate angles and focus on making it look attractive. According to neuroscientific research, doing these tasks makes it easier for learners to appreciate ambiguous situations (Immordino-Yang, 2016). Furthermore, the rigidity of math-art challenges helps people overcome worries about math topics. Thanks to math-art, students may become both skilled and flexible in their studies, ready for whatever changes the world brings. Educators can use this idea to guide activities that encourage people to value learning and progress throughout their lives.

Technology has made it possible to combine mathematics and art in many new ways. Apps that use virtual reality, such as those in the education sector, enable students to get closer to three-dimensional math objects, using models that are easy to handle and explore. In fact, with Tinkercad, learners have the tools to create 3D models, covering both math and art with accuracy. On p5.js and similar websites, students can program visual artworks designed in the spirit of pioneering artists such as Benoit Mandelbrot. Thanks to these tools, students in resource-limited areas can learn about math-art with open-source technologies. With the help of organizations such as Bridges, students and educators worldwide can share their math-art projects online with each other. Technology integration encourages teachers to introduce students to new approaches where they can think creatively and use technology when solving math problems. Because of this, students learn skills they need for architectural and VR work, as well as science visualization, so modern educators must integrate math and art into their plans.

This chapter explored the deep connection between the two by discussing how they have collaborated from time to time. It highlights cultural milestones that demonstrate how both disciplines have long informed and enriched one another, using the case study of prominent artists and figures such as M.C. Fisher and Benoit Mandelbrot. It showcases how their work has used a blend of both to dissolve disciplinary boundaries, resulting in the expansion of creative horizons. To encourage analytical creativity, this chapter further explores educational strategies that blend logic and imagination to foster innovative, adaptable learners. It advocates a unified lens to appreciate the interconnectedness of human thought and unlock innovations at the crossroads of logic and art.

THEORETICAL FRAMEWORK: HISTORICAL CONNECTIONS BETWEEN MATH AND ART

Mathematics and the arts have a historical lineage of interdependence. This relationship is based on the human endeavour to understand and express it in the form of order, beauty, and harmony. Mathematical concepts have always been a tool and an inspiration for artistic creation, and the arts provide a canvas for expressing these abstract mathematical ideas (Lipovetsky, 2022). Major examples include the golden ratio, Fibonacci sequence, symmetry, and perspective in the Renaissance, which illustrate the union and showcase how mathematical precision enhances aesthetic expression and how art has externalised mathematical thoughts.

The value of the Golden ratio, which is approximately 1.618, has been widely recognised for its aesthetically pleasing proportions. The golden ratio was first recognised in the Parthenon's architecture and was later codified by mathematicians such as Euclid, becoming a bridge between numerical harmony and visual balance (Thapa *et al.*, 2018). Paintings by Leonardo da Vinci have integrated this ratio in his famous painting, Mona Lisa, where the proportion of the face and composition subtly align with the divine proportion. Similarly, the Fibonacci series is a sequence in which each number is the sum of its preceding number (1,1,2,3,58, *etc.*), which can be found in natural spirals, such as sunflower seeds and nautilus shells. They were the inspiration behind many artists in the Middle Ages and the Renaissance (Muhamad & Azzaldeen 2021). The recursive elegance of this sequence mirrors the quest for organic unity in art.

The principle of symmetry in mathematics is a cornerstone of artistic design. In art, bilateral symmetry can be found in Greek pottery and radial symmetry in Islamic tile work; this concept has been harnessed to evoke balance and completeness (Wichmann *et al.*, 2017). During the Renaissance, the development of linear perspective revolutionised art by using geometric principles to create the illusion of depth. It was pioneered by Filippo Brunelleschi and formalised by Leon Battista Alberti. Their perspective relied on the mathematical concepts of vanishing points and proportional scaling (Alexander, 2019).

This complex interplay of both disciplines can engage multiple dimensions of human intelligence. In Howard Gardner's theory of intelligence, he identified logical mathematical intelligence as a distinct capacity for abstract reasoning and problem solving, which contributes to the discovery of mathematics. Gardner's theory also comes into play as artists use mathematical concepts to calculate proportions or design perspective, pointing to the overlap with Gardner's spatial intelligence (Vivian *et al.*, 2024). Modern neuroscience has found that creativity in both disciplines emerges from an integrated process. Other examples include Benoit Mandelbrot's fractal geometry, used to showcase a display of artistic flair in visualising complexity, and the work of M.C. Escher, which displays mathematical precision to craft impossible realities.

The link between mathematics and art is present in cultures worldwide. In Africa, fractal geometry, which shows repeating patterns at different levels, existed for a long time before Benoit Mandelbrot's explanation. For example, Ba-ila villages in Zambia are built according to a circular layout and developed with structures of the same size repeated in a mirror pattern, much like the branching pattern found in nature (Husain *et al.*, 2022). Also, Kuba textiles from the Democratic Republic of Congo include elaborate geometrical patterns that host mathematical symbols. Such patterns, usually neglected in Western writing, point out that indigenous systems combine mathematics and artistry.

In South Asia, where they began, Indian mandalas mix the concept of symmetry with spiritual beliefs. They are based on Hindu and Buddhist storytelling and use accurate measurements to build radial designs that show the harmony found in astronomy as well as symmetry and proportions in mathematics. Building mandalas requires setting up concentric designs based on Euclidean geometry's focus on balance.

Mathematical art is a central element in East Asian art. For example, making Japanese origami requires people to follow specific algorithms to transform flat paper into complex structures. Research on origami in modern times has established links to computational geometry and uses it in engineering and design. In the Song Dynasty, Guo Xi painted Chinese landscapes with proportion grids that resemble the way Renaissance artists would eventually use perspective. They prove that including math and art in education is a common practice everywhere and helps disprove the idea that it originated in Europe. Creativity in mathematics takes the form of a proof or theorem that is based on imagination but must follow strict rules (Bruhn & Lüken, 2023). Imitation and emotion are examples of creativity leveraged in art, using mathematics as a basis. The framework shows that maths and art go hand in hand and both contribute to our creative thinking.

Dimensions of Integration

• Mathematics & Art

Mathematics and the arts have extended to multiple domains, such as visual arts, music, architecture, and digital creativity, revealing a unique dimension where precision and aesthetic expression converge. The synthesis of artistic and mathematical outputs also deepens our understanding of mathematical concepts through tangible, sensory experiences. By exploring geometry in visual art, ratio in music, spatial reasoning in architecture, and algorithms in digital design, one can uncover an interplay that can break disciplinary boundaries.

• Visual Art & Geometry

The concept of the visual has a significant link with mathematics, providing a structure that artists have used to manipulate and evoke wonder and harmony. Various concepts, such as tessellation, have been explored by M. C. Escher, exemplifying the integration of both subjects. His work on the circle limit series is rooted in the mathematical study of plane tiling. His art transforms rigid rules into surreal interlocking forms, a perfect blend of logic and imagination. Benoit Mandelbrot's fractal geometry, in which self-similar patterns repeating at different scales mirror nature's complexity, like coastlines and fern leaves. They have inspired artists to depict infinite depths within finite boundaries. In addition, many Islamic buildings have used geometric patterns to create a blend of spiritual aesthetics where order symbolises the divine (Glaeser, 2020).

• Music & Mathematics

Mathematics is woven into the fabric of sound. Any Rhythm, Harmony, and scale are based on the numerical relationship of ratio and frequency (Roberts, 2016). A perfect mix of the two can result in auditory confusion. The discovery of the octave (2:1 frequency ratio), a musical interval proposed by Pythagoras, and the fifth (3:2) has been the foundation of Western music theory (Parncutt *et al.*, 2018). These ratios translate vibrations into consonant sounds, a principle later refined in the equal-tempered scale, where the octave is divided into 12 semitones with a frequency ratio. Johann Sebastian Bach, a famous composer integrated mathematical precision into emotional resonance, exploiting these relationships in both counterpoint and fugues (Junod, 2017). In modern music, minimalist composers such as Philip Glass have used repetitive, additive, and rhythmic patterns that involve arithmetical progression. To compose harmonic textures, spectral music is analysed (Winterson, 2024). Thus, music becomes an art form in which mathematics orchestrates the invisible.

- **Architecture & Spatial Reasoning**

In architecture, mathematics has a significant influence through spatial reasoning to create functional and aesthetically pleasing structures. Concepts such as proportion, structure, and symmetry guide architectural design and contribute to the building of structural beauty. Gothic cathedrals have used geometric engineering to distribute weight and achieve soaring heights, as seen in the symmetrical façade (Bork, 2023). Modern architects, such as Zaha Hadid, have used parametric design and mathematical models that involve curvature and scale to craft fluid organic forms. All these structures demonstrate how spatial reasoning can be grounded in mathematical calculation, which translates abstract concepts into a lived environment.

- **Mathematical Modelling in Digital Art**

Technological advancements have introduced a new field in the arts driven by digital technology. Mathematics in modelling in the arts and generative arts is based on algorithms that allow artists to create complex visuals through coded instructions. Processing a programming language provides artists with the opportunity to define rules, such as recursive functions or random walks, to create an evolving pattern, such as John Maeda's abstract designs. Fractal-based software generates an infinite zoom into Mandelbrot sets, assimilating mathematical exploration with visual miracles.

Mathematics is intertwined as a complementary force in this dimension. This Integration reflects a profound harmony of mathematics, offering tools to measure and model, while art provides these tools with meaning, emotion, and imagination. When both disciplines are combined, they reveal a shared human value to bring meaning to this catastrophic world through precision and beauty (Kugler *et al.*, 2023).

Math-Art in Emerging Fields

Both mathematics and art are merged in these new disciplines, enabling their application to today's issues and concerns. One example of how this works is data visualization, which is expanding rapidly. Artists partner with scientists to represent complex datasets as attractive and engaging interactive maps and models of epidemics. To display the information, the visualizations use mathematical methods, but the use of colours, shapes, and interactions makes the data clearer and more appealing (Hagler *et al.*, 2024). To illustrate, "Wind Map" by Fernanda Viegas and Martin Wattenberg combines the beauty of graphic design with the usefulness of scientific methods to present real-time wind patterns in the U.S.

Bio art is another field that mixes mathematical theories with biology to develop new artworks. Simulating how slime molds grow is one of the uses Marta de Menezes has found for algorithms. Many artists use these installations to help viewers learn about the math behind life and to combine science with beautiful art. They expand the limits of art and treat it as a form of scientific communication.

Immersive virtual worlds in games are created when mathematics and art are combined. With Unreal Engine and other game engines, linear algebra and trigonometry allow developers to render 3D settings, and artists add textures and stories to make the games feel emotional. No Man's Sky, along with other games, uses programming to randomly produce big and unique worlds, demonstrating the use of math in imagination. Thanks to the connection between math and art, there are now new routes for education and careers.

Looking at data and combining it with biology and digital methods, these fields broaden our opportunities to be creative. Through these scenarios, students can use math skills and collaborate to solve problems, which are important for a successful career in the 21st century.

PEDAGOGICAL IMPLICATIONS:

Enhancing Learning Through Interdisciplinary Math and Art Integration

Mathematics and arts education provide opportunities to use transformative implications to enrich pedagogical engagement to foster critical and creative thinking and abstract concepts with tangible experiences. Interdisciplinary teaching leverages the inherent connections between these fields to create a holistic learning environment. Moving beyond siloed disciplines to cultivate curiosity and innovation. By integrating STEAM (Science, Technology, Engineering, Arts, and Mathematics) approaches, arts-infused projects, and project-based learning, educators can harness students' potential to think critically and creatively, preparing them for a world that demands adaptability and synthesis.

• Enhancing Student Engagement Through Interdisciplinary Teaching

Interdisciplinary teaching helps connect abstract concepts of mathematics to the sensory and expressive realm of art, making learning more relatable and dynamic. When students observe how geometry shapes a tessellation, or how ratios explain musical harmony with maths as a living creative force in place of rote procedure (Fathauer, 2020). Researchers have suggested that approaches have been taken to increase motivation, particularly for students who may struggle with traditional math instruction (Sujatha & Vinayakan, 2023). One of the prime examples of student engagement through interdisciplinary teaching is Gardner's theory of multiple intelligences theory, which combines logical-mathematical intelligence with spatial and musical intelligence. Teachers can create diverse learning styles that foster inclusivity. By providing an opportunity to participate in hands-on activities, passive learners can be transformed into active creators.

• Arts Integration Projects in Schools

Art integration projects are the best example of combining math and the arts co-existing in the classroom. In a primary setting, students can explore the concept of the Fibonacci series by creating spiral artwork inspired by sunflowers. By linking numerical patterns to visual outcomes, middle schoolers can analyse symmetry in Islamic tile design, thus applying geometric transformation while discussing cultural context. These types of projects not only reinforce mathematical concepts but also build cultural awareness and communication skills as students articulate their processes and inspiration (Penerosa & Pischnotte, 2021).

• Fostering Critical and Creative Thinking

Bringing together maths and arts, both critical and creative thinking are important for problem-solving and innovation. As students analyse the patterns in Escher's tessellations or Bach's fugues, they develop creative thinking, which requires them to deconstruct and justify the relationship between the elements. When concepts learned in the classroom are implemented in practical settings, such as designing a mathematically sound yet visually striking structure, critical thinking flourishes. This integration of arts and mathematics challenges the misconception that maths is logical and art is intuitive, challenging the notion of right-brain and left-brain dichotomy (Abraham, 2024). Students learn that imagination plays a huge role within constraints, which builds their flexibility and resilience.

Project-Based Learning: Combining Math with Visual Art or Music

Project-based learning (PBL) has emerged as a powerful framework to combine mathematics with the arts, encouraging students to work and explore real-world applications collaboratively. Some examples in this context are students working on a project that might involve creating an aesthetically scaled model of a structure by calculating proportions and angles. Another interesting example is the "Math in motion" project, where students can be encouraged to compose a short piece of music using fractional notes and harmonic ratios to perform in the classroom. These learning experiences can be made engaging and conceptually understood. A study by Ozkan suggests that students should be trained by involving these practices more intuitively when they hear rhythms and foster teamwork as they negotiate design choices and refine their creations iteratively (Ozkan, 2023).

Teachers must be given the skills and confidence to connect math and art in education. Many teachers are only skilled in one field, as it's uncommon for math teachers to focus on art and for art teachers to emphasize math (Vale & Barbosa, 2025). Addressing the gap requires professional development that teaches how to collaborate and use new teaching methods.

Workshops from the Bridges Organization give teachers an example of how to collaborate with mathematicians and artists to make tessellations, fractal sculptures or music. These activities break down the link between math and art, helping teachers plan enjoyable lessons for their students. A workshop might teach a math teacher how to use GeoGebra to generate symmetrical designs that can be used in lessons.

Such programs help teachers better prepare for various situations. Math and art teachers may join forces to plan classes, including a unit on Islamic art that explores patterns and their cultural backgrounds. Partnering on activities encourages both groups to learn from one another and to practice ways of thinking that involve several disciplines. Moreover, regional education boards can work with universities to organize summer institutes where teachers receive help in designing STEAM lessons.

It is important for policymakers to prioritize funding so that teacher training includes modules on math and art in certification courses. Teachers with limited access to further training can find lesson plans on the National Council of Teachers of Mathematics website to support their teaching. Empowered Educators can help students develop analytical and creative skills, aligning with the aims of STEAM education (Escudeiro *et al.*, 2024).

The implication of integrating maths and arts has a profound impact on enhancing engagement by making learning experimental and integrating curricula like STREM to nurture well-rounded thinkers, and using projects to mix critical analysis with creative expression. By integrating multiple interdisciplinary approaches, educators can reinforce mathematical mastery and cultivate students' ability to see a connection between different domain skills, which is vital for navigating an increasingly complex world. Education has become a space where numbers dance, spaces sing, and students thrive as both analysts and artists. The convergence of mathematics and art is evident in real-world projects, and the work of iconic figures and institutions can spotlight this interdisciplinary fusion.

Challenges and Considerations: Integrating Math and Art in Education

These are major roadblocks in the process of implementing interdisciplinary integration of mathematics and art in education. Resistance from current educators, evaluating combined outcomes, the need for specialised training, and resource limitations are some of the barriers (Kucukkaragoz & Meylani, 2025). These issues highlight the complexities involved in shifting entrenched educational paradigms and the practical obstacles in effectively fostering this fusion.

Traditional educators often shy away from blending mathematics and the arts, rooted in the belief that the two disciplines are distant. There is a worry among maths educators that both disciplines are inherently distinct. Math instructors may worry that artistic elements may detract from the rigour of computation and problem-solving; on the other hand, it is believed by the art education community that rigour in mathematics kills the creative freedom needed for learning the arts (Sánchez *et al.*, 2022). This has been the case for a long time due to the inclination of educators to teach in the comfort of their teaching methods.

Assessment is yet another factor that acts as a hurdle. Traditional or standardised tools fail to evaluate interdisciplinary work (Shuler *et al.*, 2016). In grading a factually inspired artwork, one needs to balance mathematical precision, such as accurate recursive equitation, with artistic-like composition. Making a clear assessment that is dual-focused is of vital importance for the development of both the skills

The implementation of this method depends on teachers' expertise in both subjects, although many instructors lack dual-field expertise. A math educator typically understands ratios as a concept but struggles to apply this knowledge to musical scales, whereas artistic teachers struggle with the geometric basis of perspective. A lack of professional development programs exists throughout normal educational systems to teach teachers how to develop and deliver such lessons. Educators may find empowerment through workshops that teach generative art coding, yet tools and access determine their ability to be prepared for success.

Interdisciplinary tasks become more challenging due to budget restrictions, as these projects require materials such as art supplies and software, along with extra planning time that many schools currently lack. Educational tools, such as 3D printers designed for generating golden ratio models, remain out of reach for students in rural areas, whereas urban schools must manage overcrowded schedules. Well-funded school institutions receive an unfair advantage through funding practices that abandon unsupported ones. The solution requires the development of novel methods that combine free digital resources with specific funding streams to establish a level playing field.

Mathematics and art integration requires the development of specific solutions to address four significant obstacles within the educational system. An effective solution to these obstacles requires teachers to experience cultural changes, along with improved assessment systems, comprehensive professional growth initiatives, and equal resource distribution strategies. This interdisciplinary vision becomes achievable when organisations pursue solutions that create an education system that combines precise teaching methods with creative approaches.

Future Directions

Due to technological advancements, it is possible to integrate mathematics and art in the classroom, and augmented reality platforms enable users to access immersive experiences that connect the two. The combination of augmented reality and virtual reality lets students view geometric patterns on real artworks through their devices and provides them with the chance to explore symmetry in real time and then explore fractal landscapes or architectural designs in virtual space (Guazzaroni *et al.*, 2019). The AI algorithm, Generative Adversarial Networks (GANs), allows creators to create visual art based on mathematical models by producing computer-generated artworks like the works of M.C. Escher. Through these technologies, people gain enhanced visual understanding through interactive tools that make abstract concepts accessible to all learners.

Research across disciplines is essential to advance this combination because it unites mathematicians, artists, teachers, and technologists. Cross-disciplinary research should analyse the connection between fractal geometry in ecological art and musical ratios in data sonification to achieve greater value for both fields. When authorities employ this methodology, they create innovative outcomes by crossing organizational boundaries and developing total problem solutions that resolve practical challenges, such as eco-friendly planning and design.

Educators who embrace these tools should use them to support project-based learning by designing virtual galleries with AR applications. The incorporation of math-art relations within educational standards should result in modular programs that unite algebra education with digital design instruction. Policy leaders need to dedicate financial resources to develop technological infrastructure and conduct training for teachers while establishing universal technology access for all districts (Liu *et al.*, 2025). The implementation of preliminary STEAM curriculum programs with VR technology would generate data for the large-scale deployment of these initiatives. Multiple strategic actions will promote the convergence of mathematics and art while developing students for the merging future of analytical and creative skill applications.

The fusion of math and art shapes the world around us and helps people prepare for diverse jobs. The pursuit of creative ideas is very important, helping students in game design, animation, urban planning, and data science, since these require knowledge of math and art. Thus, Pixar animators often use differential geometry when animating characters, and urban planners use geometric algorithms to incorporate appealing and beneficial features into city buildings (Wagner & Tamayo, 2024). They show that people who can work with math and art are highly valued today.

Mixing math and art invites people to become more understanding and familiar with other cultures. Using graphics and anecdotes, students can see social issues related to migration through both data and personal experience. By showing the link between mathematics and art, math-art encourages individuals to take part in public affairs.

Using math-art in schools can help avoid educational disparity. Math lessons are offered to those exploring public parks through works of art, and people are encouraged to learn coding in community centres. Programs like UNESCO's Arts Education help people continue to respect and appreciate the cultures around them.

Being involved in new areas such as mathematics and art, schools give students a renewed understanding of education. Community programs and schools should highlight the benefits of this synthesis at the request of the government.

CONCLUDING REMARKS

Research into mathematical-artistic combinations unveils several centuries of significant historical elements that show how these fields stem from human mental capabilities. Interdisciplinary projects that merge visual art forms, music, architecture, and digital media improve educational outcomes by stimulating critical thought and student interest. Fundamental research and historical examples, such as classroom tessellation designs and Leonardo da Vinci's work, demonstrate practical application, and the Bridges Organization builds cultural significance through its initiatives. Various obstacles force educators to resist and test assessment capacities, while educators need training and insufficient resources to prevent the wide-scale adoption of this approach.

The combination of analytical thinking and creative pursuits creates substantial worth, which eliminates stock divisions to establish a single intellectual environment. Students learn adaptable abilities through this integration because it combines reasoning with creative thinking to solve complicated issues. This relationship between mathematics and art produces transformative effects by transforming classes into connected explorations between numbers, beauty, and order in art. The momentum behind this digital tool innovation continues to grow through transdisciplinary research support, which creates new possibilities for seeing, creating, and understanding.

Academic integration represents an inherent drive for humans to combine knowledge and expression fundamentals. Our approach to accepting this combination will produce learners who familiarise themselves with both mathematical formulas and artistry, so that innovative breakthroughs emerge from the union of precision and wonder. Math-art serves as an instrument that permits the design of educational systems and lifestyle principles from a fresh perspective.

The transformative power of math-art integration extends beyond the classroom, reshaping societal values and fostering global cultural dialogue. By drawing on diverse traditions, such as the geometric precision of Mesoamerican Maya calendars or the symmetrical patterns in Persian miniature paintings, this synthesis highlights the universality of human creativity and intellect.

These cultural artifacts, rooted in mathematical principles, serve as bridges across time and geography, reminding us that the quest for beauty and order transcends borders. For instance, the Maya's use of cyclical mathematics in their calendar systems not only facilitated astronomical predictions but also inspired intricate artistic motifs that adorned their temples, blending utility with aesthetic reverence. Similarly, Persian miniatures employed proportional grids to achieve balanced compositions, reflecting a mathematical harmony that resonated with spiritual themes. By integrating such global examples into education, teachers can cultivate a sense of shared heritage, encouraging students to see themselves as part of a worldwide continuum of analytical and artistic innovation. This approach fosters cultural empathy and global citizenship, equipping learners to address complex global challenges, such as sustainable development or intercultural collaboration, with both rigor and creativity. As educators embrace these diverse perspectives, they create inclusive learning environments that celebrate the multiplicity of human expression while grounding it in universal mathematical truths.

The societal impacts of math-art integration are profound, preparing individuals for a future where interdisciplinary skills drive progress in technology, culture, and civic life. In fields like urban planning, professionals use mathematical algorithms to design efficient city layouts while incorporating artistic elements to enhance liveability, as seen in projects like Singapore's Gardens by the Bay, where geometric modelling meets ecological aesthetics (Tan, 2017). In the creative industries, animators at studios like Pixar rely on differential geometry to craft realistic character movements, blending technical precision with narrative artistry. Beyond professional applications, math-art integration fosters public engagement through initiatives like community art installations. For example, large-scale sculptures inspired by polyhedral geometry, such as those displayed in public parks, invite passersby to explore mathematical concepts intuitively, democratizing access to knowledge. These societal applications underscore the economic and cultural value of math-art skills, positioning them as essential for addressing 21st-century challenges like climate change or digital equity. By nurturing students who can navigate both analytical and creative domains, education systems can contribute to a workforce and citizenry capable of innovating responsibly and inclusively. Policymakers should prioritize funding for public programs that showcase math-art integration, ensuring its benefits reach diverse communities and amplify its role in shaping a cohesive, forward-thinking society.

Education lies at the heart of this transformation, and the integration of mathematics and art offers a blueprint for reimagining pedagogical paradigms. Traditional disciplinary silos often stifle curiosity, but interdisciplinary approaches like STEAM (Science, Technology, Engineering, Arts, and Mathematics) empower students to forge connections across fields, fostering holistic thinkers. For instance, a classroom project where students design a virtual reality (VR) model of a mathematically inspired sculpture not only teaches geometry and coding but also encourages artistic experimentation and collaborative problem-solving. Such projects mirror real-world applications, like the design of parametric architecture by firms like Zaha Hadid Architects, where mathematical models create fluid, visually striking structures (Schumacher *et al.*, 2017). Community-based learning further amplifies this impact, with initiatives such as math-art workshops in libraries or museums that engage learners of all ages. The Bridges Organization's annual conferences, which bring together educators, artists, and mathematicians, exemplify how collaborative platforms can inspire innovative curricula, such as units on symmetry in Aboriginal Australian art or polyhedral modelling in 3D printing. To overcome barriers such as educator resistance and resource constraints, professional development programs must equip teachers with cross-disciplinary skills, drawing on models like UNESCO's Arts Education framework. By embedding math-art integration in national standards and providing equitable access to tools like open-source software, education systems can nurture students who excel as both analysts and creators, ready to tackle complex, interconnected challenges with ingenuity.

Looking to the future, the convergence of mathematics and art promises to redefine how we create, learn, and connect in an increasingly digital and globalized world. Emerging technologies, such as artificial intelligence (AI) and augmented reality (AR), are expanding the possibilities of this integration. AI-driven tools, like those used in algorithmic music composition, allow students to explore harmonic ratios through interactive platforms, creating soundscapes that blend mathematical precision with emotional resonance. AR applications enable learners to visualize mathematical concepts in three-dimensional space, such as exploring the geometry of a dodecahedron or designing virtual art installations, making abstract ideas tangible and engaging. Transdisciplinary research, uniting mathematicians, artists, and technologists, will further unlock innovative applications, such as using mathematical models to create sustainable urban designs or signifying data to communicate scientific findings through music. Global initiatives, like the Art Science Museum in Singapore, showcase these possibilities through exhibits that merge mathematical splendor with artistic expression, inspiring educators and students alike. To realize this vision, stakeholders must invest in technological infrastructure, teacher training, and equitable access to resources, ensuring that all learners can participate in this transformative synthesis. By embracing math-art integration, we can cultivate a generation of thinkers who navigate formulas and artistry with equal fluency, driving breakthroughs that harmonize precision and wonder to shape a more innovative, inclusive, and beautiful world.

REFERENCES

Abraham, A. (2024). *The creative brain: Myths and truths..* MIT Press. [http://dx.doi.org/10.7551/mitpress/14313.001.0001]

Ahmed, H. (2023). Art and the Human Experience: A Journey of Discovery and Understanding. *Journal of Religion and Society, 1*(1), 50-62.

Alexander, A.R. (2019). *Proof! How the world became geometrical.* Scientific American; Farrar, Straus & Giroux..

Barbin, E., Menghini, M., Volkert, K. (2019). *Descriptive geometry, the spread of a polytechnic art.* Cham: Springer.
[http://dx.doi.org/10.1007/978-3-030-14808-9]

Bork, R. (2023). Paper Thin? The Evidence for 12th-Century Gothic Design Drawings. *Arts, 12*(6), 220.

[http://dx.doi.org/10.3390/arts12060220]

Bruhn, S., Lüken, M.M. (2023). A framework to characterize young school children's individual mathematical creativity—an integrative review. *Asian J. Math. Edu., 2*(1), 116-144. [http://dx.doi.org/10.1177/27527263231163267]

Buckley, J., Seery, N., Canty, D. (2019). Investigating the use of spatial reasoning strategies in geometric problem solving. *Int. J. Technol. Des. Educ., 29*(2), 341-362. [http://dx.doi.org/10.1007/s10798-018-9446-3]

Escudeiro, P., Escudeiro, N., & Campos, M. (2024). Empowering education: The STEAME project's vision for teacher competence and innovation in STEM learning environments. *INTED2024 Proceedings*, 3256–3261.
[http://dx.doi.org/10.21125/inted.2024.0870]

Fathauer, R. (2020). Tessellations: Mathematics, art, and recreation. Boca Raton: *A K Peters/CRC Press.* [http://dx.doi.org/10.1201/9780429197123]

Gardner, H., Gardner, E. (2008). *Art, mind, and brain: A cognitive approach to creativity.*. Basic Books.

Glaeser, G. (2020). Geometry and its applications in arts, nature and technology. Springer Nature.

Grey, A. (2018). *The mission of art (20th anniversary ed.).* Shambhala Publications.

Guazzaroni, G., & Pillai, A. S. (Eds.). (2019). *Virtual and augmented reality in education, art, and museums.* IGI Global.

Hagler, J., Baur, C., & Ogawa, E. (2024). Staging Ars Electronica Archive: Alternative concept for exploring media art in education focusing on new animation art. *In 29th International Symposium on Electronic Art, ISEA* 2024 (pp. 1–7). ISEA International.

Husain, A., Nanda, M. N., Chowdary, M.S., Sajid, M. (2022). Fractals: an eclectic survey, part-I. *Fractal Fract., 6*(2), 89.
[http://dx.doi.org/10.3390/fractalfract6020089]

Immordino-Yang, M.H. (2016). Emotion, sociality, and the brain's default mode network. *Policy Insights Behav. Brain Sci., 3*(2), 211-219.
[http://dx.doi.org/10.1177/2372732216656869]

Junod, P. (2017). *Counterpoints: Dialogues between music and the visual arts.* Reaktion Books.

Kucukkaragoz, H., Meylani, R. (2025). Resistance to Learning: Reasons and Remedies via a Qualitative Research Synthesis. *J. Edu. Learn., 19*(1), 294-313.

Kugler, E., Kárpáti, A. (2023). Teaching mathematics through art. *Routledge eBooks.* Routledge. [http://dx.doi.org/10.4324/9781003376927-6]

Lipovetsky, S. (2022). [Review of the book Math and art: An introduction to visual mathematics, by S. Kalajdzievski]. *Technometrics, 64*(3), 425–426..
[http://dx.doi.org/10.1080/00401706.2022.2091873]

Liu, T., Luo, Y.T., Pang, P.C.I., Kan, H.Y. (2025). Exploring the impact of information and communication technology on educational administration: A systematic scoping review. *Educ. Sci., 15*(9), 1114. [http://dx.doi.org/10.3390/educsci15091114]

Meng, X., Zhang, L., Meng, P., Yu, Z., Diao, Z. (2024). Harmony in Complexity: Unveiling the Symbiotic Dance of Mathematics and Visual Art Through the Lens of Order. *J. Knowl. Econ., 15*(3), 13654-13672. [http://dx.doi.org/10.1007/s13132-023-01667-1]

Muhamad, A.H., Azzaldeen, A. (2021). The concept of the golden proportion in dentistry. *J Dent Med Sci, 20*, 37-45.

Ozkan, Z.C. (2023). The effect of project-based learning in visual arts lesson on lesson outcomes and attitudes. *International Journal on Social and Education Sciences, 5*(2), 367-380.

Parncutt, R., Hair, G. (2018). A psychocultural theory of musical interval. *Music Percept., 35*(4), 475-501. [http://dx.doi.org/10.1525/mp.2018.35.4.475]

Penerosa, R., & Pischnotte, A. (2021). Statewide Arts Integration Programming: A closer look at successes and challenges for elementary students, classroom teachers, and arts educators. *J. Learn. Through Arts*, 17(1). [http://dx.doi.org/10.21977/D917151495]

Pleasants, K. (2023). Integrative Approaches in Education: Bridging STEM and the Arts. *Xpertno International Journal of Interdisciplinary Research, 1*(1), 30-42. [http://dx.doi.org/10.5281/ZENODO.10053569]

Posy, C., & Ben-Menahem, Y. (Eds.). (2023). Mathematical knowledge, objects and applications: Essays in memory of Mark Steiner. Springer Nature..

Roberts, G. (2016). *From music to mathematics..* Johns Hopkins University Press. [http://dx.doi.org/10.56021/9781421419183]

Sánchez, A., Font, V., Breda, A. (2022). Significance of creativity and its development in mathematics classes for preservice teachers who are not trained to develop students' creativity. *Math. Educ. Res. J., 34*(4), 863-885. [http://dx.doi.org/10.1007/s13394-021-00367-w]

Schumacher, R.F., Zumeta Edmonds, R., Arden, S.V. (2017). Examining implementation of intensive intervention in mathematics. *Learn. Disabil. Res. Pract., 32*(3), 189-199. [http://dx.doi.org/10.1111/ldrp.12141]

Shuler, S.C., Brophy, T.S., Sabol, F.R. (2016). Arts assessment in an age of accountability: Challenges and opportunities in implementation, design, and measurement. *Meeting the Challenges to Measurement in an Era of Accountability.* Routledge. [http://dx.doi.org/10.4324/9780203781302-10]

Šprajc, I., Inomata, T., Aveni, A.F. (2023). Origins of Mesoamerican astronomy and calendar: Evidence from the Olmec and Maya regions. *Science Advances, 9*(1), eabq7675. [http://dx.doi.org/10.1126/sciadv.abq7675]

Sujatha, S., Vinayakan, K. (2023). Engaging Students with Mathematics: A Review of Motivation and Engagement Strategies. *Int. J. Interdiscip Res. Arts Humanit., 8*(2), 55-60.

Tan, P. (2017). Advancing inclusive mathematics education: Strategies and resources for effective IEP practices. *International Journal of Whole Schooling, 13*(3), 28-38.

Thapa, G.B., Thapa, R. (2018). The relation of golden ratio, mathematics and aesthetics. *Journal of the Institute of Engineering, 14*(1), 188-199. [http://dx.doi.org/10.3126/jie.v14i1.20084]

Vale, I., Barbosa, A. (2025). Mathematics and art: Ideas to use in teacher education. *Arts Commun., 3*(4), 6487. [http://dx.doi.org/10.36922/ac.6487]

Vivian Maanu, F.O.B., Ebenezer Appiagyei, , Hatsu Edo, , Klu kwasi, T. (2024). Connectivity and application of the principle of multiple intelligences in the three domains of learning mathematics. Journal of Mathematics Instruction. *Journal of Mathematics Instruction, Social Research and Opinion, 3*(2), 113-128. [http://dx.doi.org/10.58421/misro.v3i2.200]

Wagner, D., Tamayo, C. (2024). Invisibilization and Intersectionality in Mathematics Education: A Panoramic View: Invisibilização e Interseccionalidade na Educação Matemática: Uma Visão Panorâmica. In *Fourth International Handbook of Mathematics Education* (pp. 379-411). Cham: Springer Nature Switzerland. [http://dx.doi.org/10.1007/978-3-031-51474-6_16]

Wichmann, B., Wade, D. (2017). Islamic design: A mathematical approach. *Mathematics and the Built Environment.* Springer.

[http://dx.doi.org/10.1007/978-3-319-69977-6]

Winterson, J. (2024). *Maths & Music.* University of Huddersfield Press. [http://dx.doi.org/10.5920/mathsAndmusic.fulltext]

CHAPTER 11

Environmental Studies: An Interdisciplinary Approach

Saman Raza[1,*], **Tahmeena Khan**[2] and **Manisha Singh**[3]

[1] *Department of Chemistry, Isabella Thoburn College, Lucknow, Uttar Pradesh 226007, India*

[2] *Department of Chemistry, Integral University, Lucknow, Uttar Pradesh 226006, India*

[3] *Department of Education, Integral University, Lucknow, Uttar Pradesh 226006, India*

Abstract: Environmental pollution and degradation have become a looming crisis in recent years and threaten to endanger the Earth and all life on it. Climate change, pollution, natural disasters, extinction of species, behavioural changes in living things, *etc*., are all pointing towards a huge environmental catastrophe. The only solution is the combined effort of people, governments, and nations to protect the environment, and this needs mass mobilization, which can be achieved through educating people. Environmental studies is a subject that plays an important role in spreading awareness about and finding solutions for environmental issues. This is an interdisciplinary subject that integrates disciplines like biology, chemistry, physics, geography, economics, political science, engineering, sociology, *etc*. Together, learners and professionals of these disciplines contribute towards bringing about positive changes in the environment. This chapter explores the interdisciplinary nature of Environmental Studies and highlights its importance.

Keywords: Education, Environmental Studies, Humanities, Interdisciplinary, Science, Teaching.

INTRODUCTION

Environmental Studies is an extensive and dynamic field that explores the intricate and evolving relationship between humans and the natural world. Human survival, health, economies, and cultures are all intimately linked to the ecological systems that support life on Earth (Fang *et al*., 2022). This basic understanding supports the urgent need to pay attention to environmental issues such as climate change, biodiversity loss, air and water pollution, deforestation, land degradation, and resource depletion, and to respond to these challenges in a timely manner.

[*] **Corresponding Author Saman Raza:** Department of Chemistry, Isabella Thoburn College, Lucknow, Uttar Pradesh 226007, India; Email: samanmahek@gmail.com

These environmental concerns are complex and interwoven, frequently including several contributing aspects from biological, ecological, technical, political, economic, and cultural domains. For example, the issue of deforestation is more than just a biological or ecological matter; it also involves agricultural policy, economic incentives, population pressures, and so on. Similarly, global climate change affects all life on Earth, involving not only scientists dealing with pollution, but also industries, politics, economics, and the law.

Understanding these complexities highlights the importance of Environmental Studies. The study of the subject equips individuals, societies, and nations with the knowledge and skills needed to identify and evaluate human impact on the natural world and to make informed decisions that promote sustainability. Environmental Studies fosters environmental literacy, encouraging people to recognize their role in the protection of the Earth that gives us life and sustenance. By integrating science, policy, culture, and ethics, Environmental Studies has evolved into an interdisciplinary subject (Abe *et al.*, 2020). It enhances our ability to tackle persistent global challenges such as climate change (Sobti *et al.*, 2022), biodiversity loss (Dubey *et al.*, 2023), and environmental pollution and degradation (Gaur *et al.*, 2023) through interdisciplinary and hands-on methods. Also, by linking environmental issues with economic and social welfare, it helps in sustainable development.

Environmental Studies seeks to integrate diverse perspectives, methodologies, and remedies from different disciplines, such as:

- Natural and physical sciences, such as chemistry, physics, biology, oceanography, ecology, geology, *etc.*, that offer basic and extensive knowledge of environmental processes,
- Social sciences such as sociology, economics, and political science, that observe human behaviour and societal response,
- Humanities, such as philosophy, ethics, and history, provide an insightful understanding of our relationship with the environment.
- Applied fields such as law, public health, and education help in formulating viable solutions for sustainability (Fig. **1**) (Fang *et al.*, 2022).

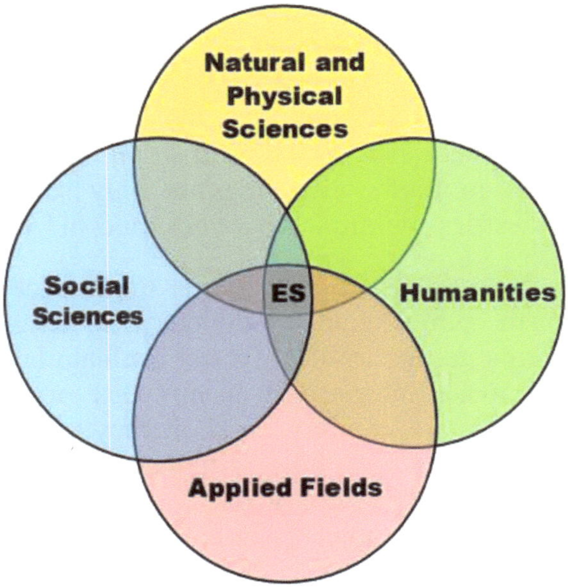

Fig. (1). Interdisciplinary nature of Environmental Studies (ES).

The value of Environmental Studies resides in its comprehensive approach. It is not simply an academic field, but also a practical guide for ethical living, decision-making, and leadership that aids in the resolution of rising environmental crises.

The Concept of Interdisciplinary Studies

Interdisciplinary studies combine concepts, theories, techniques, and methodologies from several academic subjects in a holistic manner. This technique allows for a more comprehensive grasp of numerous issues, ideas, and possibilities by bridging gaps between disciplines and generating new insights that single-discipline approaches cannot.

For example, a student of chemistry would have a better understanding of biomolecules if he studies biology as well, while a student of literature would benefit from studying history and philosophy. It is particularly useful when dealing with multifaceted and challenging issues like environmental degradation, public health, or development, that are too complex to be fully addressed by the study of any one field alone.

In the case of Environmental Studies, interdisciplinarity is its core or foundation (MacLeod et al. 2018). As mentioned earlier, it brings together several disciplines like natural and physical sciences, social sciences, humanities, and professional fields. This amalgamation helps in a proactive understanding of environmental issues at a local, national, and global level. For instance, while studying climate change, scientists use data from climatology, meteorology, physics, *etc.*; economists study its financial impact; policy makers evaluate climate policies; and sociologists explore its implications on society.

This enables a comprehensive and effective understanding of environmental issues. For instance, the resolution of an issue like air pollution involves not just a knowledge of chemistry but also climatology, meteorology, economics, political science, sociology, and health sciences, which would help in formulating better remedies, policies, laws, and health management related to air pollution. This chapter aims to explore how Environmental Studies functions as an interdisciplinary subject and why such an approach is crucial for environmental health and sustainability.

KEY DISCIPLINES INTEGRATED INTO ENVIRONMENTAL STUDIES

Natural and Physical Sciences

The basic sciences are the major components of Environmental Studies. The role of these is elaborated below, along with examples of their application:

Biology:

Biology, the study of microorganisms, plants, and animals, is crucial in understanding environmental aspects like changes in the behaviour of organisms with climate change, loss of habitat, and pollution, extinction of species, the genetic modification of crops, *etc*. For example, studies on flower pollinators like bees have emphasised their role in food security and the dangers posed by pesticides and habitat loss. In a study by Klein *et al.*, it was found that approximately 35% of global crop production depends on animal pollination, mainly bees, emphasizing their ecological and economic importance (Klein *et al.*, 2006). The loss of bee populations owing to habitat degradation, pesticide exposure, and diseases such as Colony Collapse Disorder has prompted serious concerns about food security and biodiversity conservation (Potts *et al.*, 2010). Another example where biology becomes a part of environmental studies is the study of genetic modification in crops, which has increased agricultural productivity and pest resistance in crops but has also raised concerns about the environmental risks and socioeconomic impacts of such crops (Ngongolo *et al.*, 2025). These examples show how the study of biology helps in the conservation of species and the environment, and sustainable agricultural practices.

Chemistry:

By studying the composition of polluted air, water, and soil, as well as the reactions and fate of chemicals in natural systems, Chemistry plays a crucial role in understanding environmental pollution (Hussain *et al.*, 2019). The phenomenon of acid rain is a good example. It was extensively studied through environmental chemistry, wherein it was discovered that sulfur dioxide (SO_2) and Nitrogen Oxides (NO_X), which are released from fossil fuel combustion, react with water vapor in the atmosphere to form sulfuric acid and nitric acid. This acidic rainfall causes widespread damage to forests, fields, waterbodies, and affects the ecosystems and environment (Prakash *et al.*, 2023). The knowledge acquired from these studies led to international agreements such as the 1979 Convention on Long-Range Transboundary Air Pollution (Fig. **2**), where it was agreed upon by 32 European countries to limit SO_2 and NO_X emissions and reduce acid rain effects (UNECE, 2021).

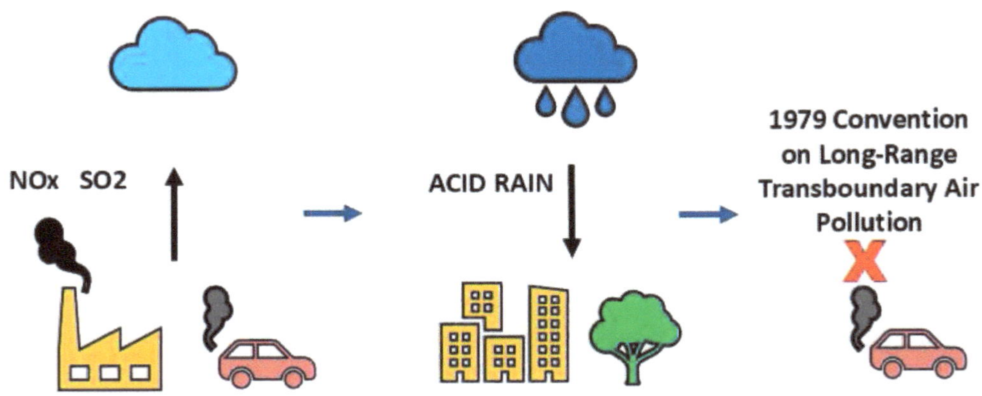

Fig. (2). Acid rain leading to an international agreement.

The detection and control of Persistent Organic Pollutants (POPs), such as DDT and PCBs, is another important example. Following public awareness of these compounds' harmful effects and bioaccumulation brought about by Rachel Carson's 1962 book Silent Spring, international efforts were made to restrict or outlaw them through the 2001 Stockholm Convention on Persistent Organic Pollutants (UNEP, 2001). Thus, environmental chemistry serves as the scientific foundation and driving force behind important legislative and regulatory actions aimed at safeguarding ecosystems and public health.

Physics:

As physics explains energy flows, thermodynamics, and climatic dynamics, it makes a substantial contribution to environmental studies (Monteith *et al.*, 2013).

The greenhouse effect, which forms the basis of the science of global warming, is a good illustration. When infrared radiation from the Earth's surface is trapped by gases in the atmosphere, such as carbon dioxide (CO_2), methane (CH_4), and water vapor, it cannot escape into space and causes the planet to warm. This phenomenon is known as the greenhouse effect. Principles of radiation and energy balance, which examine the absorption, emission, and transfer of energy in the atmosphere, provide an explanation for this process (Bardos and Pironneau 2022).

Some other examples include climate models that use physics to simulate how energy moves through the atmosphere and oceans, helping predict future climate scenarios based on current greenhouse gas emissions (IPCC, 2021).

Geology:

Geology is the study of the Earth's structure, history, and processes such as soil formation, plate tectonics, earthquakes, *etc*. Understanding resource distribution, environmental changes throughout geological time, and natural risks are all facilitated by it.

For instance, knowledge of soil composition and erosion mechanisms helps in land management and sustainable agriculture strategies (Montgomery, 2007); this is a very important aspect, as sustainable agriculture is of extreme importance in the age of climate change. Geology also helps in infrastructure development, aiding in avoiding mishaps and disasters. Furthermore, building rules in seismic zones are guided by earthquake geology knowledge to lower the chance of disaster (Dutta 2024).

Meteorology

Meteorology is the study of weather and atmospheric phenomena. It is essential for climate modelling, weather forecasting, and comprehending the effects of air pollution. For example, early warning systems for cyclones, such as Tropical Cyclone Forecasting [https://wmo.int/content/tropical-cyclone-forecasting], have been made possible by meteorological advancements, saving thousands of lives every year. Furthermore, climate models inform international policy by simulating potential future global warming scenarios (IPCC, 2021). For example, the Indian Meteorological Department's monsoon forecasting helps farmers plan crops, reducing agricultural losses, while monitoring of urban heat islands in megacities like Delhi informs urban planning to mitigate heat stress (Kumar *et al.*, 2019; Sikka, 2018).

Ecology

This branch concentrates on the relationships between organisms and their environment. For instance, the study of coral reef ecosystems enables ecologists to learn how biodiversity sustains marine food webs and how the climate-induced bleaching of corals poses perilous threats to entire marine ecosystems, such as in the Great Barrier Reef (Hughes *et al.*, 2017). In the Amazon rainforest, a terrestrial ecosystem, case studies illustrate how deforestation alters the relationship between species, carbon cycling, climate moderation, and ecosystem balance, revealing the need to maintain biodiversity for ecological equilibrium (Laurance *et al.*, 2016). These illustrations exemplify the role of ecology in managing and mitigating environmental impacts through the destruction of nature and helping the ecosystem through conservation efforts.

Oceanography

This branch studies ocean currents, marine life, and the composition of seawater. Oceanographic research has contributed to understanding phenomena such as El Niño, its effects on world weather, agriculture, and fisheries (NOAA, 2024).

These scientific disciplines, in combination, help track and monitor changes over time, evaluate potential ecological impacts and risks, and develop plans for mitigation and adaptation. The combination with Environmental Studies provides a comprehensive, empirically supported grasp of natural systems and their feedback to human activities.

Social Sciences

Apart from the natural and physical sciences that are the major components, Environmental Studies also benefits from social studies. Individuals and social systems are key contributors to both the destruction and protection of the environment. Contributing disciplines from social sciences include:

Sociology

Sociology is the study of cultures, communities, and social behaviour of humans. It comes in handy in environmental studies when assessing how people relate and interact with nature, perceive risks concerning the environment, and practice sustainability. For instance, environmental justice studies demonstrate how underrepresented populations are often the most impacted by pollution and climate crises (Chowkwanyun, 2023). Knowing social perceptions and attitudes helps design effective environmental campaigns and policies. For example, research on the indigenous people of the Amazon has documented the application of traditional ecological knowledge in aiding the conservation efforts of the forest (Nascimento *et al.*, 2024).

Economics

Environmental economics analyzes the effect on the economy of policies concerning the environment, resource utilization, and external negative factors like pollution [Smith, 2001]. For example, it targets unsustainable practices by creating markets for carbon trading. For example, the European Union Emission Trading Scheme (EU ETS) is a market-driven cap and trade system where emission allowances for greenhouse gases are allocated to firms for trading purposes (Bryant, 2019).

Cost and benefit evaluation of the renewable energy projects enables governments to effectively utilize resources. These principles are the basis for environmental taxation, which aims at reducing unhealthy activities and provides means for protection at the same time (Ashiabor *et al.*, 2021).

Political Science

The study of political science focuses on political structures, laws, policies, *etc.*, and their influence on environmental governance and policymaking (Brown, 2016). It examines international agreements, environmental legislation, and the function of NGOs and governments in promoting sustainability. An example of international collaboration in climate policy is the historic 2015 Paris Agreement, which committed nations to reducing global warming (Van Calster *et al.*, 2021).

Geography:

Geography studies how ecosystems, human activity, and environmental events are distributed physically. In order to comprehend land use, resource management, and environmental change, it combines human and physical geography (Fu, 2020). For instance, research on expansion in urban geography informs sustainable city design to prevent pollution and habitat loss (Seto, 2012), while GIS (Geographic Information Systems) technology aids in mapping deforestation in the Amazon, directing conservation efforts (Moutinho, 2021).

Management

Environmental management focuses on applying scientific knowledge, policy, and technology to sustainably manage natural resources and mitigate environmental impacts (Burger, 2008). For example, Integrated Water Resources Management (IWRM) approaches balance social, economic, and ecological needs for sustainable water use (Wang, 2025).

Humanities

Humanities enrich Environmental Studies by offering ethical, historical, and cultural perspectives:

- **Philosophy**: Raises questions about human responsibility towards nature (Mathews, 2014).
- **Ethics**: Develops frameworks for environmental responsibility and justice (Ott, 2020).
- **History**: Provides context for how environmental practices have evolved (Roy, 2019).
- **Literature**: Helps in spreading awareness and shaping environmental consciousness (Buell *et al.*, 2011).
- **Cultural Studies**: Highlights indigenous and traditional knowledge systems [Phaedra, 2015].

Law and Public Policy

Legal frameworks are essential for enforcing environmental protection:

- **Environmental Law**: Governs the use and protection of natural resources.
- **International Law**: Addresses global environmental challenges like climate change through treaties (Louka, 2006).
- **Policy Studies**: Evaluates the effectiveness of environmental policies and programs.

For example, the Environment Protection Act of 1986 [https://www.indiacode.nic.in/bitstream/123456789/4316/1/ep_act_1986.pdf] in India provides a legal foundation for regulating pollutants and protecting ecosystems.

Health Sciences and Epidemiology

Environmental factors significantly influence human health:

- **Public Health**: Studies the impact of environmental pollution on health (Karak *et al.*, 2025).
- **Toxicology**: Analyses the effects of chemicals on organisms (Sauvé, 2024).
- **Epidemiology**: Tracks the spread of diseases linked to environmental conditions, like the link between air pollution and respiratory diseases like asthma, which is studied using epidemiological methods (Dharmage *et al.*, 2019).

The role of these disciplines and their interdisciplinarity came to the fore during the COVID-19 pandemic. Epidemiologists tracked the spread of the SARS-CoV-2 virus by collecting and analysing data on infection rates, transmission patterns, and mortality. Virologists studied the virus structure, helping develop vaccines. Public health experts designed interventions such as social distancing and mask mandates. Behavioural scientists worked on communication strategies to encourage vaccine acceptance. This combined effort led to global vaccination campaigns and health policies that mitigated the pandemic's impact (Chowdhury *et al.*, 2020; Krpan *et al.*, 2020; Naser *et al.*, 2020; Samudrala *et al.*, 2020).

BENEFITS OF INTERDISCIPLINARY ENVIRONMENTAL STUDIES

The benefits of Interdisciplinary Environmental Studies are summarized in Fig. (**3**).

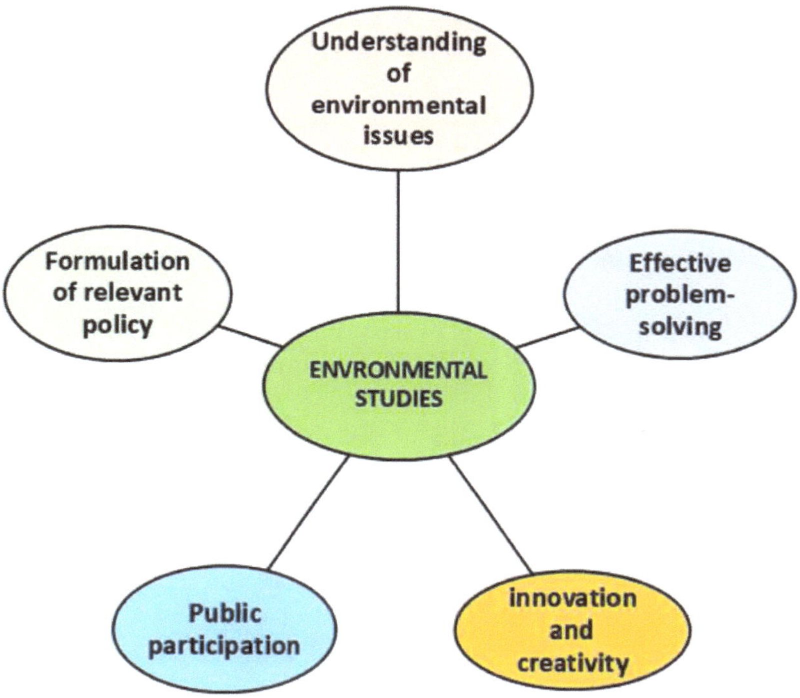

Fig. (3). Benefits of environmental studies.

- **Comprehensive Understanding**: Interdisciplinary environmental studies provide a deeper and more complex understanding of environmental issues by drawing on a variety of fields, including science, sociology, law, the humanities, and more. Students and professionals can better understand the socio-ecological complexity of environmental challenges, recognize underlying causes, and see links across systems with the aid of this holistic approach. For instance, understanding ecology, economics, legislation, and indigenous knowledge is necessary to combat deforestation (Fisher *et al.*, 2011; Murmu *et al.*, 2018).

- **Effective Solutions**: Environmental Studies facilitates the development of more efficient and contextually aware solutions by combining many disciplinary perspectives. It enables problem-solving that focuses on the underlying causes rather than just treating the symptoms. For example, tackling urban air pollution requires a combination of public health statistics, regulations to control emissions, engineering for cleaner transportation, and public awareness efforts (Jafari *et al.*, 2021).
- **Innovation**: Collaboration across disciplines encourages innovation and creativity. Innovations that might not occur in a single area can result from the exchange of ideas across different fields. For instance, collaboration between physicists, chemists, engineers, and environmental economists frequently leads to the creation of green technologies like solar panels or bioplastics (Costa *et al.*, 2023; Ncube *et al.*, 2023; Paipa-Sanabria *et al.*, 2025; Sati *et al.*, 2023).
- **Policy Relevance**: Interdisciplinary perspectives greatly aid in the formulation of environmental policy. More solid and generally embraced policies are those based on empirical evidence, social justice concerns, economic feasibility, and moral values. Frameworks for disaster risk reduction, biodiversity policies, and climate action plans can all benefit from interdisciplinary research, which can increase their efficacy and inclusivity (Förster, 2022; Righi *et al.*, 2021).
- **Public Engagement**: Environmental Studies also enhances public communication and participation. Humanities and social sciences help frame environmental narratives in relatable and culturally sensitive ways, boosting awareness and motivating behavioural change. By integrating ethical arguments, historical context, and storytelling, public engagement becomes more empathetic and action-oriented.
- **Educational Implications:** Environmental education increases the knowledge of ecological systems by students, ensures sustainable practices, and equips the learners with future careers in environmental science and sustainability fields. It also enhances performance in academics as it ensures learning is relevant and captivating. The benefits of environmental studies and their education have been elaborated in the following sections.

Educational Implications of Environmental Studies and Its Education

Given the interdisciplinary nature of environmental studies, its teaching and learning at the school and college level is of utmost importance. Environmental education is essential in India to combat critical environmental issues such as air and water pollution, green revolution-induced deforestation, and climate change. It equips students with an understanding and the ability to become environmental stewards and ensures sustainable development. The benefits can be enumerated as follows.

- Enhanced academic performance: Studies have shown that introducing environmental education in school curricula can improve academic performance (Husin *et al.*, 2025). By relating in-class learning to actual environmental problems, the information becomes more interesting and applicable, which is more beneficial to the students. This applicability increases their interest in sciences, geography, and social studies, which consequently results in better academic performance.
- Social change: When students acquire knowledge of environmental concerns and sustainable practices, they tend to pass on this information to their families and communities. This becomes a ripple effect, influencing larger societal changes as communities embrace more environmentally friendly practices. Schools can be catalysts for these changes by developing a culture of sustainability in their local communities (Yarkandi, 2012).
- Better career opportunities: Students who receive environmental education are better prepared for the growing number of green economy professions. There is an increase in employment in environmental research, conservation, renewable energy, and sustainability consultancy. By providing students with a strong understanding of environmental challenges, schools and colleges may prepare them to pursue these types of careers as well as engage in innovative solutions to environmental problems.

Integrating Environmental Studies into the Curriculum

Schools can incorporate environmental education by updating lesson plans to reflect environmental issues, having eco-clubs and workshops, and spreading eco-friendly practices throughout the school community. Joining forces with local environmental agencies can also be beneficial. Incorporating environmental education into school and college curricula is not only an educational improvement but a long-term strategy for future generations. By creating environmental awareness and sustainable skills among students, schools have a crucial role to play in developing a greener and more sustainable future. Incorporating environmental studies as an interdisciplinary subject into the curriculum is the most important step that needs a lot of planning and strategy (Winther *et al.*, 2010). The curriculum should integrate science, humanities, and social studies. Institutes can create interdisciplinary lesson plans that incorporate environmental themes into subjects such as science, geography, history, and social studies. For instance, an ecosystems lesson in science can involve coverage of biodiversity and conservation, while a history lesson can involve the effects of industrialization on the environment. Other than this, fieldwork and project-based learning are great methods to educate students and enhance experiential understanding of the environment. For example, visits to parks, gardens, museums, and wildlife sanctuaries give students hands-on experiences of natural ecosystems. To be effective, environmental education also requires proper training of teachers to impart to them the necessary knowledge and skills to incorporate environmental issues into their lesson plans. and resources. Making teaching resources like lesson plans, activity guides, and multimedia materials, and interdisciplinary research centres accessible to teachers can also help them deliver effective environmental education.

Challenges and Solutions for Environmental Education Implementation

Environmental Studies, as an interdisciplinary subject, while offering numerous benefits, also faces several significant challenges and barriers that hinder its full potential. Each of these challenges is rooted in issues within educational institutes, funding structures, policy frameworks, and educational practices. Below is an elaboration on each of these key barriers:

- **Institutional Framework**: Academic and organizational structures are often divided into distinct departments with little interaction or collaboration. These frameworks discourage interdisciplinary dialogue and restrict resource sharing, which hinders the development of integrated environmental programs. Faculty members and researchers may face difficulties in coordinating cross-departmental initiatives due to bureaucratic hurdles or disciplinary territoriality.
- **Communication Gaps**: Each discipline has its own technical jargon, methodologies, and epistemologies, which can create barriers to effective communication and collaboration. For example, a climate scientist and an economist may approach the same environmental problem from entirely different perspectives and use terminology that is unfamiliar to the other. This can lead to misunderstandings and inefficient teamwork unless deliberate efforts are made to foster cross-disciplinary fluency.
- **Funding Constraints**: Limited resources often hamper the introduction of environmental education. Interdisciplinary research and education often struggle to secure funding because grant agencies may prioritize traditional, single-discipline research proposals. Many funding bodies are structured around disciplinary criteria, making it difficult for interdisciplinary teams to demonstrate alignment with specific funding priorities. Additionally, evaluation committees may lack the interdisciplinary expertise needed to fairly assess such proposals.
- **Assessment Difficulties**: Standard methods of evaluating academic performance and student learning are not always suitable for interdisciplinary approaches. For

instance, standardized tests may not capture the integrated thinking and problem-solving skills that are central to Environmental Studies. Similarly, academic institutions may struggle to design curricula and metrics that fairly assess contributions from multiple disciplines.

Potential Solutions:

Addressing these barriers requires systemic reform and innovation. Educational institutes can establish interdisciplinary departments or research centres that foster collaboration across disciplines. Joint appointments and team-teaching initiatives can encourage faculty from different backgrounds to work together. The problem of limited resources can be overcome through alliances with local environmental groups, government institutions, and NGOs. These alliances can open windows of opportunity for schools to access resources, expertise, and financial support for environmental education initiatives. Funding agencies can develop special grants for interdisciplinary research, and educational institutions can adopt more flexible and holistic assessment strategies. (Fig. **4**).

Fig. (4). Challenges and solutions in interdisciplinary studies.

Overcoming these challenges is essential to harnessing the full potential of Environmental Studies. Only through genuine interdisciplinary integration can we develop sustainable solutions to the complex environmental issues of our time.

Through government support, several environmental education initiatives have been undertaken in India that provide resources to facilitate interdisciplinary teaching-learning. Some of the effective environmental education programs in India are:

- Paryavaran Mitra is a Centre for Environment Education (CEE) initiative in India that aims to create environmental awareness and action amongst students. Through classroom lessons, excursions, and community projects, Paryavaran Mitra [https://www.ceeindia.org/programmes/paryavaran-mitra-educati-n-children-rajasthan.php] promotes students to be environmental ambassadors and has impacted thousands of schools in India.
- The Green Schools Programme or GSP [https://www.cseindia.org/green-schools-program-253] has been initiated by the Centre for Science and Environment (CSE) of India to enhance environmental education in schools by adopting green practices. Under GSP, students are educated on waste management, water conservation, energy efficiency, and biodiversity.
- Eco-Schools [https://www.ecoschools.in/]is an international program that has been adopted in several schools in India. The program offers a curriculum and operational framework for schools to weave sustainability into their day-to-day activities. Schools engage in student-led activities such as waste reduction, energy efficiency, and biodiversity, which makes them feel responsible for their contribution to the environment.

Besides this, several educational institutes have curated interdisciplinary Environmental Studies curricula, such as the TERI School of Advanced Studies in India, which offers interdisciplinary environmental programs combining science, policy, and management.

CONCLUDING REMARKS

In today's world, environmental studies is a crucial interdisciplinary subject. Its interdisciplinary approach promotes creative and sustainable solutions as well as a deep awareness of environmental issues. Interdisciplinary approaches in education, research, and policy are becoming essential as the globe struggles with ecological issues, such as resource shortages, biodiversity loss, and climate change.

Environmental Studies promotes a comprehensive worldview that recognizes the interdependence of ecological and human systems by combining knowledge from the natural sciences, social sciences, humanities, and practical areas. This all-encompassing viewpoint enables people, groups, and organizations to confront environmental issues as complex issues that call for cooperation, ingenuity, and critical thinking rather than as isolated issues.

The ability of present and future generations to think across boundaries, bridge knowledge systems, and implement context-sensitive solutions is critical to the future of sustainable development. In order to foster environmental literacy, policy innovation, and global leadership, it is imperative that multidisciplinary environmental studies be promoted in educational institutions. In conclusion, environmental studies' interdisciplinary character gives society the knowledge and moral insight it needs to protect the environment and create a resilient, stable, fair, and sustainable future. Adopting this strategy is essential to turning environmental awareness into substantial action.

REFERENCES

Adamson, J., Gleason, W. & Pellow, D. (2016). *Keywords for Environmental Studies*. New York, USA: New York University Press.
[http://dx.doi.org/10.18574/nyu/9780814760741.001.0001]

Ashiabor, H., Milne, J.E., Mikael, S.A. (2021). *Environmental taxation in the pandemic era: Opportunities and Challenges.*. Cheltenham, UK: Edward Elgar Publishing.
[http://dx.doi.org/10.4337/9781800888517]

Bardos, C., Pironneau, O. (2022) The radiative transfer model for the greenhouse effect. *SeMA 79*, 489–525.
[http://dx.doi.org/10.1007/s40324-021-00265-y]

Brown, M. (2016). Environmental science and politics. In T. Gabrielson, C. Hall, J. M. Meyer, & D. Schlosberg (Eds.), *The Oxford handbook of environmental political theory*. Oxford University Press.
[http://dx.doi.org/10.1093/oxfordhb/9780199685271.013.1]

Bryant, G. (2019). *Carbon Markets in a Climate-Changing capitalism.*. Cambridge: Cambridge University Press.
[http://dx.doi.org/10.1017/9781108377591]

Buell, L., Heise, U.K., Thornber, K. (2011). Literature and Environment. *Annu. Rev. Environ. Resour., 36*(1), 417-440.
[http://dx.doi.org/10.1146/annurev-environ-111109-144855]

Burger J. (2008) Environmental management: integrating ecological evaluation, remediation, restoration, natural resource damage assessment and long-term stewardship on contaminated lands. Sci Total Environ 400(1-3), 6-19.
[http://dx.doi.org/10.1016/j.scitotenv.2008.06.041] [PMID: 18687455]

Centre for Environment Education. (2025). Paryavaran Mitra. *CEE India.* Available from: https://www.ceeindia.org/programmes/paryavaran-mitra-education-children-rajasthan.php

Centre for Science and Environment. (2025). Green Schools Programme. *CSE India.* Available from: https://www.cseindia.org/green-schools-program-253

Chowdhury, S. D., Oommen, A. M. (2020). Epidemiology of COVID-19. *J. Dig. Endosc., 11*(01), 03-07.
[http://dx.doi.org/10.1055/s-0040-1712187]

Chowkwanyun M. (2023) Environmental Justice: Where It Has Been, and Where It Might Be Going. *Annu Rev Public Health, 3*; 44: 93-111.
[http://dx.doi.org/10.1146/annurev-publhealth-071621-064925] [PMID: 36623927]

Costa, A., Encarnação, T., Tavares, R., Todo Bom, T., Mateus, A. (2023). Bioplastics: Innovation for green

transition. *Polymers (Basel), 15*(3), 517.
[http://dx.doi.org/10.3390/polym15030517] [PMID: 36771817]

Dharmage, S.C., Perret, J.L., Custovic, A. (2019). Epidemiology of asthma in children and adults. *Front Pediatr., 7*, 246.
[http://dx.doi.org/10.3389/fped.2019.00246] [PMID: 31275909]

Dubey, A., Mainani, R., Shah, P., Solanki, H. (2023). Biodiversity Loss and its Impacts on Human Well-being. *International Journal of Plant and Environment, 8*(3), 47-51.
[http://dx.doi.org/10.18811/ijpen.v8i03.06]

Dutta. (2024). Seismic design considerations for buildings in earthquake-prone areas. *ResearchGate.* https://www.researchgate.net/publication/379986342

Environment Protection Act (1986). *India Code.* Available from: https://www.indiacode.nic.in/bitstream/123456789/4316/1/ep_act_1986.pdf

Fang, WT., Hassan, A., LePage, B.A. (2023). Introduction to Environmental Education. In: *The Living Environmental Education. Sustainable Development Goals Series.* Springer, Singapore.
[http://dx.doi.org/10.1007/978-981-19-4234-1_1]

Fisher, B., Lewis, S.L., Burgess, N.D., Malimbwi, R.E., Munishi, P.K., Swetnam, R.D., Kerry Turner, R., Willcock, S., Balmford, A. (2011). Implementation and opportunity costs of reducing deforestation and forest degradation in Tanzania. *Nat. Clim. Chang., 1*(3), 161-164.
[http://dx.doi.org/10.1038/nclimate1119]

Förster, J Förster, J. (2022). *Linkages between biodiversity and climate change and the role of science-policy practice interfaces for ensuring coherent policies and actions. Deutsche Gesellschaft für Internationale Zusammenarbeit (GIZ)*; International Institute for Sustainable Development (IISD); Helmholtz Centre for Environmental Research (UFZ).

Fu, B. (2020). Promoting geography for sustainability. *Geography and Sustainability, 1*(1), 1-7.
[http://dx.doi.org/10.1016/j.geosus.2020.02.003]

Gaur, N., Sharma, S., & Yadav, N. (2024). Environmental pollution. In *Green chemistry approaches to environmental sustainability: Status, challenges and prospective* (pp. 23–41). Elsevier: Academic Press.
[http://dx.doi.org/10.1016/b978-0-443-18959-3.00010-0]

Hughes, T.P., Kerry, J.T., Álvarez-Noriega, M., Álvarez-Romero, J.G., Anderson, K.D., Baird, A.H., Babcock, R.C., Beger, M., Bellwood, D.R., Berkelmans, R., Bridge, T.C., Butler, I.R., Byrne, M., Cantin, N.E., Comeau, S., Connolly, S.R., Cumming, G.S., Dalton, S.J., Diaz-Pulido, G., Eakin, C.M., Figueira, W.F., Gilmour, J.P., Harrison, H.B., Heron, S.F., Hoey, A.S., Hobbs, J.P.A., Hoogenboom, M.O., Kennedy, E.V., Kuo, C., Lough, J.M., Lowe, R.J., Liu, G., McCulloch, M.T., Malcolm, H.A., McWilliam, M.J., Pandolfi, J.M., Pears, R.J., Pratchett, M.S., Schoepf, V., Simpson, T., Skirving, W.J., Sommer, B., Torda, G., Wachenfeld, D.R., Willis, B.L., Wilson, S.K. (2017). Global warming and recurrent mass bleaching of corals. *Nature, 543*(7645), 373-377.
[http://dx.doi.org/10.1038/nature21707] [PMID: 28300113]

Husin, A., Helmi, H., Nengsih, Y.K., Rendana, M. (2025). Environmental education in schools: sustainability and hope. *Discov. Sustain., 6*(1), 41.
[http://dx.doi.org/10.1007/s43621-025-00837-2]

Hussain, C. M., & Keçili, R. (2020). Environmental pollution and environmental analysis. In C. M. Hussain (Ed.), *Modern environmental analysis techniques for pollutants* (pp. 1–36). Elsevier: Academic Press.
[http://dx.doi.org/10.1016/b978-0-12-816934-6.00001-1]

Intergovernmental Panel on Climate Change IPCC. (2021). Climate Change 2021: The Physical Science Basis. *Cambridge University Press.*https://www.ipcc.ch/report/ar6/wg1/

Jonidi Jafari, A., Charkhloo, E., Pasalari, H. (2021). Urban air pollution control policies and strategies: a systematic review. *J. Environ. Health Sci. Eng., 19*(2), 1911-1940.
[http://dx.doi.org/10.1007/s40201-021-00744-4] [PMID: 34900316]

Karak, P., Parveen, A., Modak, A., Adhikari, A., Chakrabortty, S. (2025). Microplastic Pollution: A Global Environmental Crisis Impacting Marine Life, Human Health, and Potential Innovative Sustainable Solutions. *Int. J. Environ. Res. Public Health, 22*(6), 889.
[http://dx.doi.org/10.3390/ijerph22060889] [PMID: 40566316]

Klein, A.M., Vaissière, B.E., Cane, J.H., Steffan-Dewenter, I., Cunningham, S.A., Kremen, C., Tscharntke, T. (2007). Importance of pollinators in changing landscapes for world crops. *Proc. Biol. Sci., 274*(1608), 303-313.
[http://dx.doi.org/10.1098/rspb.2006.3721] [PMID: 17164193]

Krpan, D., Makki, F., Saleh, N., Brink, S.I., Klauznicer, H.V. (2021). When behavioural science can make a difference in times of COVID-19. *Behav. Public Policy, 5*(2), 153-179.
[http://dx.doi.org/10.1017/bpp.2020.48]

Kumar, P., Robins, A., Vardoulakis, S., Norford, L. (2019). Urban heat islands: Causes, consequences, and mitigation. *Environ. Pollut., 247*, 1313-1326.
[http://dx.doi.org/10.1016/j.envpol.2019.01.057]

Laurance, W.F. *et al.* (2016). An Amazonian Forest and Its Fragments as a Laboratory of Global Change. In: Nagy, L., Forsberg, B., Artaxo, P. (eds) *Interactions Between Biosphere, Atmosphere and Human Land Use in the Amazon Basin*. Springer, Berlin, Heidelberg.
[http://dx.doi.org/10.1007/978-3-662-49902-3_17]

Louka, E. (2006). *International Environmental Law: Fairness, Effectiveness, and World Order*. Cambridge: Cambridge University Press.
[http://dx.doi.org/10.1017/CBO9780511618109]

MacLeod, M., & Nagatsu, M. (2018). What does interdisciplinarity look like in practice: Mapping interdisciplinarity and its limits in the environmental sciences. *Studies in History and Philosophy of Science Part A, 67*, 74-84.
[http://dx.doi.org/10.1016/j.shpsa.2018.01.001]

Mathews, F. (2014). Environmental Philosophy. In: Oppy, G., Trakakis, N. (eds) *History of Philosophy in Australia and New Zealand.* Springer, Dordrecht.
[http://dx.doi.org/10.1007/978-94-007-6958-8_22]

Monteith, J., & Unsworth, M. (2008). *Principles of environmental physics* (3rd ed.). Amsterdam, Netherlands: Academic Press.

Moutinho, S. (2021, August 16). *Brazil's first homemade satellite will put an extra eye on dwindling Amazon forests.* Science. https://www.science.org/content/article/brazil-s-first-homemade-satellite-will-put-e-tra-eye-dwindling-amazon-forests

Murmu, S. C., & Bhattacharya, S. (2019). Indigenous knowledge on forest management: An approach towards sustainable development. *Journal of Culture, Society and Anthropology, 1*(4), 1–18.

Naser, N., Masic, I., Zildzic, M. (2020). Public Health Aspects of COVID19 Infection with Focus on Cardiovascular Diseases. *Mater. Sociomed., 32*(1), 71-76.
[http://dx.doi.org/10.5455/msm.2020.32.71-76] [PMID: 32410896]

Nascimento, M. N., N. Aukes, T. F., & H. McMichael, C. N. (2024). Indigenous and colonial influences on Amazonian forests. *Plants, People, Planet, 6*(4), 803-823.
[http://dx.doi.org/10.1002/ppp3.10515]

National Oceanic and Atmospheric Administration. (2024). What are El Niño and La Niña?. *NOAA Ocean Service.* Available from: https://oceanservice.noaa.gov/facts/ninonina.html

Ncube, A., Mtetwa, S., Bukhari, M., Fiorentino, G., Passaro, R. (2023). Circular Economy and green chemistry: the need for radical innovative approaches in the design for new products. *Energies, 16*(4), 1752.
[http://dx.doi.org/10.3390/en16041752]

Ngongolo, K., Mmbando, G.S. (2025) Necessities, environmental impact, and ecological sustainability of

genetically modified (GM) crops. *Discov Agric, 3*(29), 1-11.
[http://dx.doi.org/10.1007/s44279-025-00180-0]

Ott, K. (2020). Environmental ethics. *Online Encyclopedia Philosophy of Nature / Online Lexikon Naturphilosophie.*
[http://dx.doi.org/10.11588/oepn.2020.0.71420]

Paipa-Sanabria, E., González-Montoya, D., Coronado-Hernández, J.R. (2025). Green Technologies for Environmental Air and Water Impact Reduction in Ships: A Systematic Literature Review. *J. Mar. Sci. Eng., 13*(5), 839.
[http://dx.doi.org/10.3390/jmse13050839]

Pezzullo, P. C. (Ed.). (2015). *Cultural studies and environment, revisited.* New York, NY: Routledge.

Potts, S.G., Biesmeijer, J.C., Kremen, C., Neumann, P., Schweiger, O., Kunin, W.E. (2010). Global pollinator declines: trends, impacts and drivers. *Trends Ecol. Evol., 25*(6), 345-353.
[http://dx.doi.org/10.1016/j.tree.2010.01.007] [PMID: 20188434]

Prakash J, Agrawal SB, Agrawal M. (2023) Global Trends of Acidity in Rainfall and Its Impact on Plants and Soil. *J Soil Sci Plant Nutr. 23*(1): 398-419.
[http://dx.doi.org/10.1007/s42729-022-01051-z] [PMID: 36415481]

Righi, E., Lauriola, P., Ghinoi, A., Giovannetti, E., Soldati, M. (2021). Disaster risk reduction and interdisciplinary education and training. *Prog. Disaster Sci., 10*, 100165.
[http://dx.doi.org/10.1016/j.pdisas.2021.100165]

Roy, D. (2019) Concept of Environmental History. *International Journal of Humanities & Social Science Studies (IJHSSS)*, 4(2), 42-47. https://www.ijhsss.com/files/6.-Dr.-Debashish-Roy.pdf

Samudrala, P.K., Kumar, P., Choudhary, K., Thakur, N., Wadekar, G.S., Dayaramani, R., Agrawal, M., Alexander, A. (2020). Virology, pathogenesis, diagnosis and in-line treatment of COVID-19. *Eur. J. Pharmacol., 883*, 173375.
[http://dx.doi.org/10.1016/j.ejphar.2020.173375] [PMID: 32682788]

Sati, K. A., Kumar, S. (2023). Solar Energy Advancements and Their Environmental Impacts. *Futuristic Trends in Renewable & Sustainable Energy, IIP Proceedings, 2*(29), 62-76.

Sauvé, S. (2024). Toxicology, environmental chemistry, ecotoxicology, and One Health: definitions and paths for future research. *Front. Environ. Sci., 12*, 1303705.
[http://dx.doi.org/10.3389/fenvs.2024.1303705]

Seto, K.C., Güneralp, B., Hutyra, L.R. (2012). Global forecasts of urban expansion to 2030 and direct impacts on biodiversity and carbon pools. *Proc. Natl. Acad. Sci. USA, 109*(40), 16083-16088.
[http://dx.doi.org/10.1073/pnas.1211658109] [PMID: 22988086]

Sikka, D.R. (2018). Forecasting Indian monsoons: Advances and challenges. *Curr. Sci., 115*(6), 1021-1028.

Smith, V. K. (2001). Environmental economics. In N. J. Smelser & P. B. Baltes (Eds.), *International encyclopedia of the social & behavioral sciences* (pp. 4611–4617). Elsevier: Pergamon.
[http://dx.doi.org/10.1016/b0-08-043076-7/02314-7]

Sobti, R.C., Malhotra, S.K., Jaiswal, K., Puri, S. (2022). *Environmental Studies and Climate Change..* CRC Press.
[http://dx.doi.org/10.1201/9781003220824]

UNEP (2001). United Nations Environment Programme. *Stockholm Convention on Persistent Organic Pollutants*. Geneva, Switzerland.

United Nations Economic Commission for Europe. (2021). *1979 Convention on long-range transboundary air pollution.* https://unece.org/sites/default/files/2021-05/1979%20CLRTAP.e.pdf.

Van Calster, G., Reins, L. (2021). *The Paris Agreement on climate change: A Commentary..* Edward Elgar Publishing.

[http://dx.doi.org/10.4337/9781788979191]

Wang. (2025). Sustainable Water Management: Balancing Economic, Social, and Environmental Needs. *Journal of Lifestyle and SDGs Review, 5*(5), e06606.
[http://dx.doi.org/10.47172/2965-730X.SDGsReview.v5.n05.pe06606]

Winther, A.A., Sadler, K.C., Saunders, G.W. (2010). Approaches to environmental education. In: *The inclusion of environmental education in science teacher education.* Springer Dordrecht.
[http://dx.doi.org/10.1007/978-90-481-9222-9_3]

World Meteorological Organization. (2023). Tropical cyclone forecasting [Fact sheet]. *WMO.* Available from: https://wmo.int/content/tropical-cyclone-forecasting

Yarkandi, A.H. (2012). Strengthening environmental education in school curricula. *Journal of Education and Vocational Research, 3*(8), 264-270.

CHAPTER 12

Spaces of Interaction: Architecture as a Conduit for Cross-disciplinary Collaboration in Education

Zeba Nisar[1,*] and **Khurram Ashraf**[1]

[1] *Department of Architecture, Integral University, Lucknow, Uttar Pradesh 226006, India*

Abstract: This chapter examines the transformative role of architecture in fostering cross-disciplinary collaboration within contemporary educational settings. As traditional, discipline-specific pedagogies give way to integrated and interactive learning models, the spatial design of educational environments has become a critical enabler of innovation, knowledge exchange, and collaborative problem-solving. Drawing on theories from environmental psychology, spatial syntax, and educational design, the chapter explores how architectural elements such as openness, modularity, technological integration, and spatial fluidity can actively shape interdisciplinary engagement. Through detailed case studies of institutions like Stanford School, MIT Media Lab, and the University of Melbourne Learning Hub, the chapter demonstrates how intentional spatial configurations can dismantle academic silos and cultivate dynamic ecosystems for learning. It also addresses key challenges, including institutional inertia, financial constraints, functional conflicts, and accessibility concerns, while proposing actionable design strategies and policy recommendations. By rethinking educational spaces as adaptive, inclusive, and responsive systems, the chapter argues that architecture can serve not only as a backdrop but as a catalyst for educational innovation in an increasingly interconnected academic landscape.

Keywords: Adaptive spaces, Educational architecture, Interdisciplinary collaboration, Learning environments, Pedagogy, Spatial design, Technology integration.

INTRODUCTION

Architecture plays a pivotal role in shaping educational environments by influencing how individuals learn, interact, and collaborate. Beyond its functional role in providing shelter and infrastructure, educational architecture actively contributes to cognitive engagement, social interaction, and interdisciplinary knowledge-sharing. The spatial organization of educational settings, from formal

* **Corresponding Author Zeba Nisar:** Department of Architecture, Integral University, Lucknow, Uttar Pradesh 226006, India; Email: deanarch@iul.ac.in

Tahmeena Khan, Manisha Singh, & Saman Raza (Eds.)

classrooms, auditoriums, and libraries to informal collaborative hubs, maker spaces, and digital learning zones, has a profound impact on the effectiveness of learning experiences. Thoughtfully designed spaces can stimulate creativity, encourage interaction, and foster an interdisciplinary learning culture, while poorly designed environments may reinforce disciplinary silos, limit engagement, and inhibit collaboration (Oblinger, 2006).

As contemporary education shifts towards collaborative, technology-driven, and experiential learning models, traditional, rigid, and compartmentalized spatial layouts are increasingly seen as inadequate. Modern pedagogical approaches emphasize active learning, interdisciplinary engagement, and student-centred instruction, necessitating an architectural response that supports these evolving paradigms. This transition calls for flexible, adaptable, and interactive spaces that blur disciplinary boundaries, allowing seamless collaboration between students, educators, and researchers across various fields. In this chapter, we examine how architecture can act as a conduit for cross-disciplinary collaboration in education by enabling dynamic learning spaces that foster interaction across disciplines. By analysing theoretical foundations, design strategies, and case studies, this chapter aims to highlight the role of spatial configurations in breaking disciplinary boundaries and promoting holistic learning experiences.

The Shift Toward Interactive and Interdisciplinary Learning Environments

The 21st century has witnessed a fundamental shift in education, moving away from passive knowledge transmission toward active, problem-solving, and interdisciplinary learning. Cross-disciplinary collaboration fosters innovation by combining knowledge from multiple domains, enabling students to develop creative problem-solving skills and adaptability (Repko *et al.*, 2016). This shift has been driven by several factors shown in Fig. (**1**), including:

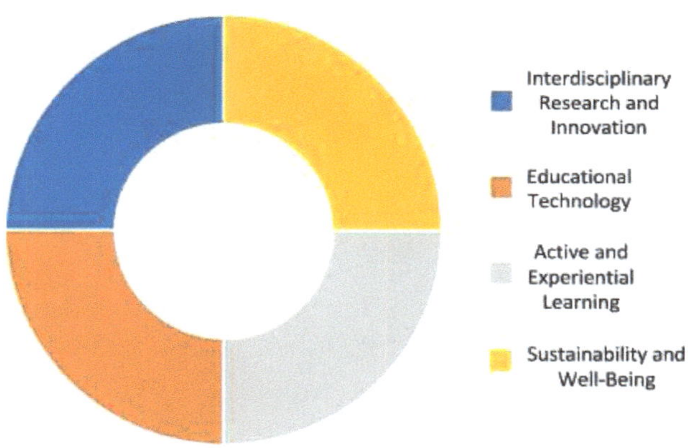

Fig. (1). Shift toward interactive and interdisciplinary learning environments.

- *The Rise of Interdisciplinary Research and Innovation:* Fields such as STEM (Science, Technology, Engineering, Mathematics), humanities, arts, and business increasingly intersect, requiring educational environments that facilitate cross-disciplinary engagement. Institutions such as the MIT Media Lab and Stanford School exemplify how spatial design can integrate multiple disciplines, fostering creativity and innovation.
- *Advancements in Educational Technology:* The integration of smart classrooms, Virtual Reality (VR), Augmented Reality (AR), and AI-driven learning systems requires spaces that are both technology-ready and adaptable. This necessitates architectural flexibility to accommodate hybrid learning models, digital collaboration platforms, and immersive learning environments (Oblinger, 2006).
- *The Need for Active and Experiential Learning:* Traditional lecture-based learning environments are giving way to problem-based learning (PBL), hands-on workshops, and interactive group projects. Architectural interventions must, therefore, prioritize multi-functional spaces that support both structured learning and informal peer collaboration (Malhotra and Aulakh 2024). *The Influence of Sustainability and Well-Being*: Research suggests that natural light, ventilation, biophilic design, and open spatial configurations enhance student performance, engagement, and mental well-being (Levengood, 2017). The design of educational spaces should therefore integrate sustainability principles, flexible use of materials, and climate-responsive strategies to promote holistic learning experiences.

Architectural spaces play a crucial role in supporting these pedagogical shifts. Traditional classroom designs often reinforce hierarchical, teacher-centred learning models that limit spontaneous interaction. Conversely, open and adaptable learning environments, such as maker spaces, co-learning hubs, and multidisciplinary labs, encourage fluid exchange of ideas among students and educators from diverse disciplines (Cutieru, 2024). Through intentional architectural design, educational institutions can cultivate ecosystems that blur disciplinary boundaries, allowing students to engage in cross-disciplinary dialogue organically.

THEORETICAL FRAMEWORK

The role of architecture in education extends beyond providing physical shelter; it actively shapes learning experiences, social interactions, and interdisciplinary collaboration. Educational spaces are not passive containers but dynamic environments that facilitate or constrain knowledge exchange. From the spatial arrangement of classrooms to the design of collaborative hubs, architecture plays a crucial role in determining whether disciplines remain siloed or integrate into holistic learning ecosystems.

The Concept of Space as a Facilitator of Interaction

Space is not merely a passive backdrop but an active agent in shaping human interaction and learning. Theories in environmental psychology, sociology, and architectural design suggest that the physical environment significantly influences behaviour, cognition, and collaboration (Charehjoo, *et al.*, 2018). In the context of education, spaces designed for interaction can:

- Foster spontaneous interdisciplinary dialogue through strategic spatial configurations.
- Support collaborative learning methodologies, such as Problem-Based Learning (PBL), design thinking, and experiential education.
- Enhance social constructivist learning, wherein knowledge is co-constructed through discourse, negotiation, and shared experiences.

One key framework that explains this relationship is space syntax theory, which examines how spatial configurations impact movement, interaction, and knowledge exchange (Hillier *et al.*, 1984). According to this theory:

- *Highly integrated spaces* (*e.g.*, open-plan co-learning hubs, research incubators, and interdisciplinary innovation centres) foster greater cross-disciplinary engagement, as they promote informal interactions and knowledge exchange.
- *Segregated spatial layouts* (*e.g.*, discipline-specific faculty buildings with restricted access) can inhibit collaboration and reinforce intellectual silos.
- *Threshold spaces* (*e.g.*, corridors, atriums, breakout areas, and shared lounges) act as informal learning conduits, facilitating serendipitous encounters and interdisciplinary dialogue.

By applying these spatial principles, educational institutions can design environments that enhance interdisciplinary engagement rather than reinforce disciplinary silos.

Architectural Design Principles Supporting Interdisciplinary Engagement

Historically, educational architecture has been shaped by functional requirements and traditional pedagogical models. However, the increasing emphasis on cross-disciplinary collaboration necessitates architectural spaces that are adaptive, inclusive, and interaction-driven. Well-designed learning environments must transcend conventional spatial hierarchies, fostering knowledge exchange, creative problem-solving, and interdisciplinary dialogue. Several architectural principles support this transformation:

Openness and Transparency

- Spatial transparency, achieved through glass partitions, open-plan workspaces, and shared research zones, enhances visual and physical connectivity across disciplines.
- Such configurations encourage serendipitous encounters, unstructured conversations, and interdisciplinary collaboration, reinforcing the notion that knowledge production is inherently social and dynamic (Sailer, 2011) (Fig. **2**).

Fig. (2). (a) Bromley Central Library (*Source: www.thedesignconcept.co.uk),* and (b) Interior view of Thapar University (*Source: www.archdaily.com).*

The Harvard Innovation Lab (i-Lab) integrates spatial transparency to foster intellectual exchanges between students, faculty, and industry professionals, promoting a culture of interdisciplinary innovation.

Spatial Flexibility and Modularity

- Flexible learning environments incorporate movable partitions, multipurpose furniture, and adaptable layouts, allowing reconfiguration based on evolving academic and research needs.
- These spaces accommodate diverse learning styles, from lecture-based instruction to project-based collaboration, enhancing multi-use, cross-disciplinary engagement (Brown *et al.*, 2006). Stanford School's collaborative studios employ modular, reconfigurable spaces, enabling students to seamlessly transition between design prototyping, group discussions, and interdisciplinary research projects as illustrated in (Fig. **3**).

Fig. (3). Stanford School's collaborative spaces (Source: www.worksight.jp/en/issues/181.html).

Common Spaces and Interaction Hubs

Informal learning zones, including atriums, lounges, shared workspaces, and maker labs, serve as neutral grounds for interdisciplinary interactions. These spaces stimulate spontaneous knowledge-sharing, fostering peer learning and cross-domain problem-solving. The MIT Media Lab incorporates strategically placed social spaces, ensuring that students, researchers, and faculty from diverse disciplines engage in casual, yet intellectually stimulating exchanges, driving innovation.

Technology Integration

- Smart classrooms, digital collaboration platforms, VR labs, and AI-enhanced environments facilitate virtual interdisciplinary engagement across geographical boundaries.
- Hybrid learning environments, which blend physical-digital interactions, are pivotal in expanding cross-disciplinary research networks (Oblinger, 2006).

By integrating these architectural strategies, institutions can establish ecosystems that facilitate knowledge co-production, foster innovation, and support experiential, interdisciplinary learning.

Evolution of Educational Spaces and Their Influence on Collaboration

The architectural design of educational spaces has undergone significant transformation, reflecting shifts in pedagogical models, technological advancements, and research methodologies. This evolution can be categorized into four key phases, each demonstrating a progressive movement toward dynamic, interactive, and cross-disciplinary learning environments.

The Industrial-Age Classroom (19th–Early 20th Century)

- Characterized by rigid, hierarchical, and discipline-specific learning environments, with fixed seating arrangements and compartmentalized faculty buildings.
- Limited interaction due to compartmentalized faculty buildings. Education was teacher-centred, with minimal emphasis on collaborative learning.

The old college system, where faculties operated in disciplinary isolation before.

The Progressive Education Movement (Mid-20th Century)

- Influenced by John Dewey's experiential learning theory, educational spaces began shifting toward student-centred, inquiry-driven models (Dewey, 1938).
- Introduction of laboratory schools, studio-based learning, and hands-on experimental environments that encouraged active learning.

The Bauhaus School of Design integrated arts, crafts, and technology, creating an interdisciplinary architectural model where students engaged in collaborative, practice-oriented learning.

The Open-Plan Learning Movement (Late 20th Century–Early 21st Century)

- The rise of constructivist educational paradigms led to spatial openness, promoting fluid learning arrangements and multi-functional collaboration zones.
- Libraries and traditional study halls evolved into learning commons, facilitating interdisciplinary discourse and project-based learning.
- The Learning Commons Model, pioneered by North American universities, transformed traditional academic libraries into interactive, technology-integrated research hubs (Marques, 2018).

The Digital and Post-Pandemic Era (21st Century–Present)

- The emergence of digital education, AI-driven adaptive learning, and remote collaboration has redefined educational space, emphasizing hybrid, immersive, and technologically enhanced environments.
- Smart campuses, cloud-based collaborative research platforms, and AI-integrated spatial designs have expanded the boundaries of interdisciplinary engagement.

The University of Melbourne's Learning Hub integrates hybrid digital-physical learning environments, allowing students and researchers to collaborate across global networks in real-time. The historical evolution of educational spaces highlights a paradigm shift toward fluid, interactive, and interdisciplinary learning environments, aligning with contemporary pedagogical needs.

The historical transformation of educational architecture illustrates a paradigm shift from static, discipline-segregated models to dynamic, interdisciplinary ecosystems. As pedagogical frameworks continue to evolve, the design of learning spaces must be reimagined to accommodate emerging knowledge domains, collaborative methodologies, and global research networks.

ARCHITECTURE AS A MEDIUM FOR LEARNING AND COLLABORATION

The architectural design of educational spaces plays a crucial role in shaping pedagogical methodologies, cognitive engagement, and interdisciplinary collaboration. As contemporary education transitions towards integrative, interactive, and research-driven learning models, architectural interventions must evolve to create dynamic, flexible, and adaptable environments that facilitate interdisciplinary knowledge exchange. The configuration of learning environments directly influences communication, engagement, and collaborative problem-solving. While traditional classroom layouts often reinforce linear, discipline-specific learning models, open and interactive spatial typologies promote cross-disciplinary engagement by facilitating knowledge sharing, peer learning, and innovative research (Oblinger, 2006). Key spatial configurations include:

Open-Plan Learning Spaces

- Encourage fluid movement and spontaneous intellectual exchanges across disciplines.
- Facilitate multifunctional usage, where spaces can transition between lectures, group discussions, and hands-on research.

Flexible and Adaptable Spaces

- Incorporate modular and reconfigurable furniture, allowing students and faculty to customize learning environments based on evolving project needs.
- Enable multi-modal learning experiences, supporting both structured classroom instruction and unstructured collaborative work.

Makerspaces and Innovation Labs

- Provide shared technological resources, such as 3D printers, fabrication labs, and digital workstations, to support interdisciplinary experimentation.

- Bridge engineering, design, and applied sciences, fostering a culture of iterative prototyping and innovation.

Hybrid Learning Environments

- Combine physical and digital learning spaces, allowing for global interdisciplinary collaboration.
- Utilize smart classrooms, virtual reality (VR) zones, and cloud-based research networks to facilitate seamless cross-disciplinary engagement.

Harvard Innovation Labs incorporates integrated digital collaboration tools to extend cross-disciplinary discussions beyond physical boundaries. By implementing these architectural strategies, educational institutions can dismantle disciplinary silos, creating fluid and interconnected ecosystems that facilitate collaborative innovation across diverse academic fields.

Case Studies of Educational Institutions Fostering Cross-Disciplinary Collaboration

To understand the role of architecture in enabling cross-disciplinary collaboration, we examine three leading educational institutions—Stanford School, MIT Media Lab, and the University of Melbourne's Learning Hub. Each of these institutions has implemented innovative spatial configurations that facilitate interaction among students and faculty from diverse disciplines.

Case Study 1: Stanford School (Hasso Plattner Institute of Design, USA)

The Stanford School is a globally recognized institution specializing in design thinking and interdisciplinary problem-solving. The facility is structured to support collaborative ideation, rapid prototyping, and hands-on experimentation. The architectural design features are:

Open and Reconfigurable Spaces

- Movable walls, modular furniture, and adaptable workstations allow students to reshape their environment based on project needs.
- Rooms can be easily transformed from lecture halls to group discussion areas, fostering spontaneous collaboration.

Transparency and Connectivity

- The use of glass partitions and open-plan layouts enhances visibility, encouraging students to interact across disciplines.

- Central atriums and shared workspaces create natural collision points where diverse ideas intersect.

Maker Spaces and Prototyping Labs

- Equipped with 3D printers, fabrication labs, and digital tools, these spaces facilitate hands-on learning and rapid prototyping of ideas.
- Encourages iterative design thinking by providing access to resources for students of different disciplines.

Technology-Integrated Learning Hubs

- Smart boards, digital collaboration tools, and video conferencing rooms support remote interdisciplinary collaboration.

Impact on Learning and Collaboration

- Facilitates real-world problem-solving, integrating engineering, business, and social sciences.
- Enhances collaborative decision-making, fostering multi-disciplinary research on global challenges such as healthcare innovation, urban design, and sustainability, as shown in Fig. (**4**).

Fig. (4). Stanford School (Source: https://alchetron.com/Hasso-Plattner-Institute-of-Design).

Case Study 2: MIT Media Lab, USA

The MIT Media Lab, established in 1985, is a leading interdisciplinary research hub at the intersection of technology, design, and applied sciences. Its architectural design embodies fluidity, openness, and transparency, reinforcing the lab's anti-disciplinary ethos. The architecture of the MIT Media Lab, designed by Fumihiko Maki, is a transparent, open, and interconnected structure that embodies the lab's ethos of interdisciplinary collaboration, illustrated in Fig. (**5**).

Fig. (5). MIT Media Lab, USA (Source: www.architecturalrecord.com/articles/8235-mit-media-lab).

Open Laboratory Spaces

- Unlike traditional university settings with segregated departments, the Media Lab features large, open-plan workspaces that encourage cross-disciplinary projects.
- There are no assigned offices, allowing researchers and students to move freely and interact across teams.

Glass Walls and Visual Connectivity

- Transparency is a major architectural theme- glass-walled offices and research spaces allow students to see what others are working on.
- This visual accessibility encourages spontaneous conversations and knowledge exchange.

Flexible and Multi-Use Areas

- Spaces can be repurposed dynamically - a classroom may turn into a workshop or event space as needed.
- Movable partitions and furniture support rapid reconfiguration for different projects.

Collaborative Prototyping Studios

- The lab houses robotics labs, digital fabrication studios, and AI research hubs, where students from different disciplines (computer science, design, and neuroscience) can experiment and build together.

Impact on Learning and Collaboration

- The lack of physical boundaries creates a culture of shared exploration and collective problem-solving.
- Students and faculty from varied backgrounds (architecture, music, AI, biology) collaborate on breakthrough innovations, such as wearable tech and responsive environments.

- The cross-pollination of ideas between technology, design, and the arts has led to advancements in fields like smart cities, human-computer interaction, and bioengineering.

Case Study 3: The Learning Hub, University of Melbourne, Australia

The Learning Hub at the University of Melbourne is an interactive educational space designed to support cross-disciplinary engagement, which is evident in Fig. (**6**). It integrates formal classrooms, informal study areas, and collaborative zones to create a holistic learning environment. The architectural design features are:

Fig. (6). The Learning Hub, University of Melbourne, Australia.

Multi-Layered Spatial Design

- The Hub is divided into different zones, each catering to a unique mode of learning- individual study, group collaboration, and social interaction.
- Spaces range from quiet reflection areas to vibrant discussion pods.

Flexible and Adaptive Learning Environments

- Movable seating, writable walls, and reconfigurable rooms allow students to customize their workspace.
- The library is designed with fluid circulation patterns, ensuring smooth transitions between disciplines.

Technology-Rich Learning Spaces

- Features immersive digital screens, VR labs, and interactive whiteboards for blended learning.
- Digital collaboration tools allow students from different faculties to engage in virtual discussions and research projects.

Social and Collaborative Nodes:

- Includes cafe-style workspaces, outdoor learning terraces, and interdisciplinary studios, creating a dynamic knowledge-sharing culture.

Impact on Learning and Collaboration

- The diverse spatial configurations support various learning styles, accommodating both independent thinkers and collaborative teams.
- The blurring of boundaries between faculties has led to cross-disciplinary research initiatives, particularly in areas like urban planning, public health, and digital humanities.
- The emphasis on flexibility and inclusivity ensures that students from diverse academic backgrounds feel encouraged to engage in interdisciplinary dialogues.

Key Design Strategies for Enhancing Interdisciplinary Interaction

One of the fundamental principles in designing interactive educational environments is the strategic organization of spaces to support diverse learning activities. Zoning helps in balancing individual, collaborative, and interdisciplinary learning experiences. Key strategies include:

Spatial Planning and Zoning for Diverse Learning Needs

Open and Flexible Layouts

- Traditional classrooms with rigid seating arrangements limit interaction. Open-plan spaces with reconfigurable furniture allow for dynamic learning (Open-Plan Classrooms 2024).
- Movable partitions, modular furniture, and multi-use spaces accommodate different pedagogical styles, from lectures to hands-on workshops (Oblinger, 2006).

Blended Formal and Informal Learning Spaces

- Semi-open spaces such as atriums, lounges, and co-working areas encourage spontaneous interactions among students from different disciplines. The University of Pennsylvania's Weitzman School of Design integrates informal workspaces into its academic buildings, fostering impromptu discussions.

Distributed vs. Centralized Collaboration Zones

- A decentralized approach places smaller collaborative spaces across campuses, making interaction accessible at multiple points (Brown *et al.*, 2006). A centralized hub model, such as MIT's Media Lab, creates a singular, high-energy space for cross-disciplinary innovation.

The Significance of Common Spaces, Hubs, and Informal Learning Areas

Common areas serve as catalysts for social and intellectual exchanges that transcend disciplinary silos. Thoughtfully designed spaces promote interaction between students, faculty, and researchers.

Learning Commons and Innovation Hubs

- Learning Commons integrate libraries, co-working spaces, and maker labs to facilitate cross-disciplinary engagement (Brown *et al.*, 2006).

Collaborative Workspaces and Maker Spaces

- Hands-on, project-based learning spaces, such as fab labs and maker spaces, allow students from different disciplines to prototype ideas together (Abdessadak, 2025).

Georgia Tech's Invention Studio is an open-access maker space for engineering, architecture, and art students, as can be seen in Fig. (7).

Fig. (7). Georgia Tech - Makerspace at HLG studio. Source: www.hlgstudio.com/.

Courtyards, Atriums, and Outdoor Learning Spaces

- Open-air collaboration zones—such as courtyards and terraces—enhance informal peer-to-peer learning.
- The University of Sydney's Abercrombie Building integrates indoor-outdoor learning spaces.

Integration of Technology in Architectural Design to Enhance Interaction

Technology is a crucial component in shaping future-ready educational spaces. Digitally-enhanced architecture can support real-time collaboration across disciplines, campuses, and even geographical locations.

Smart Classrooms and Interactive Technologies

- Smartboards, AR/VR tools, and collaborative digital platforms transform physical spaces into interactive learning environments (Almufarreh, 2023; Kazendi, 2017). Harvard's Innovation Lab (i-Lab) integrates digital collaboration tools to facilitate interdisciplinary learning.

Virtual Learning Environments and Hybrid Spaces

- Mixed-reality and hybrid learning spaces extend the boundaries of physical classrooms by integrating telepresence technologies (Verma *et al.*, 2023). The University of Helsinki's Think Corner, which combines digital and physical learning environments, is an example.

Adaptive Lighting, Acoustics, and Ergonomic Design

- Dynamic lighting and acoustic design enhance focus and collaborative engagement (Yu, 2021). The University of Nottingham's George Green Library integrates acoustically responsive design for quiet and group learning.

Sustainable and Inclusive Design Approaches in Educational Environments

Educational architecture should not only facilitate interaction but also be sustainable, accessible, and inclusive.

Sustainable Design for Cross-Disciplinary Learning

- Green buildings that incorporate biophilic design principles enhance cognitive function and well-being (Kellert *et al.*, 2013). The Singapore University of Technology and Design (SUTD) incorporates green walls and passive cooling systems in its collaborative spaces.

Universal Design for Inclusive Learning

- Spaces should be physically and digitally accessible to students of all abilities (Burgstahler, 2015). Gallaudet University's Deaf Space Project optimizes architecture for visual and spatial accessibility.

Adaptive Reuse of Buildings for Learning Environments

- Repurposing historic buildings for educational use encourages cultural and interdisciplinary engagement. The Cambridge Judge Business School, housed in a repurposed 19th-century hospital, fosters multidisciplinary collaboration.

By integrating flexible zoning, informal interaction spaces, technology, and sustainable design, educational institutions can create dynamic and interdisciplinary learning ecosystems. The success of institutions like Stanford School, MIT Media Lab, and the University of Melbourne illustrates how spatial design serves as an active facilitator of collaboration, innovation, and transformative learning.

CHALLENGES AND LIMITATIONS

Despite the growing recognition of architecture's role in facilitating interdisciplinary collaboration, several institutional, financial, spatial, and cultural barriers hinder the successful implementation of interactive learning environments. These challenges include rigid institutional structures, financial limitations, functional conflicts, cultural resistance, accessibility concerns, and sustainability constraints.

Institutional and Financial Constraints

One of the most significant challenges in designing interdisciplinary learning spaces is institutional resistance to change. Many educational institutions are structured around rigid, discipline-specific models that reinforce departmental silos. This hierarchical organization makes it difficult to implement spatial designs that encourage cross-disciplinary collaboration (Oblinger, 2006).

Additionally, financial limitations often restrict the ability of institutions to retrofit existing buildings or construct new, interactive spaces. Designing innovative learning environments requires significant investment in:

- Flexible and adaptive infrastructure (*e.g.*, movable partitions, multifunctional furniture).
- Technological integration (*e.g.*, smart classrooms, digital collaboration tools).
- Sustainable materials and energy-efficient designs to ensure long-term viability.

While funding is often allocated for discipline-specific infrastructure, interdisciplinary spaces tend to be undervalued in budget considerations. Public institutions, in particular, may struggle with securing the necessary resources for large-scale architectural interventions (Boys, 2011).

Potential Solutions:

- Encouraging Public-Private Partnerships (PPP) to fund architectural innovations in education.
- Implementing low-cost modular solutions that allow flexibility within existing spaces.
- Advocating for policy-level reforms that emphasize interdisciplinary learning environments in national and institutional educational frameworks.

Balancing Flexibility with Functional Needs

Interdisciplinary learning spaces require a high degree of spatial flexibility, yet this can sometimes conflict with the functional needs of different disciplines. For example:

- Science and engineering labs require specialized equipment and safety measures, which may not align with open, flexible spaces.
- Art and design studios need large workspaces with proper lighting, which may differ from traditional classrooms.
- Library spaces must balance quiet study zones with collaborative discussion areas, leading to design conflicts (Dugdale, 2009).

If flexibility is prioritized excessively, there is a risk that spaces may become too generalized, making them unsuitable for specific pedagogical activities. Conversely, if spaces are designed with strict, predefined functions, they may hinder interdisciplinary interactions.

Potential Solutions:

- Zoning and hybrid models: Creating multi-purpose spaces that include designated areas for different forms of learning (*e.g.*, quiet zones, active collaboration zones).
- Adaptive furniture and spatial configurations: Using movable partitions, reconfigurable seating, and smart walls to allow customization based on changing needs.
- User-driven design: Involving students and faculty in the design process to create spaces that are both functional and adaptable (Manca *et al.*, 2020).

Cultural and Pedagogical Barriers to Spatial Transformation

Even with well-designed architectural spaces, cultural and pedagogical mindsets often present barriers to cross-disciplinary collaboration. Many faculty members and students are accustomed to traditional learning environments, where interaction occurs within discipline-specific boundaries. Some key issues include:

- Resistance to open learning environments: Faculty and students may feel uncomfortable with spaces that lack clear hierarchical divisions between disciplines.
- Pedagogical misalignment: Many education models still rely on lecture-based, individual learning, rather than collaborative, interdisciplinary approaches.
- Assessment challenges: Traditional grading and evaluation systems often do not account for interdisciplinary projects, making faculty hesitant to adopt such learning methods (Chandramohan *et al.*, 2009).

Potential Solutions:

- Professional development for educators: Training faculty in collaborative teaching methods and interdisciplinary pedagogy.
- Pilot projects and incremental changes: Introducing small-scale interdisciplinary projects within existing curricula before large-scale spatial redesigns.
- Flexible scheduling: Implementing cross-departmental courses and shared teaching spaces to encourage faculty collaboration.

Accessibility and Inclusivity Concerns

Creating truly inclusive learning environments remains a significant challenge. Many architectural designs fail to address the diverse needs of students with different abilities, cultural backgrounds, and learning styles. Some common accessibility issues include:

- Physical barriers: Lack of ramps, elevators, and tactile pathways for students with mobility impairments.
- Acoustic challenges: Poor sound insulation and echo-heavy spaces, which impact students with auditory disabilities.
- Neurodiversity considerations: Spaces that are overstimulating (*e.g.*, excessive glass walls, bright lighting) can be challenging for students with autism or ADHD (Sharma, 2020).

Potential Solutions:

- Universal design principles: Ensuring all spaces are physically accessible, well-lit, and acoustically optimized.
- Sensory-friendly learning environments: Providing quiet study areas, flexible lighting options, and adaptable classroom layouts.

- Cultural inclusivity: Designing spaces that reflect diverse cultural identities and learning traditions, fostering a sense of belonging.

Sustainability Challenges in Educational Spaces

Sustainability is a growing concern in architectural design, yet many educational institutions struggle to integrate eco-friendly materials, energy-efficient technologies, and biophilic design elements (Biancardi *et al.*, 2020).

Some common sustainability challenges include:

- Budget constraints are limiting the adoption of green building technologies.
- Retrofitting issues in older buildings, where energy efficiency upgrades may be expensive.
- Lack of institutional policies that mandate sustainable construction practices.

Potential Solutions:

- Implementing passive design strategies (*e.g.*, natural ventilation, daylighting) to reduce energy consumption.
- Using local and recycled materials to reduce the carbon footprint of new constructions.
- Encouraging student-led sustainability initiatives to create awareness and demand for eco-friendly educational spaces (Chopra, 2024).

By implementing these recommendations, institutions can bridge the gap between spatial design and interdisciplinary pedagogy, ensuring that architecture actively fosters educational innovation.

CONCLUDING REMARKS

The exploration of architecture as a catalyst for interdisciplinary collaboration in education underscores its pivotal role in reshaping pedagogical practices, fostering innovation, and enhancing knowledge exchange. The spatial organization of educational environments significantly influences how disciplines interact, how students engage with knowledge, and how institutions adapt to emerging learning paradigms. This research has demonstrated that intentional spatial design through flexible configurations, interactive zones, and technology-enhanced spaces can effectively facilitate cross-disciplinary engagement and experiential learning.

A key finding of this study is that learning environments must evolve beyond conventional discipline-specific enclosures to embrace fluid, adaptable, and integrative spaces. The success of institutions such as Stanford School, MIT Media Lab, and the University of Melbourne Learning Hub illustrates how open, modular, and collaborative environments can break down academic silos and create a culture of shared inquiry and innovation. These case studies reinforce that spatial design is not a neutral backdrop to education but an active agent in shaping learning experiences, cognitive development, and research productivity (Thomas *et al.*, 2011).

While the benefits of interdisciplinary architectural design are evident, challenges persist in institutional restructuring, financial investment, functional adaptability, and pedagogical realignment. Traditional academic structures, often resistant to change, require policy-level interventions and stakeholder engagement to integrate spatial configurations that support interdisciplinary education. Addressing these barriers necessitates a multi-dimensional approach that combines evidence-based design principles, technological integration, and sustainability-focused solutions.

Looking ahead, the future of educational spaces will be defined by hybrid digital-physical environments, AI-driven spatial optimization, immersive learning technologies (VR/AR), and biophilic design strategies (Baltezarević *et al.*, 2024). The dynamic nature of global education systems demands that architectural fram-eworks remain flexible and adaptable, ensuring continuous alignment with evolving pedagogical and technological advancements.

In conclusion, architecture is a fundamental enabler of interdisciplinary education, shaping how students and educators engage, interact, and innovate. By rethinking spatial paradigms and embracing responsive, inclusive, and technology-integrated design, institutions can prepare for the next generation of education, fostering creativity, collaboration, and knowledge co-creation in ways that extend beyond traditional academic boundaries.

REFERENCES

Abdessadak, J. (2025) FabLabs, Makerspaces or Third places: Catalysts for University Students' Innovation, *Journal of Information Systems Engineering and Management, 10*(47s).

Almufarreh, A. (2023). Exploring the potential of mixed reality in enhancing student learning experience and academic performance: An empirical study. *Systems, 11*(6), 292. [http://dx.doi.org/10.3390/systems11060292]

Biancardi, A., Colasante, A., D'Adamo, I., Daraio, C., Gastaldi, M., Uricchio, A. (2020). Strategies for developing sustainable communities in higher education institutions. *Sci. Rep., 13*, 20596. [http://dx.doi.org/10.1038/s41598-023-48021-8] [PMID: 37996505]

Boys, J. (2011). *Towards creative learning spaces: Re-thinking the architecture of post-compulsory education..* New York, NY: Routledge.

Brown, M., Long, P. (2006). Trends in learning space design. In: Oblinger, D.G., (Ed.), *Learning spaces.* EDUCAUSE. Available from: https://www.educause.edu/ir/library/pdf/PUB7102.pdf

Burgstahler, S. E. (2015). *Universal design in higher education: From principles to practice (2nd ed.).* Cambridge, MA: Harvard Education Press.

Chandramohan, B., & Fallows, S. (Eds.). (2009). *Interdisciplinary learning and teaching in higher education: Theory and practice.* New York, NY: Routledge.

Charehjoo, F., Etesam, I., Rasoulpour, H. (2018) The Role of Environmental Psychology in Architecture and Urban Design. *Scientific Journal Academia Architecture and Construction, 1*(2), 11-15.

Chopra, S. (2024). Enhancing learning spaces: How architecture impacts education. *Creators Architects.* Available from: https://insights.creatorsarchitects.com/enhancing-learning-spaces-how-architecture-im-acts-education-9a87ac3c3785

Cutieru, A. (2024). The design of learning spaces: Architecture as a teaching tool. *ArchDaily.* Available from: https://www.archdaily.com/972189/the-design-of-learning-spaces-architecture-as-a-teaching-tool

Dewey, J. (1938). Experience and education. *Macmillan.* Available from: https://talkcurriculum.wordpress.com/wp-content/uploads/2014/09/dewey-j-1938-experience-and-education-pp-17-31-new-york-ny-touchstone.pdf

Dugdale, S. (2009). Space strategies for the new learning landscape. *EDUCAUSE Review, 44*(2), 51–63.

Hillier, B., Hanson, J. (1984). *The social logic of space.* Cambridge: Cambridge University Press. [http://dx.doi.org/10.1017/CBO9780511597237]

Kazendi. (2017). How mixed reality is shaping the future of education. *AR/VR Journey.* Available from: https://arvrjourney.com/how-mixed-reality-is-shaping-the-future-of-education-538d3f8b0dc4

Kellert, S.R., Heerwagen, J., Mador, M. (2013). *Biophilic design: The theory, science and practice of bringing buildings to life.* Hoboken, NJ: John Wiley & Sons.

Levengood, M. (2017). Impact of sustainable design elements on student learning (Senior honors theses & projects). *Eastern Michigan University.* Available from: https://commons.emich.edu/cgi/viewcontent.cgi?referer=&httpsredir=1&article=1521&context=honors

Malhotra, A., Aulakh, R.S. (2024). (Multi)functionality: As an Urban Development Tool for Placemaking. In: Nandineni, R.D., Ang, S., Mohd Nawawi, N.B. (eds) Sustainable Resilient Built Environments. SRBE 2022. Advances in 21st Century Human Settlements. Springer, Singapore. [http://dx.doi.org/10.1007/978-981-99-8811-2_9]

Manca, S., Cerina, V., Tobia, V., Sacchi, S., & Fornara, F. (2020). The Effect of School Design on Users' Responses: A Systematic Review (2008–2017). *Sustainability, 12*(8), 3453. [http://dx.doi.org/10.3390/su12083453]

Marques, Silas. (2018). Trends in Academic Library Space: From book boxes to learning commons. *Open Information Science, 2*, 59-74. [http://dx.doi.org/10.1515/opis-2018-0005]

Oblinger, D.G. (2006). *Learning spaces..* EDUCAUSE.

Petare, P. A., Shamim, M., Gupta, T., Verma, R., & Singh, G. (2023). Exploring the impact of virtual learning environments on student engagement and academic achievement. *Journal of Survey in Fisheries Sciences, 10*(1S), 5912–5923.

Psychology Town. (2024). *Open-plan classrooms: Revolutionizing traditional learning spaces.* https://psychology.town/environmental/open-plan-classrooms-revolutionizing-learning/

Radcliffe, D., Wilson, H., Powell, D. (2009). *Designing next generation learning spaces. The University of Queensland Press.* Available from: https://www.researchgate.net/publication/237249564

Repko, A.F., Szostak, R., Buchberger, M.P. (2016). *Introduction to interdisciplinary studies..* SAGE

Publications.

Sailer, K. (2011). Creativity as social and spatial process. *Facilities, 29*(1/2), 6-18. [http://dx.doi.org/10.1108/02632771111101296]

Sharma, M. (2020). Neurodiversity in education: Celebrating differences and embracing strengths. *Global International Research Thoughts, 8*(1), 16-19. [GIRT].

Stojanović, J., & Miletić, A. (2024). The importance of artificial intelligence in the modern education system. *Baština, 34*(62), 471–479.

Thomas, D., Brown, J. (2011). *A new culture of learning: Cultivating the imagination for a world of constant change.*. CreateSpace Independent Publishing.

Yu, J. (2021). Exploring the relationships between learning space and student learning in higher education: A comparative case study in China. In: Imms, W., Kvan, T., (Eds.), *Teacher transition into innovative learning environments.* Springer. [http://dx.doi.org/10.1007/978-981-15-7497-9_18]

CHAPTER 13

Creative Teaching Tools for Enhanced Learning: Promoting Student Engagement through Blended Learning in Higher Education

Shahla Tanveer[1,*] and **Mohammad Imran Ahmad**[1]

[1] *Department of Chemistry, Integral University, Lucknow, Uttar Pradesh 226006, India*

Abstract: The continuously evolving landscape of higher education necessitates the adoption of innovative pedagogical strategies that address the multifaceted needs of diverse learners as well as the dynamic nature of the digital environment. This chapter critically investigates the incorporation of inventive teaching tools aimed at augmenting learning outcomes within the realm of higher education. It scrutinizes the emerging technologies and methodologies that have surfaced in recent years, which include immersive technologies such as Virtual and Augmented Reality, generative Artificial Intelligence, microlearning, gamification, wearable technologies, blockchain applications for credentialing, and collaborative learning platforms. These educational tools not only enhance the instructional experience but also foster heightened levels of student engagement, motivation, and individualized learning opportunities. Through rigorous analysis and the presentation of contemporary case studies, the chapter elucidates how these innovative tools can effectively reconcile the disparities between conventional academic delivery methods and the expectations of modern learners. The results underscore the critical significance of faculty development, digital literacy, and institutional support as prerequisites for the successful implementation of these pioneering innovations. The chapter concludes by offering recommendations for future research endeavours and best practices for the effective integration of creative teaching tools, aimed at cultivating inclusive, adaptable, and efficient learning environments within higher education.

Keywords: Creative teaching tools, Enhanced learning, Game components, Higher education, Interactive learning, Learning based on games, Learning based on simulation.

INTRODUCTION

The incorporation of inventive pedagogical instruments within tertiary education has evolved as a remarkable emphasis for teachers aspiring to augment student

* **Corresponding Author Shahla Tanveer:** Department of Chemistry, Integral University, Lucknow, Uttar Pradesh 226006, India; E-mail: shahla@iul.ac.in

Tahmeena Khan, Manisha Singh, & Saman Raza (Eds.)

participation, retention rates, and academic achievement. Conventional teaching approaches frequently inadequately address the multifaceted requirements of contemporary learners who necessitate engaging, interactive, and tailored educational experiences. This discourse examines an array of groundbreaking instructional tools and their influence on student outcomes, utilizing insights derived from pertinent scholarly articles.

Higher education currently finds itself at a pivotal crossroads, confronted with the dual imperative of equipping students with enduring knowledge and state-of-the art competencies necessary for navigating an increasingly intricate and swiftly evolving global landscape. The conventional lecture-based pedagogical model, although instrumental in imparting fundamental concepts, frequently proves inadequate in fostering the dynamic skill sets requisite for modern societal demands. Such skill sets encompass critical analysis, intricate problem-solving, proficient collaboration, and the agility for lifelong learning. In light of this exigency, a significant paradigm shift is occurring, characterized by an escalating focus on the development and application of innovative educational tools. These tools are intended not solely to improve learning outcomes but also to cultivate a deeper and more meaningful engagement with the educational experience. The article highlights the value of student-centred learning and the application of knowledge in real-world contexts. By implementing these strategies, educators can create dynamic learning environments that cater to diverse student needs, fostering deeper understanding and essential skills for success in academia and future careers. The article advocates for a shift from traditional teaching methods to more interactive and application-based learning experiences in order to better prepare students for the demands of the professional world. The article examines a number of innovative teaching and learning approaches in higher education, including gamification, flipped learning, project-based learning, and technology-enhanced learning.

TEACHING METHODOLOGIES FOR ACTIVE LEARNING

To engage students as well as to improve learning outcomes, active learning methodologies have evolved as a powerful approach. These methodologies focus on the participation of students, collaboration, and hands-on experiences, so as to make learning more meaningful and effective (Konopka *et al.*, 2015).

The goal of active learning approaches is to teach with emotion as well as to evoke feelings. To this degree, learning is reinforced and solidified through the employment of various teaching approaches that enhance one another. Several active learning approaches, such as (i) flipped teaching, (ii) game-based learning, (iii) Game-based learning (iv) learning based on projects, (v) interactive storytelling, and (vi) process modelling, are reviewed in this section and can be used in university classrooms (Fig. **1**).

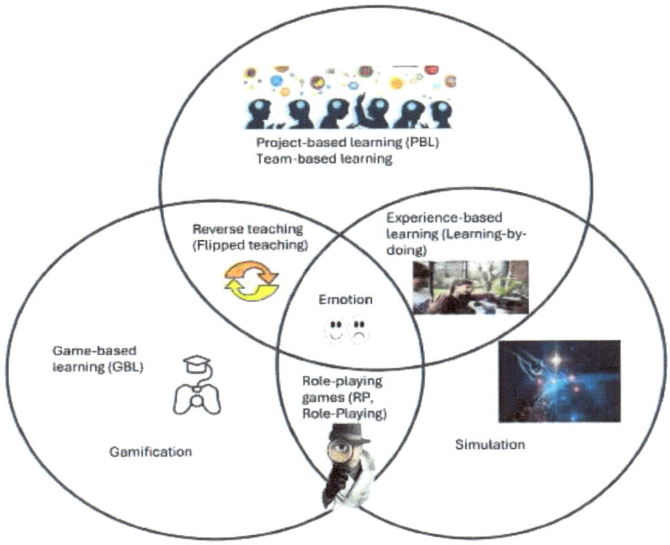

Fig. (1). Methodologies for active learning.

Reverse Teaching/ Flipped Classroom

The flipped classroom model, which can be referred to as reverse teaching, is a student-focused teaching tool aiming to enhance the best of learning during classroom hours (Ozdamli *et al*., 2016). Complicated and higher-order thinking skill-based outcomes are developed with the assistance and direction of the teacher (typically during in-class activity), while the simpler or more tangible learning outcomes are completed independently by the students (generally at home). The purpose behind this teaching model is to develop multimedia content for the educator, usually, and let students better understand the concept that will be taught in class. The aim of multimedia content extends beyond helping students visualize information; simultaneously, it should integrate formative tasks and assessments to enhance effective learning (Milman, 2015). Based on what students have learnt at home can be taken up to a higher level or more complex contents can be created later in the classroom or activities can be carried out on the concepts that have been already learnt, keeping in mind that the activities to be conducted within the cohort should have an increased value to it, being not able to do them autonomously at home. After the class, pupils could engage in independent activities at home to consider and assimilate the concepts they had studied.

Among the benefits of flipped teaching are the following: dispositions (positive student evaluations, positive student-teacher impressions, and positive attitudes); interactions (student-student, general, and instructor-student); educational contributions (easy learning, enabling individualised learning, enhances happiness, well prepared before class, enhanced autonomy, offers collaboration opportunities, more feedback, *etc.*); and enhanced learning performance, satisfaction, throughout engagement, self-motivation, and creativity, among others (Naik, 2023).

Gamification:

Gamification leverages game design to heighten motivation and engagement. It involves two main approaches: **game-based learning** and **gamified elements**.

Game-Based Learning (GBL):

The GBL teaching tool uses the " Learn while you play" technique to acquire knowledge and abilities, by incorporating digital games/games during teaching. This learning is further assessed by educators. According to Alkhafaf (2010), this approach is built on games with programs, plans, and supplies that are intended to serve as advice to help pupils reach particular goals. One of GBL's main features is the utilisation of a secure environment to examine and explore various choices and behaviours while engaging students in aspects connected to competition and enjoyment. Unlike other traditional learning methods, which take an impersonal approach, game-based learning incorporates an emotional component (Greipl *et al.*, 2020).

It is important to create game-based activities that can inspire and involve every student. According to (Laine *et al.*, 2020), the basic elements of any educational game must include challenge, competitiveness, accomplishments and goals, control, feedback, and innovation.

(Plass *et al.*, 2015) articulate the fundamental supportive rationale for the integration of game-based activities within educational contexts: (i) The impact that game mechanisms and rewards have on motivation (ii) the engagement level of the participant; (iii) the capacity to tailor the game in accordance with the learners' requirements (regarding feedback and learning challenges) and aspirations (such as personal interests or the minimization of learning duration);

and (iv) the facilitation of secure environments through digital games wherein errors and failures are permissible, thus mitigating the adverse effects of failure while augmenting motivational prospects. Among the principal benefits of adopting this pedagogical approach, the following are particularly noteworthy: it enhances the language acquisition process (specifically, skills in listening, writing, and reading) (Eltahir *et al.*, 2021), it yields advantages for attentional and visual perceptual processes (Green *et al.*, 2012), it aids in the development of creative and unconventional problem-solving methodologies (Liu and Israel 2022), it positively influences self-regulation and emotional control, (Gabbiadini, 2017) and it proves beneficial for the acquisition of transversal skills associated with collaborative efforts, creativity, complex cognitive processes, and information technology (Binkley *et al.*, 2012).

Gamified Elements:

In place of comprehensive game experiences, educators may incorporate "game elements" (such as points, badges, leaderboards, and narratives) into conventional tasks. The objective is to render activities "appealing and inspiring." For example, awarding badges for the mastery of specific subjects, or conceptualizing a project as an adventurous quest. These components engage both intrinsic and extrinsic motivation: progress indicators, incentives, and feedback mechanisms promote sustained involvement. As noted in a particular review, "the implementation of gamification is likely to enhance learner motivation," thereby rendering even vocabulary exercises captivating. Educators have observed that when student work is disseminated (for instance, through social media platforms), motivation and engagement increase, as learners recognize that their efforts are being acknowledged by a genuine audience. Essentially, gamification transforms the educational experience into an interactive "game" wherein learners are actively engaged and pursue defined objectives.

The educational approach revolves around incorporating elements of games in serious settings to craft an opportunity to learn that can exist independently of the enjoyable aspects. However, when developed in alignment with this methodology's principles, these experiences become engaging and inspiring for pupils. Games are a helpful tool, and their design should support the development of transversal skills and the specified learning objectives (Latorre-Cosculluela, *et al.*, 2025). To accomplish this, the game-based activity should encompass both extrinsic motivational factors (like rewards or accolades) and intrinsic motivational factors (such as learning progress and personal fulfilment).

Ramakrishnan (2019) highlights key elements that need to be part of a learning game activity for it to be both enjoyable and beneficial: It should (i) test each student's unique abilities and knowledge; (ii) have clear, concise goals that give teachers and students feedback; and (iii) be flexible and adaptive over time. Consequently, game-based elements serve as an effective method not only for meeting learning goals from the student's standpoint but also from the teachers' viewpoint. Some benefits of implementing gamification include assisting in classroom management, enhancing the classroom atmosphere, and shortening learning durations. This is a primary reason for its increasing adoption in non-STEM disciplines when teaching STEM subjects, which are often perceived as more challenging (Buckley, 2017). Research within higher education supports the assertion that gamification promotes a more engaged attitude among students, fostering a more relaxed setting and encouraging peer collaboration (De la Torre and Berbegal-Mirabent, 2020). Other studies, such as Su (2016), assert that gamification is a clear and necessary path for university education, as it empowers students to take charge of their own learning while simultaneously boosting motivation and active involvement.

Project-Based Learning (PBL):

This strategy allows students to take up a diligent role, thus motivating them academically. The students develop a project that aims to address a real-world problem, typically working in teams and collaborating to create an outcome that effectively resolves the issue at hand. Before this, the teacher examines the project to ensure that students possess all the necessary resources to tackle it and that the process of solving it will enable them to develop the targeted skills. As noted by Kokotsaki *et al.* (2016), this method is grounded in three elements of constructivism: learning is tailored to the context, students are proactive in their learning journey, and they achieve their goals through socialising and the sharing of knowledge.

The fundamental characteristics of PBL involve starting with a fundamental question that pupils should solve using a critical-thinking approach akin to that utilized by a team of skilled professionals within the subject area, aided by education tools that allow them to extend their capacity (Yu and Zin 2023). To allow the students to mirror the intricacies of actual problem-solving skills, they are put in teams to brainstorm answers to their fundamental questions. To finalize their learning, the students produce a collection of concrete items that answer their fundamental query.

The main benefits of PBL include the identification of individual learning strengths and useful technology use, as well as the improvement of skills and competencies like teamwork, project management, effective communication, problem-solving, decision-making, and time management. Additionally, it increases students' self-esteem and motivation, bridges the gap between university education and real-world applications, fosters collaborative skills for knowledge construction, and creates integrative connections across various disciplines (Zhang & Ma 2023).

As noted by Guo *et al.* (2020), the implementation of PBL in higher learning facilitates the achievement of four types of learning results: logical, affective, behavioural, and artefact performance. Logical results pertain to the knowledge gained and the logical techniques utilized by pupils within PBL. Affective results relate to students' views on the advantages of PBL and their overall educational experience with this approach. Behavioural results are connected to students' abilities and their level of commitment. Lastly, the evaluation of student-made documents, artefacts, and multimedia is known as artefact performance.

Role-Playing Games:

A role-playing game involves players adopting certain business "roles" and exchanging words while outlining their acts within a set narrative. According to Cheville (2016), a role-playing game is defined as "a textual system comprising rules for game mechanics, narratives (modules) that provide significance and context to character actions, and avenues for social interaction through which a story is collaboratively created" (p. 810). The game mechanics encompass the guidelines that dictate and direct the actions of the player, along with the game's reactions to those actions. Thus, the mechanics of a game clearly delineate how the game operates for its players. Each player, assuming their designated role, enhances their skills and knowledge through the narrative. This narrative features a sequence of scenarios that progressively increase in difficulty (Prager, 2019). These challenges aim to develop higher-level skills and offer various solutions that might include tactical, social, moral, or strategic aspects, unlike drill-and practice games, which focus on building memory retention (Bergström, 2012; An and Cao, 2017). As Prager (2019) notes, the main difference between games and role-playing games is that the former emphasises the player's development within the story. Therefore, the gamer's decisions, based on their understanding, are vital for the progression of the story, which can take different paths. Throughout the game, the player's skills and knowledge improve as the narrative unfolds. In this respect, the player significantly impacts the game's evolution. There are various kinds of role-playing games, such as (i) tabletop role-playing games, where players decide their characters' movements based on their roles, with success or failure determined by a formal set of norms and procedures. Within these norms, gamers have the liberty to devise, and their choices influence the direction and result of the game; (ii) live-action role-playing games, where participants physically embody characters within a fictional context; and (iii) role-playing video games, which adapt tabletop role-playing games into electronic formats. Some scholars categorize digital role-playing games into two interrelated groups: single-player games that utilize RPG mechanics and multiplayer games that involve social interaction. In the realm of business role-playing games, a key benefit is that students can physically engage with the scenarios, narratives, challenges, *etc*., found in a real-world business environment. However, the scenario, whether it be a tabletop game or a video game, must be meticulously designed and clearly defined to avoid any ambiguity regarding the teachers' expectations for what the students should learn or experience. Numerous applications can be found in the literature; for instance, interested readers can explore the works of Innes *et al*. (2015), Cuiñas *et al*. (2022), and Karmakar *et al*. (2022), among others.

Simulation-Based Learning:

Simulation-Based Learning (SBL) is utilized in higher education settings (universities) to prepare aspiring modellers, particularly in technical fields like Industrial Engineering, Operations Research, Management Science, and Business Analytics, among others (McHaney, 2018). Simulation methods are employed to assess both ongoing methods (for enhancement and maximization) and future methods (to predict results) aimed at achieving the most effective design across various purposes (Prado, 2020).

SBL is a pedagogical approach that departs from conventional teaching methods, allowing students to receive prompt feedback on topics such as the reception of raw materials, server configuration, and supply chain design. This experimental freedom and quick feedback enable students to test various designs (grounded in theoretical concepts) and experiences that would be difficult to replicate in real life due to potential dangers, high costs, or ethical concerns. Consequently, this method permits students to apprehend their shortcomings without facing the actual repercussions.

From the standpoint of the teacher, SBL provides an opportunity for endless experimentation, which expands and improves the educational experience. Furthermore, it instantly creates an electronic record of data that may be used to personalise learning experiences or improve the educational process according to student profiles (*e.g.*, according to their interests, talents, and abilities). Replicating the results of these trials also makes it easier to compare different SBL settings, which enhances the experience's accuracy and authenticity.

The primary advantages of SBL linked to the circular economy and sustainability practices across several areas are summed up in the study of De la Torre *et al*. (2021). Chemical engineering (*e.g.*, for identifying processes); architecture and construction engineering (*e.g.*, for estimating material energy consumption, or for evaluating the efficiency of new designs); and business and management (*e.g.*, for evaluating customer acceptance, choosing managers, or showcasing the possible benefits of new policies).

Artificial Intelligence Tools in Education

Artificial Intelligence has been seamlessly integrated into a multitude of instruments employed for diverse objectives within academic environments. Intelligent tutoring systems, chatbots, automated assessment systems, and dashboards are the four AI-driven tools that are the main focus of the current study. According to Celik (2023) and Chen *et al*. (2022), these tools have been recognised because they are the most widely used AI-based tools in the field of education. Luckin *et al*. (2016) divided the current state of AI in education into three categories: intelligent virtual reality, intelligent support for collaborative learning, and personal tutoring systems. Artificial intelligence tools used in educational settings were examined in five primary areas in this investigation. These include chatbots, automated assessment systems, Intelligent Virtual Reality (IVR), learning platforms, and Intelligent Tutoring Systems (ITS).

Moreover, supplementary instruments that leverage artificial intelligence in educational settings have been assessed. (Fig. **2**) represents AI as a teaching-learning tool in education.

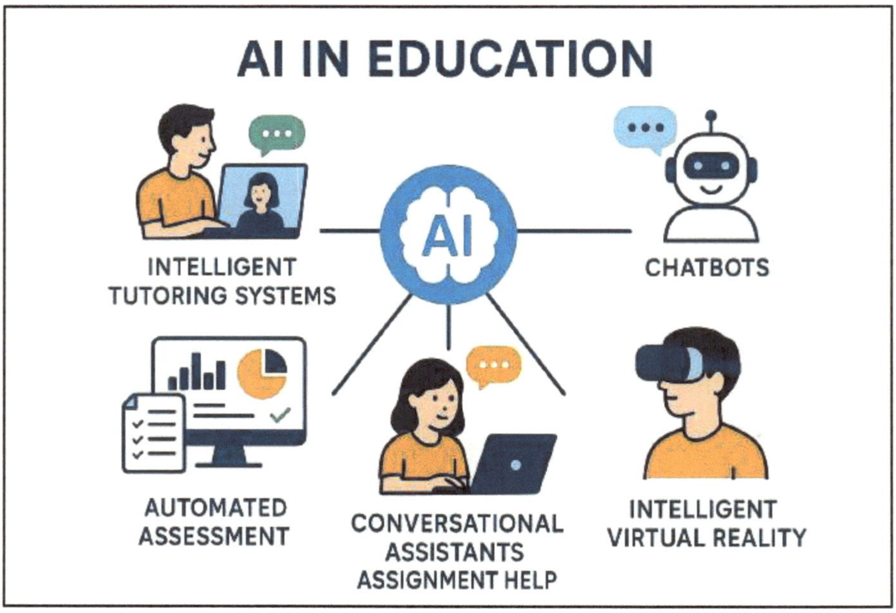

Fig. (2). AI applications in education.

Educational Platform:

An online platform that is enabled with a virtual tutoring system has the potential to enhance professional development courses. A multitude of free and accessible courses is presently offered, presenting a diverse array of captivating and systematically organized features and content. Numerous courses, exemplified by those available on platforms such as Udemy, MOOCs, Google AI, Khan Academy, Alison, edX, Coursera, and Udacity, have incorporated artificial intelligence technology into their curricula (Zhang, 2021). Examples of learning platforms include Altitude Learning, Carnegie Learning, Knewton's Alta, and Cognii Virtual TA.

- Altitude Learning represents a digital learning platform that emphasizes an educational paradigm centred around the student, aiming fundamentally to empower learners to navigate their own educational trajectories through a tailored academic experience. By offering choices for solitary or collaborative assignments, the artificial intelligence tool promotes a student-centric model of education that integrates contemporary research on efficacious pedagogical strategies.
- Leveraging artificial intelligence and machine learning, Carnegie Learning provides technology-driven curriculum solutions for high school and college students in subjects like math, world languages, and literacy. Notably, their MATHia software for mathematics was crafted by experts from Carnegie Mellon University.
- Alta by Knewton is a prominent AI-driven platform for higher education, offering complete courseware that blends open educational resources with adaptive learning. This creates a highly personalized learning journey for students. Alta includes all necessary materials like text, videos, examples, and assessments for a course and covers various subjects such as math, statistics, chemistry, and economics, making it an impactful tool for individualized and productive learning in higher education.
- Cognii Virtual TA is an AI-driven virtual tutoring platform that personalizes learning for students. Through sophisticated machine learning, it identifies individual capabilities and shortcomings, delivering focused instruction and immediate feedback to foster academic improvement and achievement. This AI-powered tool effectively enhances educational quality by giving students tailored help to maximize their capacity.

Chatbots:

Chatbots act as digital helpers for students and teachers, facilitating communication through text or voice. Educators can interact with these chatbots using speech or typing (Luo *et al.*, 2019) and receive information on student advancements via chatbot notifications (Chocarro *et al.*, 2021). Their natural conversational abilities make chatbots valuable tools, exemplified by "Pounce" at Georgia State University, which helps students with administrative tasks like registration, admission, and financial aid (Greenhow *et al.*, 2020). Other AI-powered chatbot models include ChatGPT, Perplexity, and Ivy.

- ChatGPT is an example of an AI-powered chatbot used in education. This generative AI is designed to have natural, human-like conversations with users. It has been educated on a massive dataset from the internet and refined by human input to communicate authentically.
- A chatbot called Perplexity uses generative AI to have conversations with people. A special thing about it is that it shows where its information comes from by including clickable links in its answers, so users can find out more.
- Ivy offers a suite of AI chatbot tools designed to assist universities and colleges with various functions, such as application forms, enrollment, tuition costs, deadlines, and other administrative duties. A notable feature of Ivy is its ability to develop recruitment campaigns using data analysis. This AI tool can also provide students with crucial information about loans, scholarships, grants, tuition payments, and more. Additionally, its capacity to generate customized chatbots for individual departments makes it a versatile tool for the entire institution.

AI-Powered Virtual Reality:

Virtual reality is understood as a human-computer interface that simulates real environments for user interaction, a precise three-dimensional digital copy of the physical world, or simply cyberspace (Angra *et al.*, 2025). An "intelligent" virtual reality system is produced by combining artificial intelligence with augmented reality. Virtual Reality (VR), Mixed Reality (MR), Augmented Reality (AR), and haptic technologies have all become increasingly popular in dental education recently (Saghiri *et al.*, 2022). Engaging and guiding students in realistic, game-like virtual learning environments is one use for Intelligent Virtual Reality (IVR). Virtual agents can act as peers, professors, or aides for students in remote or virtual labs (Perez *et al.*, 2017). Virtual reality's highly interactive software greatly enhances the educational experience in bioscience education by encouraging active engagement and self-directed learning. These software are very useful in the medical and biosciences field where students not only study organs but also gain hands-on experience in working with them, by independently investigating their structures (Fabris, 2019). A variety of VR tools, including Nearpod, CBC VR, Unimersive, Sesqui VR, Veative, YouTube 360 videos, EON Reality education, and NYT VR, have been created for use in educational settings.

FURTHER APPLICATIONS OF AI IN EDUCATION:

By incorporating AI-backed systems and applications, education gains the advantage of quick and interactive student support. AI enables instant answers to student questions and concerns, facilitating quick problem resolution. Furthermore, AI offers personalized conversational assistants for assignment help, promoting flexible, self-paced learning. Ultimately, AI integration significantly enhances the learning experience and fosters academic achievement. The following lists other AIEd (Artificial Intelligence in Education) tools used in higher education or individually.

- Quillbot helps users rephrase and rewrite text by examining a wide range of synonyms. It proposes alternative word choices to achieve the best fit for fluency, vocabulary, tone, and style.
- Curipod quickly generates interactive slide presentations on any topic. It offers built-in features like polls, word clouds, open-ended questions, and a drawing tool.
- Grammarly is an AI-powered online writing assistant that helps users produce clear and error-free text. It integrates with various applications like email, productivity software, and social media, using AI to check grammar, spelling, style, tone, and more to enhance writing quality.
- SciSpace Copilot is a powerful AI application designed to help scientists and researchers navigate extensive scientific literature. Utilizing advanced natural language processing, it analyses publications and extracts key information, enabling researchers to quickly identify significant and relevant papers, thus saving time and effort in their literature reviews (Aytaç, 2024). The following (Table **1**) summarises all the teaching tools discussed in this chapter for enhanced learning.

Table 1. Creative teaching tools for enhanced learning.

Teaching Tools	Main Features	Student's Approach		Benefits	Challenges
		Autonomous	Collaborative		
Reverse Teaching/Flipped Teaching	Students work autonomously on concrete learning results at home, and more complex learning results are worked on in class	Home assignments simpler tasks educational outcomes	Value-added activities during teaching hours	•Academic achievement •Student-teacher engagement •Efficient use of time •Student independence •Tailored and adaptable learning experiences	Students' resistance to the new educational approach, where a portion of the tasks is completed at home.
GBL(Game Based Learning)	Utilizing games and/or video games in the educational setting to acquire knowledge and abilities.		Elements related to competition and entertainment.	Formulates innovative and unconventional approaches to problem-solving. Engages with the acquisition of transversal skills (teamwork, creativity, critical thinking, information processing, technology)	Encourage and engage the students in the activity. Adjust the game based on the requirements of the students.
Gamified Elements	Educational activities that incorporate game elements. They serve as a supportive resource, and their conception/ design aligns with defined learning goals.	Inspirational aspect associated with growth in education and individual fulfilment.	Competing and leisure related factors	Oversee class, Enhance classroom atmosphere and shorten learning durations. • Encouraging engaged approaches during lessons, promoting comfortable settings, and fostering peer assistance.	It should be adjustable / versatile as time goes on.

(Table 1) cont.....

Teaching Tools	Main Features	Student's Approach		Benefits	Challenges
		Autonomous	Collaborative		
Project-Based Learning	Learners create a project aimed at addressing a real issue through teamwork, resulting in a final product that tackles and resolves the problem being examined.	Personal learning abilities and the hands-on experience to utilize technology. Enhance self-confidence.	Respond to a particular challenge, typically in groups. Engage in social interactions and exchange knowledge. Utilize learning technologies to recognize their potential beyond personal limitations.	Working together, project organization, interaction, troubleshooting, making choices	• The instructor evaluates the project beforehand to confirm that students possess all the essential resources to address it and will cultivate all the skills required. • Connections that integrate various disciplines.
Role Playing Games	•A text-based system provides guidelines for gameplay mechanics. •Narrative modules give meaning and context to character actions. •Social engagement methods enable collaborative storytelling. •Role-playing games differ from traditional games by emphasizing player growth within the narrative.	• The actions of the player • The roles of the students • The game's mechanics, therefore, clearly outline how it will function for its participants • Each player, in their respective role, enhances their abilities and understanding through the narrative • Games designed for single-player, utilizing	Multiplayer games featuring social engagement.	The abilities and understanding of the player grow as the game progresses, paralleling the unfolding story. In this sense, the player significantly impacts the evolution of the game. •Learners can engage physically with the scenes, narratives, challenges, and other aspects that take place in the real world.	•As the game progresses, the player's skills and understanding develop with the narrative, allowing them to influence the game's evolution. •The game must be carefully planned and clearly explained to ensure teachers' expectations and students' learning goals are unambiguous.

(Table 1) cont.....

Teaching Tools	Main Features	Student's Approach		Benefits	Challenges
		Autonomous	Collaborative		
Simulation	Simulation methods are utilized to assess existing processes (enhancement and optimization) as well as prospective processes (forecasting solutions) in order to achieve the most effective design with various goals.	A student obtains prompt responses regarding the topic being discussed.	Learners obtain prompt responses regarding the topic being discussed.	•Students can experiment freely •Prompt feedback helps reinforce learning. •Mistakes can be made safely without real-world consequences. •Quick response time allows exploration of theoretical designs hard to test in practice. •Instant electronic data records enhance training and support personalized learning. •The simulation must be thoroughly organized and clearly structured.	The simulation must be thoroughly organized and clearly outlined.
AI Tools	AI tools feature intelligent tutoring systems, chatbots, automated assessment systems, dashboards, and IVR. These enable both collaborative and autonomous learning support for students.	Intelligent tutoring and VR promote self-directed learning. Personalised AI support (e.g, conversational assistants) enables flexible, self-paced study. These systems offer immediate answers and resources.	Intelligent support for collaborative learning. Virtual agents in VR act as peers or assistants in group settings. Chatbots enhance student-teacher communication and group updates.	•Real-time feedback and instant help •Enhances engagement through interactive, game-like environments •Promotes deeper understanding in practical fields (*e.g.*, bioscience) •Expands access to structured and diverse online courses	•Requires significant digital infrastructure and investment •Limited by current capabilities in natural language understanding and emotional intelligence •Risk of reduced human interaction and overreliance on automation

CONCLUDING REMARKS

A new approach to teaching needs to be integrated into universities, particularly since these institutions prepare students for the emerging roles created by Industry 5.0. Active learning methodologies used as teaching tools focus on student engagement and are essential for developing the skills required for new job roles in Industry 5.0. In active learning environments, students take on the role of active participants, with self-directed learning and assessment being crucial elements. Implementing active learning techniques in classrooms enhances students' problem-solving abilities and boosts critical thinking skills. Various teaching tools are examined, establishing that they work in tandem to strengthen and reinforce learning. This work primarily aims to explore the different active learning methodologies applicable in university settings, such as: (i) flipped classroom, (ii) game-based learning, (iii) gamified elements, (iv) project-based learning, (v) role-playing games, and (vi) process simulation. While active learning approaches can enhance student upliftment through self-directed tasks and collective efforts, there are numerous obstacles to address within the teaching context, including (i) reducing the number of pupils in each cohort to tailor education to individual needs; (ii) creating suitable environments to foster active learning experiences; (iii) developing the necessary materials and resources for this teaching method; (iv) leveraging technology to support learning practices; (v) establishing effective planning and evaluation methods, as traditional memory-based exams are outdated (vi) securing institutional backing and (vii) providing training for educators to effectively implement active methodologies in their courses. Future research initiatives will investigate the current state of active learning methodologies, focusing on the integration of sustainability as an important topic area and how it can be taught through active learning strategies. Additionally, we will explore innovative approaches to imparting sustainability principles using active learning methodologies. By enabling personalized learning, streamlined grading, and immediate feedback, AI has the potential to make education significantly more efficient and create more dynamic learning environments for students. AI is perfectly suited for enhancing personalized learning by offering customized materials that cater to diverse learning styles. Machine learning enables a highly individualized approach, creating unique student profiles that break down lessons into digestible "smart notes" for improved understanding and academic success.

REFERENCES

Alkhafaf, I. (2010). *Games are modern teaching strategies*. Amman, Jordan: Dar Almanahij for Publishing and Distribution.

An, Y.J., Cao, L. (2017). The effects of game design experience on teachers' attitudes and perceptions regarding the use of digital games in the classroom. *TechTrends, 61*(2), 162-170.

[http://dx.doi.org/10.1007/s11528-016-0122-8]

Angra, S., Jangra, S., Gulzar, Y., Sharma, B., Singh, G. and Onn C.W. (2025) Twenty-two years of advancements in augmented and virtual reality: a bibliometric and systematic review. *Front. Comput. Sci, 7*, 1470038.
[http://dx.doi.org/10.3389/fcomp.2025.1470038]

Aytaç, Z. (2024). Using artificial intelligence tools in higher education. In: *Innovation in the University 4.0 System based on Smart Technologies.* Chapman and Hall/CRC.

Bergström, K. (2012). Creativity rules: How rules impact player creativity in three tabletop role-playing games. *International Journal of Role-Playing,* (3), 4-17.
[http://dx.doi.org/10.33063/ijrp.vi3.221]

Binkley, M., Erstad, O., Herman, J., Raizen, S., Ripley, M., Miller-Ricci, M., & Rumble, M. (2012). Defining Twenty-First Century Skills. In P. Griffin, B. McGaw, & E. Care (Eds.), *Assessment and Teaching of 21st Century Skills.* (pp. 17-66). Dordrecht: Springer.

Buckley, P., Doyle, E., Doyle, S. (2017). Game on! Students' perceptions of gamified learning. *J. Educ. Technol. Soc., 20*(3), 1-10.

Celik, I. (2023). Towards Intelligent-TPACK: An empirical study on teachers' professional knowledge to ethically integrate artificial intelligence (AI)-based tools into education. *Comput. Human Behav, 138*, 107468.
[http://dx.doi.org/10.1016/j.chb.2022.107468]

Chen, X., Zou, D., Xie, H., Cheng, G., Liu, C. (2022). Two decades of artificial intelligence in education. *J. Educ. Technol. Soc., 25*(1), 28-47.

Cheville, R.A. (2016). Linking capabilities to functionings: adapting narrative forms from role-playing games to education. *High. Educ., 71*(6), 805-818.
[http://dx.doi.org/10.1007/s10734-015-9957-8]

Chocarro, R., Cortiñas, M., Marcos-Matás, G. (2021). Teachers' attitudes towards chat-bots in education: A technology acceptance model approach considering the effect of social language, bot proactiveness, and users' characteristics. *Educational Studies, 49*(2), 1-19.
[http://dx.doi.org/10.1080/03055698.2020.1850426]

Cuiñas, I., Sánchez, M.G. (2022). Role-playing games as a strategy to incorporate social skills in electrical engineering programs. *IEEE Antennas Propag. Mag., 64*(1), 123-146.
[http://dx.doi.org/10.1109/MAP.2021.3129689]

de la Torre, R., Berbegal-Mirabent, J. (2020). Using game-based principles to empower students in non-STEM academic programmes. *Innov. Educ. Teach. Int., 57*(5), 511-520.
[http://dx.doi.org/10.1080/14703297.2020.1727352]

de la Torre, R., Onggo, B.S., Corlu, C.G., Nogal, M., Juan, .A.A. (2021). The role of simulation and serious games in teaching concepts on circular economy and sustainable energy. *Energies, 14*(4), 1138.
[http://dx.doi.org/10.3390/en14041138]

Eltahir, M.E., Alsalhi, N.R., Al-Qatawneh, S., AlQudah, H.A., Jaradat, M. (2021). The impact of game-based learning (GBL) on students' motivation, engagement and academic performance on an Arabic language grammar course in higher education. *Educ. Inf. Technol., 26*(3), 3251-3278.
[http://dx.doi.org/10.1007/s10639-020-10396-w]

Fabris, C.P., Rathner, J.A., Fong, A.Y., Sevigny, C.P. (2019). Virtual reality in higher education. *International Journal of Innovation in Science and Mathematics Education, 27*(8), 69-80.
[http://dx.doi.org/10.30722/IJISME.27.08.006]

Gabbiadini, A., Greitemeyer, T. (2017). Uncovering the association between strategy video games and self-regulation: A correlational study. *Pers. Individ. Dif., 104*, 129-136.
[http://dx.doi.org/10.1016/j.paid.2016.07.041]

Green, C.S., Bavelier, D. (2012). Learning, attentional control, and action video games. *Curr. Biol., 22*(6), R197-R206.
[http://dx.doi.org/10.1016/j.cub.2012.02.012] [PMID: 22440805]

Greenhow, C., Galvin, S.M., Brandon, D.L., Askari, E. (2020). A decade of research on K–12 teaching and teacher learning with social media: Insights on the state of the field. *Teach. Coll. Rec., 122*(6), 1-72.
[http://dx.doi.org/10.1177/016146812012200602]

Greipl, S., Moeller, K., Ninaus, M. (2020). Potential and limits of game-based learning. *International Journal of Technology Enhanced Learning, 12*(4), 363-389.
[http://dx.doi.org/10.1504/IJTEL.2020.110047]

Guo, P., Saab, N., Post, L.S., Admiraal, W. (2020). A review of project-based learning in higher education: Student outcomes and measures. *Int. J. Educ. Res., 102*, 101586.
[http://dx.doi.org/10.1016/j.ijer.2020.101586]

Innes, J.E., Booher, D.E. (2015). A turning point for planning theory? Overcoming dividing discourses. *Plann. Theory, 14*(2), 195-213.
[http://dx.doi.org/10.1177/1473095213519356]

Karmakar, B., Liu, P., Mukherjee, G., Che, H., Dutta, S. (2022). Improved retention analysis in freemium role-playing games by jointly modelling players' motivation, progression and churn. *J. R. Stat. Soc. Ser. A Stat. Soc., 185*(1), 102-133.
[http://dx.doi.org/10.1111/rssa.12730]

Kokotsaki, D., Menzies, V., Wiggins, A. (2016). Project-based learning: A review of the literature. *Improving Schools, 19*(3), 267-277.
[http://dx.doi.org/10.1177/1365480216659733]

Konopka, C.L., Adaime, M.B., Mosele, P.H. (2015). Active teaching and learning methodologies: Some considerations. *Creat. Educ., 6*(14), 1536-1545.
[http://dx.doi.org/10.4236/ce.2015.614154]

Laine, T.H., Lindberg, R.S.N. (2020). Designing engaging games for education: A systematic literature review on game motivators and design principles. *IEEE Trans. Learn. Technol., 13*(4), 804-821.
[http://dx.doi.org/10.1109/TLT.2020.3018503]

Latorre-Cosculluela, C., Sierra-Sánchez, V. & Vázquez-Toledo, S. (2025) Gamification, collaborative learning and transversal competences: analysis of academic performance and students' perceptions. *Smart Learn. Environ, 12*, 2.
[http://dx.doi.org/10.1186/s40561-024-00361-2]

Liu, T. and Israel, M. (2022) Uncovering students' problem-solving processes in game-based learning environments, *Computers & Education, 182*, 104462.
[http://dx.doi.org/10.1016/j.compedu.2022.104462]

Luckin, R., Holmes, W., Griffiths, M., Forcier, L.B. (2016). *Intelligence unleashed: An argument for AI in education*London, England: Pearson. Available from: http://discovery.ucl.ac.uk/1475756

Luo, X., Tong, S., Fang, Z., Qu, Z. (2019). Frontiers: Machines vs. humans: The impact of artificial intelligence chatbot disclosure on customer purchases. *Mark. Sci., 38*(6), 913-1084.
[http://dx.doi.org/10.1287/mksc.2019.1192]

McHaney, R., Reiter, L., Reychav, I. (2018). Immersive simulation in constructivist-based classroom e-learning. *Int. J. E-Learn., 17*(1), 39-64.

Milman NB (2014), "The Flipped Classroom Strategy: What Is it and How Can it Best be Used?". *Distance Learning, 11*(4), 9–11.

Naik, M. (2023). Assessing The Effectiveness of Flipped Classroom Strategy on Student Performance. *European Chemical Bulletin, 12*, 2883-2896.

Ozdamli, F., Asiksoy, G. (2016). Flipped classroom approach. *World Journal on Educational Technology:*

Current Issues, 8(2), 98-105.

Perez, S., Massey-Allard, J., Butler, D., Ives, J., Bonn, D., Yee, N., & Roll, I. (2017). Identifying Productive Inquiry in Virtual Labs Using Sequence Mining. In: André, E., Baker, R., Hu, X., Rodrigo, M., du Boulay, B. (eds) *Artificial Intelligence in Education*. Springer, Cham. [http://dx.doi.org/10.1007/978-3-319-61425-0_24]

Plass, J.L., Homer, B.D., Kinzer, C.K. (2015). Foundations of game-based learning. *Educ. Psychol., 50*(4), 258-283. [http://dx.doi.org/10.1080/00461520.2015.1122533]

Prado, A.M., Arce, R., Lopez, L.E., García, J., Pearson, A.A. (2020). Simulations versus case studies: Effectively teaching the premises of sustainable development in the classroom. *J. Bus. Ethics, 161*(2), 303-327. [http://dx.doi.org/10.1007/s10551-019-04217-5]

Prager, R.H.P. (2019). *Exploring the use of role-playing games in education.*. The MT Review.

Ramakrisnan, P. (2019). Gamifying online knowledge sharing environment: A motivating user interface. In: *Advances in visual informatics: 6th International Visual Informatics Conference, IVIC 2019,* Cham, Switzerland: Springer Nature.123-134. [http://dx.doi.org/10.1007/978-3-030-34032-2_12]

Saghiri, M.A., Vakhnovetsky, J., Nadershahi, N. (2022). Scoping review of artificial intelligence and immersive digital tools in dental education. *J. Dent. Educ., 86*(6), 736-750. [http://dx.doi.org/10.1002/jdd.12856] [PMID: 34962645]

Su, C.H. (2016). The effects of students' motivation, cognitive load and learning anxiety in gamification software engineering education: a structural equation modeling study. *Multimedia Tools Appl., 75*(16), 10013-10036. [http://dx.doi.org/10.1007/s11042-015-2799-7]

Yu, L. and Zin, Z.M. (2023) The critical thinking-oriented adaptations of problem-based learning models: a systematic review. *Front. Educ, 8*, 1139987. [http://dx.doi.org/10.3389/feduc.2023.1139987]

Zhang, L., & Ma, Y. (2023). A study of the impact of project-based learning on student learning effects: a meta-analysis study. *Frontiers in psychology, 14*, 1202728. [http://dx.doi.org/10.3389/fpsyg.2023.1202728]

Zhang, Z. (2021). The impact of digital technologies on entrepreneurship education. In *Proceedings of the 2021 6th International Conference on Social Sciences and Economic Development (ICSSED 2021)* (pp. 448–452). Paris, France: Atlantis Press.

CHAPTER 14

The Role of Computational and AI-Assisted Tools in Interdisciplinary Teaching

Anushka Pandey[1], **Anukriti Pandey**[2] and **Tahmeena Khan**[1,*]

[1] *Department of Chemistry, Integral University, Lucknow, Uttar Pradesh 226006, India*

[2] *Department of English and Foreign Languages, Dr. Shakuntala Misra National Rehabilitation University, Lucknow, Uttar Pradesh 226017, India*

Abstract: In response to the growing complexity of global challenges and the evolving demands of the workforce, interdisciplinary education has emerged as a crucial pedagogical approach. This chapter explores the transformative role of computational and AI-assisted tools in facilitating interdisciplinary teaching and learning. It outlines how technologies such as intelligent tutoring systems, AI-powered simulations, natural language processing tools, and data analytics platforms foster integrative learning experiences by bridging disciplinary divides, personalizing instruction, and enhancing learner engagement. By enabling modelling, data visualization, language translation, and collaborative problem-solving, these tools support the development of critical thinking, creativity, and systems-level understanding. Real-world case studies demonstrate the application of AI in collaborative student projects across domains like public health, climate change, and smart city planning. The chapter also critically addresses ethical considerations, including algorithmic bias, data privacy, inclusivity, and digital literacy. It concludes by emphasizing the need for informed and ethical implementation of AI in education, calling for collaborative efforts among educators, institutions, and policymakers to realize the full potential of these technologies in shaping a more connected and adaptive educational paradigm.

Keywords: Artificial intelligence in education, Collaborative learning, Computational tools, Data visualization, Digital literacy, Educational technology, Ethical AI, Intelligent tutoring systems, Interdisciplinary teaching, Personalized learning.

INTRODUCTION

In recent years, education systems worldwide have been compelled to adapt to the rapidly changing demands of society, technology, and the workforce.

* **Corresponding author Tahmeena Khan:** Department of Chemistry, Integral University, Lucknow, Uttar Pradesh 226006, India; E-mail: tahminakhan30@yahoo.com

Tahmeena Khan, Manisha Singh, & Saman Raza (Eds.)

As the boundaries between academic disciplines blur and complex global challenges emerge, there is a growing recognition that traditional teaching methods must evolve. This shift calls for innovative pedagogical strategies that not only integrate diverse fields of knowledge but also leverage advanced technologies to enhance learning (Bitar & Davidovich, 2024). The role of computational and AI-assisted tools (Fig. **1**) has become central in this transformation, especially in fostering interdisciplinary teaching approaches that prepare learners for the complexities of the modern world (Mariyono & Nur Alif Hd, 2025).

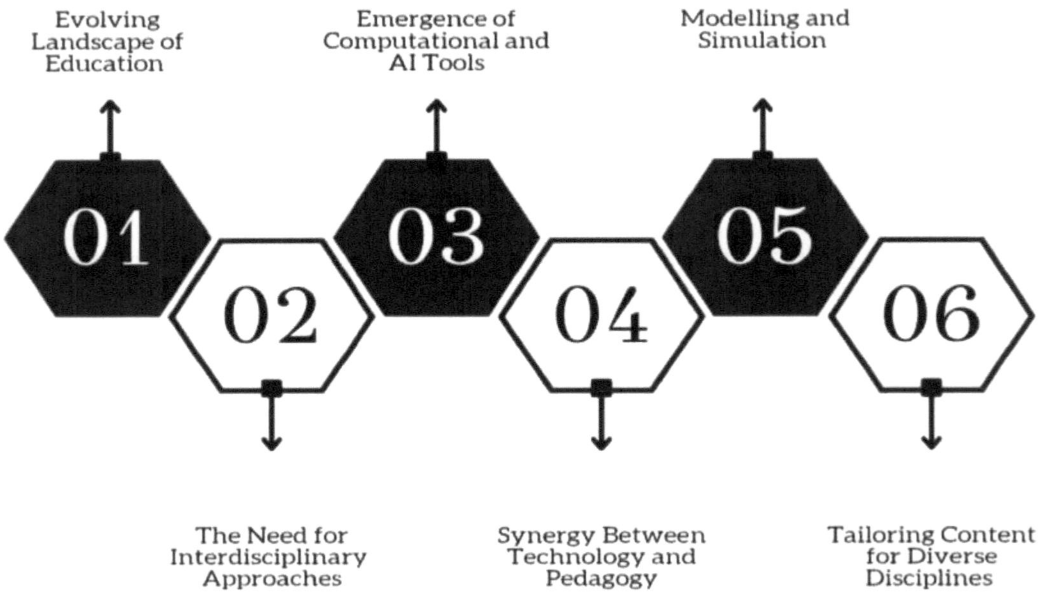

Fig. (1). Advent of AI-assisted tools in teaching.

The Evolving Landscape of Education

The educational landscape is undergoing a transformative shift, catalysed by the growing complexity of global challenges and the need for more adaptive learning systems. Traditional models, which often compartmentalize knowledge into distinct subject areas, are increasingly being re-evaluated. In their place, interdisciplinary frameworks are gaining prominence—encouraging students to synthesize knowledge from various domains to better understand and address real-world problems. This evolution reflects a broader pedagogical trend that values critical thinking, creativity, and collaborative problem-solving over rote memorization and discipline-specific isolation (Dumitru, 2024).

The Need for Interdisciplinary Approaches

While deep disciplinary knowledge remains vital, the shortcomings of isolated, single-discipline education are increasingly evident when addressing complex global challenges like climate change, public health crises, and rapid technological advancement (Anand & Singh, 2025). These problems rarely adhere to the boundaries of a single field, necessitating an educational approach that mirrors their complexity. Interdisciplinary learning fosters the integration of diverse perspectives, enabling students to form more holistic understandings and innovate at the intersections of knowledge. Such approaches not only enhance cognitive flexibility but also prepare learners to navigate an increasingly interconnected and dynamic world (Eugenijus, 2023).

Emergence of Computational and AI Tools in Education

Amidst this paradigm shift, computational and AI-assisted tools have emerged as powerful enablers of interdisciplinary education. These technologies facilitate personalized learning paths, foster collaboration across disciplines, and offer interactive, data-driven environments where theoretical knowledge can be applied in practical contexts (Chee & Sanmugam, 2023). From intelligent tutoring systems to AI-driven simulations and data visualization platforms, such tools are transforming how students engage with content and each other. Their ability to model complex systems, analyse vast datasets, and adapt to individual learner needs makes them especially well-suited to support interdisciplinary pedagogy, bridging gaps between domains and promoting deeper learning experiences (Lin *et al.*, 2023).

CONCEPTUAL FOUNDATIONS

To understand the transformative impact of computational and AI-assisted tools in interdisciplinary teaching, it is essential to first clarify the foundational concepts involved. This section outlines the core principles of interdisciplinary education, introduces the key technologies shaping modern learning environments, and discusses the pedagogical value created through the integration of technology into teaching practices (Nur Fitria, 2021).

Defining Interdisciplinary Teaching

Interdisciplinary teaching refers to an educational approach that integrates knowledge, methods, and perspectives from multiple academic disciplines to examine issues, solve problems, or explore themes that transcend traditional subject boundaries (Nikitina, 2006). Rather than viewing subjects as separate entities, interdisciplinary teaching promotes a holistic approach to learning by encouraging students to make connections, integrate ideas, and think across disciplines. This model fosters critical thinking, creativity, and adaptability—skills that are crucial in navigating complex societal and scientific challenges. In modern education, interdisciplinary teaching plays a vital role in equipping students with a deeper and more versatile understanding of the world.

Overview of Computational and AI Tools

Computational and AI tools encompass a broad range of technologies that support and enhance the learning experience. In educational contexts, these include machine learning algorithms that personalize content delivery, natural language processing (NLP) tools that enable intelligent tutoring and automated feedback, and simulation platforms that allow learners to engage with dynamic models of real-world systems (Novis-Deutsch *et al.*, 2024). Other applications include predictive analytics for early intervention, adaptive assessment systems, and data visualization tools that simplify complex concepts. These tools not only enrich the teaching and learning process but also open new possibilities for interdisciplinary exploration by enabling students to interact with data and models across diverse fields (Xiao Tan *et al.*, 2025).

The Synergy between Technology and Pedagogy

The integration of technology into pedagogy creates a powerful synergy that enhances the effectiveness of interdisciplinary education. Computational and AI tools make abstract or complex topics more tangible, enabling educators to present content in ways that are both engaging and intellectually rigorous (Alam & Mohanty, 2023). For instance, a simulation in environmental science may incorporate data from biology, chemistry, and economics, helping students understand the interconnectedness of ecological and societal systems (Anapana *et al.*, 2025). Similarly, AI-powered platforms can scaffold learning across subjects, adapt to individual needs, and facilitate collaboration among students from different disciplinary backgrounds. When thoughtfully implemented, these technologies not only support knowledge acquisition but also foster the skills needed for interdisciplinary thinking—such as problem-solving, data interpretation, and systems thinking (Anapana *et al.*, 2025).

FACILITATING INTEGRATION ACROSS DISCIPLINES

One of the most transformative contributions of computational and AI-assisted tools in education lies in their ability to bridge disciplinary divides. These technologies do not merely support learning within individual domains—they enable students and educators to work across them, fostering richer, more connected understandings of complex problems. This section explores how modelling, data analysis, and intelligent translation tools play a central role in facilitating interdisciplinary learning (Mariyono & Nur Alif Hd, 2025).

Modelling and Simulation in Multi-Disciplinary Contexts

Modelling and simulation tools offer dynamic environments where learners can experiment with real-world systems by integrating concepts from various fields. Platforms such as MATLAB and Python—along with specialized AI-driven simulators—allow students to build, manipulate, and analyze models that span engineering, biology, economics, and beyond (Constantinescu *et al.*, 2024). For example, a simulation of disease spread may involve mathematical modelling, epidemiological data, and behavioural science principles (Peddinti & Sabbani, 2024). These tools encourage systems thinking, where students recognize the interplay between components of a complex problem. By doing so, they foster a hands-on understanding of interdisciplinary relationships and prepare learners to apply theoretical knowledge to practical scenarios (Blatti *et al.*, 2019).

Data Science and Cross-Disciplinary Analysis

Data science, underpinned by AI-driven analytics, is a natural catalyst for interdisciplinary exploration. It empowers students to handle large, diverse datasets that often include variables from multiple domains—such as linking climate patterns with economic indicators or examining the social determinants of health using public health databases (Khosravi *et al.*, 2022). Tools that leverage machine learning, statistical modelling, and visualization enable learners to uncover patterns, test hypotheses, and draw informed conclusions that cut across disciplinary boundaries (Shinde & Shivthare, 2024). Through this, students not only learn to analyze data but also to ask integrative questions and frame their investigations in ways that reflect the complexity of real-world challenges.

Bridging the Gap: Tools that Translate Concepts across Fields

A significant barrier to interdisciplinary collaboration is the variation in terminologies, methodologies, and conceptual frameworks across disciplines. AI-assisted tools are increasingly addressing this challenge by functioning as mediators—translating domain-specific language and processes into more universally understandable terms (Tatipang *et al.*, 2024). Natural language processing technologies, for instance, can summarize academic content for different audiences or convert technical jargon into accessible language (Khurana *et al.*, 2023). Additionally, AI-driven knowledge graphs and ontology-based systems help map relationships between concepts in distinct fields, making it easier for learners to navigate unfamiliar academic terrain (Mohamed *et al.*, 2025). These tools play a crucial role in fostering mutual understanding and collaboration among students and educators from varied disciplinary backgrounds.

PERSONALIZED LEARNING AND ENHANCED ENGAGEMENT THROUGH AI

The integration of artificial intelligence into educational environments has redefined how learners engage with content and how educators address diverse learning needs. AI technologies not only support individualized instruction through adaptive systems but also actively promote deeper engagement, critical thinking, and problem-solving (Prajapati, 2024). This section explores how AI facilitates personalized learning experiences while simultaneously enhancing motivation and interdisciplinary inquiry through innovative tools and approaches (Yilmaz, 2023).

AI-Based Learning Analytics

AI-driven learning analytics systems collect and analyze data on student behaviour, performance, and progress in real-time. These systems use predictive models to identify knowledge gaps, learning preferences, and at-risk students, enabling timely interventions (Ravichandran, 2024). By adapting the pace, sequence, and complexity of instructional materials, AI ensures that each learner receives a tailored educational experience aligned with their strengths and challenges. In interdisciplinary settings, where students may enter with varied backgrounds, such analytics are invaluable in balancing foundational reinforcement with advanced exploration (Strielkowski, Grebennikova, Lisovskiy, Rakhimova, & Vasileva, 2024).

Intelligent Tutoring Systems

Intelligent Tutoring Systems (ITS) represent one of the most impactful applications of AI in education. These systems simulate one-on-one instruction by providing real-time feedback, scaffolding, and personalized guidance (Owusu *et al.*, 2025). Unlike static learning platforms, ITS adapts dynamically to a student's performance, offering hints, explanations, or alternative approaches based on individual needs (Liu & Yu, 2023). When applied to interdisciplinary subjects, ITS can help students navigate the complexity of integrated concepts by guiding them through the interconnections between different knowledge domains.

Tailoring Content for Diverse Disciplinary Backgrounds

In interdisciplinary classrooms, learners often come with varied levels of familiarity with each discipline involved. AI helps bridge these gaps by customizing content presentation according to a student's academic background, learning history, and prior knowledge (Martin, 2019). For example, a student with a strong foundation in mathematics but limited exposure to biology can receive adjusted modules that introduce biological concepts in a manner aligned with their existing skill set (Stanhope *et al.*, 2017). This ensures inclusivity and prevents cognitive overload, facilitating smoother integration of diverse knowledge areas.

AI-Powered Problem-Solving Environments

Tools like **WolframAlpha**, **ChatGPT**, and other AI-powered assistants support students in tackling complex, interdisciplinary problems. These platforms can perform computations, offer explanations, and even generate hypotheses based on user queries. They act as cognitive extenders, allowing learners to test ideas, visualize outcomes, and refine their thinking. When used effectively, such tools foster inquiry, experimentation, and critical analysis—key components of interdisciplinary education (Garcia *et al.*, 2025).

Scenario-Based and Inquiry-Driven Learning

AI is playing an increasingly prominent role in facilitating interactive, scenario-based learning environments. Through intelligent case studies, problem-solving simulations, and virtual labs, students are encouraged to ask questions, investigate solutions, and apply cross-disciplinary knowledge. These tools (Table **1**) simulate real-world complexity, such as public health responses or climate interventions, providing learners with rich contexts for inquiry-driven exploration (Dai, 2023). AI ensures that these experiences remain responsive to learner input, adapting to challenges and feedback in real time (Rane *et al.*, 2023).

Table 1. AI and computational tools for interdisciplinary education.

Tool / Platform	Core Function	Application in Interdisciplinary Teaching	Reference
MATLAB, Python	Modeling and simulation	Simulate biological, physical, and economic systems	(Idoko *et al.*, 2024)
R, KNIME, Orange	Data analytics and visualization	Analyze datasets across health, environment, and economics	(Pynam, Spanadna, & Srikanth, 2018)
ChatGPT, WolframAlpha	AI-assisted problem solving	Generate hypotheses, compute solutions, and explain complex concepts	(Robledo-Rella & Toh, 2024)
Intelligent Tutoring Systems	Personalized instruction and feedback	Adapt learning path based on performance across varied domains	(Mousavinasab *et al.*, 2021)

(Table 1) cont.....

Tool / Platform	Core Function	Application in Interdisciplinary Teaching	Reference
NLP-based tools	Language translation and simplification	Translate terminology across disciplines for learner understanding	(Ha & Nehm, 2023)
VR/AR Simulation Tools	Experiential learning and gamification	Immerse students in real-world scenarios combining multiple fields	(Paithankar & Khaparde)
Learning Analytics Dashboards	Progress tracking and prediction	Guide instruction and identify support needs in diverse learners	(Ramaswami, Susnjak, Mathrani, & Umer, 2023)
Knowledge Graphs/Ontologies	Conceptual linkage between disciplines	Visualize interrelations between terms from different fields	(Abu-Salih & Alotaibi, 2024)

Gamification and Simulation Tools

Game-based learning environments—powered by AI and immersive technologies like **Virtual Reality (VR)**—offer engaging, interactive platforms for exploring interdisciplinary content. AI-driven role-playing games, decision-making simulations, and educational VR experiences immerse students in complex scenarios where they must apply knowledge from multiple domains. These tools increase motivation and participation, particularly among digital-native learners, while reinforcing systems thinking and collaborative problem-solving. By turning abstract concepts into tangible experiences, gamification makes interdisciplinary learning more accessible and memorable (Papadimitriou, 2025).

SUPPORTING COLLABORATIVE AND PROJECT-BASED LEARNING

Collaborative and Project-Based Learning (PBL) has gained recognition as an effective pedagogical approach for developing interdisciplinary competencies. These approaches simulate real-world working conditions where diverse teams must apply varied skill sets to solve complex problems (Almulla, 2020). Computational and AI-assisted tools serve as critical enablers, providing platforms, automation, and intelligence to facilitate efficient collaboration across disciplinary boundaries (Patel *et al.*, 2024).

Empowering Real-Time Interdisciplinary Collaboration through Virtual Labs and Shared Workspaces

Virtual labs and **cloud-based collaborative platforms** allow geographically dispersed, cross-disciplinary teams to work synchronously and asynchronously on shared goals (Thong *et al.*, 2023). Tools like:

- **Google Colab** enables collaborative coding in Python, widely used in data science, machine learning, and simulation projects (Sukhdeve & Sukhdeve, 2023).
- **Jupyter Notebooks** offer modular documentation and live code execution, supporting reproducibility and real-time collaboration in STEM and social sciences (Mariën, 2021).
- **Overleaf** is ideal for collaborative writing and documentation, particularly for technical reports or LaTeX-based academic manuscripts (Wright, 2019).
- **Microsoft Teams and Slack** integrate messaging, file sharing, and video conferencing, offering unified communication in team-based learning environments (Morales-Gonzalez *et al.*, 2024).

These platforms facilitate the dissolution of disciplinary boundaries by enabling integrated, data-driven workflows. In a disaster response scenario, for example, civil engineers can simulate infrastructure failure models, while public health researchers concurrently perform risk assessments using interoperable datasets and shared analytical frameworks (Mühlhofer *et al.*, 2023). Features such as live commenting, version control, and embedded visuals allow all team members to contribute meaningfully, regardless of their domain (Xin *et al.*, 2024).

Structuring Effective Interdisciplinary Teams and Project Management Using AI Tools

Effective interdisciplinary collaboration hinges on building the right teams and managing their efforts smoothly. AI-driven platforms enhance both aspects:

- **Team formation algorithms** (*e.g.*, CATME, EduSourced) use learner profiles, assessment scores, personality traits, and discipline-specific skills to optimize group diversity and performance (Hastings *et al.*, 2023).
- **AI project management tools** like Trello (with Butler automation), Asana (with smart task prediction), and Monday.com help assign tasks based on strengths, set dependencies, track progress, and identify bottlenecks (Stanton, 2025).
- **Natural Language Processing (NLP)-based chatbots** in Slack or Microsoft Teams assist with reminders, task updates, and resource recommendations (Singh *et al.*, 2024).

AI can also help resolve common group challenges, such as uneven participation, by monitoring interaction patterns and recommending adjustments. These tools train students to manage interdisciplinary workflows as they would in professional environments, emphasizing communication, responsibility, and adaptability (Gomez *et al.*, 2023).

Real-World Case Studies Demonstrating AI-Supported Interdisciplinary Student Projects

Numerous institutions have implemented computational and AI tools to support interdisciplinary projects, often resulting in innovative and impactful outcomes (Kusters *et al.*, 2020). Examples include:

- **Disease Surveillance and Prediction**: In a project at a public health institute, students from data science, medicine, and environmental studies collaborated using AI-powered time-series models in Google Colab to predict dengue outbreaks. The project combined epidemiological data, weather patterns, and socio-economic indicators into an integrated machine learning framework (Ali, 2024).
- **AI-Driven Game Design for Climate Education**: A team of students in computer science, visual arts, and environmental policy used Unity 3D, NLP-based dialogue engines, and environmental simulation APIs to create an educational game that immerses users in climate-related decision-making scenarios. AI tools enabled students to simulate climate models and dynamically adjust storylines in response to user input (Dogan *et al.*, 2024).
- **Smart City Planning**: A multidisciplinary team from architecture, urban planning, and AI worked together using shared GIS platforms, traffic simulation tools, and AI-based optimization algorithms to redesign pedestrian pathways in a congested urban zone (He & Chen, 2024).

The majority of earlier research on artificial intelligence (AI) and collaborative learning recognizes human–AI collaboration as a key focus (Dwivedi *et al.*, 2023; Wang *et al.*, 2023) (Lo *et al.*, 2024; Luther *et al.*, 2024). In other terms, the user (such as a student or teacher) works together using AI tools to accomplish a task (for example, composing a report, aiding in programming, creating assessments, or resolving an issue). For instance, Elson *et al.* (2018) investigated trust and reliance on human-AI teamwork (Elson *et al.*, 2018).

The discussed studies collectively highlight the growing role and potential of AI, particularly generative AI (GenAI) like ChatGPT, in enhancing human-AI collaboration across educational and professional domains. Nah *et al.* (2023) emphasized that the design of GenAI systems should be guided by human values and needs (Fui-Hoon *et al.*, 2023). Pavlik (2023) explored the implications of human–ChatGPT collaboration in journalism and media education (Pavlik, 2023). Similarly, Luther *et al.* (2024) reported that individuals collaborating with ChatGPT to write about public alcohol bans were generally satisfied with its responses (Luther *et al.*, 2024). Students have also found ChatGPT to be a fast, helpful, and user-friendly tool for tasks such as online market research and website development (Economides & Perifanou, 2024).

Li *et al.* (2024) further examined how collaboration with GenAI influences factors like AI literacy, human agency, creativity, and copyright awareness (Li *et al.*, 2024). Numerous studies have also explored AI's role in supporting collaborative learning. These include efforts to manage collaboration—such as forming student groups or designing group tasks—as well as enhancing the collaborative process itself by analyzing group interactions, offering feedback, and evaluating group outcomes (Perifanou & Economides, 2025). For instance, Haq *et al.* (2021) and Wang and Wang (2022) proposed AI-driven classification methods based on individual characteristics (Haq *et al.*, 2021; Kasepalu *et al.*, 2022; Lewis, 2022; Ouyang & Zhang, 2024; Ponte *et al.*, 2023; Rudolph *et al.*, 2023; Zheng *et al.*, 2024), while Kasepalu *et al.* (2022) developed AI tools to assist teachers in managing group work (Kasepalu *et al.*, 2022).

Other research focused on using ChatGPT to generate collaborative problem-solving scenarios (Rudolph *et al.*, 2023) (Rudolph *et al.*, 2023), and leveraging multimodal large language models to provide learners with behaviour-based feedback (Lewis, 2022). Tan *et al.* (2022) and Ponte *et al.* (2023) used AI to analyze group interactions and provide personalized formative feedback, although only a few groups used this feedback to improve (Tan *et al.*, 2022). Ouyang and Zhang (2024) reviewed AI-supported evaluation of group performance, and Zheng *et al.* (2024) proposed using AI to deliver automated assessments and tailored recommendations to support collaborative knowledge building (Ouyang & Zhang, 2024; Zheng *et al.*, 2024).

Despite these advances, researchers such as Følstad *et al.* (2021), Gilson *et al.* (2023), and Tan *et al.* (2022) have emphasized the need for more empirical research into chatbot-supported collaborative learning. Zhu *et al.* (2023), for example, noted a lack of empirical studies on interdisciplinary collaboration and recommended adapting ChatGPT use to specific subjects and student needs (Følstad *et al.*, 2021; Gilson *et al.*, 2023; S. C. Tan *et al.*, 2022; Zhu *et al.*, 2023).

Collectively, these case studies illustrate not only the practical integration of AI tools in educational contexts but also their pedagogical benefits (Fig. **2**). They foster exposure to diverse ways of thinking, improve interdisciplinary communication, and enable students to engage more deeply with socially significant challenges.

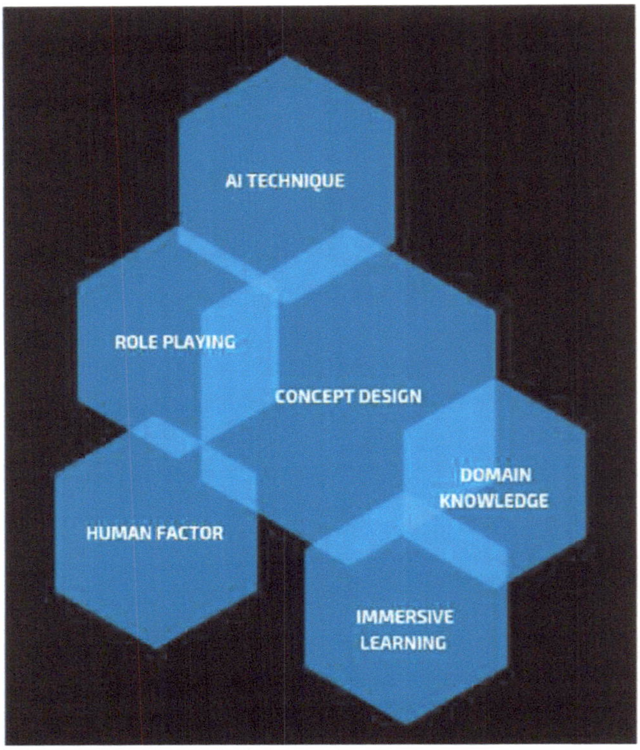

Fig. (2). Practical integration of AI tools.

VISUALIZATION AND INTERPRETATION ACROSS DOMAINS

In interdisciplinary education, visualization plays a pivotal role in bridging conceptual gaps between fields. Whether interpreting data trends, understanding biological pathways, or analyzing sociopolitical patterns, effective visual representation enhances cognitive processing and fosters deeper comprehension (Van den Beemt *et al.*, 2020). Computational and AI-assisted tools have revolutionized the way learners visualize, interact with, and interpret complex information across disciplinary boundaries (Arora *et al.*, 2025).

Role of Computational Tools in Visual Thinking

Computational tools have empowered educators and students alike to transform abstract concepts into concrete visuals that aid in conceptual understanding and retention (Jeyabal *et al.*, 2016). AI-augmented platforms such as:

- **Tableau** and **Power BI** offer intuitive dashboards and real-time visualizations, enabling learners to interactively explore trends, patterns, and correlations in data (Mishra, 2020).

- **R/Shiny** provides a customizable environment for statistical computing, making it ideal for visualizing complex models, simulations, or interdisciplinary datasets (Khedr & Hilal, 2021).
- **Python libraries** like Matplotlib, Seaborn, and Plotly enable dynamic, layered, and domain-specific visual representations—from gene expression data to economic forecasting (Bergkvist, 2023).

Such tools help cultivate *visual thinking*, a key skill for interdisciplinary problem-solving. For example, engineering and public policy students can collaborate to visualize the impact of infrastructure investment on economic indicators, enabling shared understanding through visual data exploration (Kim *et al.*, 2022).

Interpreting Complex, Multimodal Data

Interdisciplinary research often involves **multimodal data**—a mix of numerical, textual, visual, and spatial information (Dhivya *et al.*, 2025). AI tools enhance the integration and interpretation of such data through:

- **Natural Language Processing (NLP)** tools that extract sentiment, themes, and trends from unstructured text (*e.g.*, social media, literature, clinical notes) (Dreisbach *et al.*, 2019).
- **Computer vision algorithms** that interpret images and videos for applications like medical diagnostics, satellite analysis, or historical artefact recognition (Szeliski, 2022).
- **Multimodal fusion frameworks** combine data types into unified visual outputs. For instance, combining radiological scans, clinical text, and lab reports for a comprehensive medical diagnosis (Azam *et al.*, 2022).

These technologies allow learners from diverse backgrounds to collaboratively interpret rich datasets, uncover hidden insights, and construct knowledge that transcends traditional disciplinary formats.

Visual Narratives in Education

Visual storytelling harnesses the power of narrative and imagery to convey complex ideas in a compelling and relatable manner. AI and computational tools now support educators in creating **visual narratives** that enhance interdisciplinary comprehension (Lu *et al.*, 2024):

- **AI-generated infographics** and concept maps (*e.g.*, using Canva or MindMeld) help visualize systemic relationships in topics like climate change, public health, or supply chain dynamics (Hoque, 2024).

- **Story-mapping platforms** such as StoryMapJS and Knight Lab allow learners to create geographical and temporal narratives integrating maps, timelines, and multimedia content (Groshans *et al.*, 2019).
- **Generative AI models** (*e.g.*, DALL·E, Midjourney) can create illustrative images to support fictional case studies or simulate hypothetical futures, ideal for use in interdisciplinary ethics or policy discussions (Chekalina, 2024).

These tools facilitate engaging, learner-centred pedagogy where students can *construct and share meaning* visually—enabling inclusive understanding for audiences from varied academic and cultural backgrounds (Pysarchyk, 2024).

ETHICAL, SOCIAL, AND PRACTICAL CONSIDERATIONS

As AI and computational tools gain traction in interdisciplinary education, it becomes critical to evaluate the ethical, social, and practical implications of their integration (Yadav, 2024). While these technologies can drive innovation, they also raise important concerns around fairness, privacy, accessibility, and responsible usage (Table **2**). Educators and institutions must address these challenges proactively to ensure the equitable and ethical deployment of AI in teaching and learning contexts (Pedro *et al.*, 2019).

Table 2. Ethical risks of educational AI tools and mitigation strategies.

Ethical Risk	Implications in Education	Mitigation Strategies	Reference
Algorithmic Bias	- Discriminatory content recommendations- Unequal access to resources or feedback	- Use diverse and representative training data- Regular audits of AI models- Include human oversight	(Murikah *et al.*, 2024)
Data Privacy Violation	- Unauthorized sharing or misuse of student data- Violation of consent and trust	- Implement clear data governance policies- Ensure informed consent- Use encryption and anonymization	(Pina *et al.*, 2024)
Surveillance & Overmonitoring	- Student discomfort and reduced autonomy- Risk of stress or altered behavior	- Limit data tracking to essential metrics- Communicate data use clearly- Provide opt-out options	(Han *et al.*, 2023)
Lack of Transparency	- Students/educators unaware of how decisions are made- Loss of trust in educational tools	- Use explainable AI (XAI) systems- Provide algorithmic transparency reports- Train educators to interpret AI outputs	(Manna & Sett, 2024)
Exclusion of Marginalized Groups	- Tools fail to support learners with disabilities or from non-dominant cultures	- Apply universal design principles- Conduct usability testing with diverse populations- Support multi-language and adaptive interfaces	(Veytia Bucheli *et al.*, 2024)

(Table 2) cont.....

Ethical Risk	Implications in Education	Mitigation Strategies	Reference
Digital Literacy Gap	- Inequitable use of AI tools among students and faculty- Misinterpretation of AI outputs	- Offer targeted digital literacy training- Embed ethical AI literacy into curricula- Provide accessible tutorials and guides	(Jinuachan *et al.*, 2024)
Over-reliance on AI	- Erosion of critical thinking and learner autonomy- Reduced human interaction	- Encourage AI as a supplementary tool- Design reflective assignments- Promote human-AI collaboration rather than substitution	(Zhai, Wibowo, & Li, 2024)

Addressing Algorithmic Bias in Education

One of the major ethical concerns in AI-assisted education is **algorithmic bias**—systematic and unintended unfairness embedded in machine learning models or decision-making systems. Bias can arise from skewed training data, flawed assumptions, or a lack of contextual awareness, potentially disadvantaging students from underrepresented or marginalized groups (Hanna *et al.*, 2025). In interdisciplinary learning environments, biased algorithms may:

- Misrepresent disciplinary knowledge (*e.g.*, historical inaccuracies or cultural misinterpretations).
- Recommend resources or assessments inequitably based on race, gender, or language.
- Reinforce stereotypes in intelligent tutoring systems or adaptive content delivery.

Mitigating bias requires transparency in algorithm design, regular audits of AI tools, diverse datasets, and the active involvement of educators in curating and interpreting outputs. Incorporating **ethical AI literacy** into interdisciplinary curricula can also empower students to critically assess the tools they use (González-Sendino *et al.*, 2024).

Data Privacy and Digital Literacy

With the increased use of AI in education comes the collection of vast amounts of **student data**—from learning behaviour and performance metrics to biometric or behavioural cues in immersive environments (Seo *et al.*, 2021). These data raise significant concerns around:

- **Privacy**: Who owns the data? How is it stored, shared, or analyzed? Are students informed and consenting?

- **Security**: Are there protections in place against data breaches or misuse?
- **Surveillance**: Do AI tools inadvertently create a sense of constant monitoring that could impact student well-being?

In parallel, the **digital literacy gap** becomes a barrier to meaningful participation in AI-enhanced learning (Muawanah *et al.*, 2024). Students and educators need training in understanding how these tools work, what data they collect, and how to interact with them responsibly. Embedding **data ethics and digital skills training** in interdisciplinary programs can help build informed, empowered users (Atenas *et al.*, 2023).

Accessibility and Inclusivity in Interdisciplinary Learning

AI tools hold immense potential to **enhance accessibility**—from automatic captioning and translation to screen readers and voice-activated learning interfaces. For students with disabilities or diverse linguistic and cultural backgrounds, these features can significantly improve engagement and inclusion (Eziamaka *et al.*, 2024).

However, if not designed with inclusivity in mind, AI systems may **exacerbate inequalities** (Guo *et al.*, 2020), for instance:

- Interfaces that are inaccessible to visually or hearing-impaired users.
- Language models that fail to support non-dominant languages or dialects.
- Pedagogical designs that do not account for neurodiverse learners.

A truly inclusive interdisciplinary learning environment must involve **universal design principles**, **inclusive AI development practices**, and regular feedback from diverse learners (Fig. **3**). Collaborative input from educators, technologists, and accessibility advocates can guide the creation of equitable learning ecosystems that serve all students (Pagliara *et al.*, 2024).

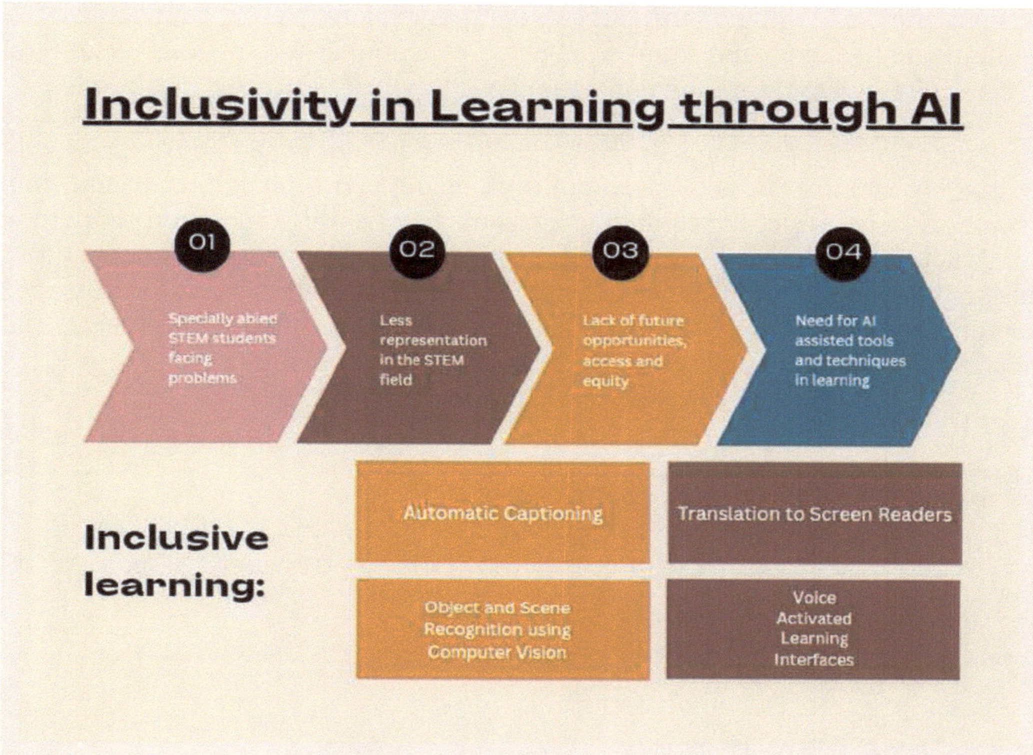

Fig. (3). Accessibility and inclusivity using AI-assisted tools.

CONCLUDING REMARKS

This chapter has examined the transformative potential of computational and AI-assisted tools in interdisciplinary teaching. As the traditional education system evolves to meet the demands of real-world complexity, the integration of AI technologies—ranging from machine learning and natural language processing to simulation platforms and adaptive learning environments—has emerged as a powerful enabler of cross-disciplinary engagement. These tools support a wide range of pedagogical goals, including enhanced visualization, personalized instruction, collaborative problem-solving, and the development of critical thinking skills. Their application across disciplines fosters not only deeper comprehension but also the kind of creative and systems-level thinking needed in today's interconnected world.

However, the implementation of AI in education also brings significant challenges. Concerns around algorithmic bias, data privacy, digital inequity, and the potential erosion of human judgment underscore the importance of ethical and thoughtful deployment. These risks must be addressed through inclusive design, transparent algorithmic decision-making, and strong data governance policies. At the same time, emerging opportunities—such as explainable AI, equitable access to intelligent systems, and more responsive educational frameworks—highlight AI's promise as a tool for reimagining how we teach and learn across disciplines.

Realizing this vision requires deliberate action. Educators must be equipped with the training and resources to meaningfully integrate AI tools into their teaching. Institutions must foster digital literacy, ensure accessibility, and build supportive infrastructures. Policymakers must update guidelines to protect learners' rights and promote innovation. With a coordinated effort, computational and AI-assisted technologies can become foundational to a more collaborative, inclusive, and impactful model of interdisciplinary education.

REFERENCES

Abu-Salih, B., Alotaibi, S. (2024). A systematic literature review of knowledge graph construction and application in education. *Heliyon, 10*(3), e25383.
[http://dx.doi.org/10.1016/j.heliyon.2024.e25383] [PMID: 38327411]

Alam, A., Mohanty, A. (2023). Educational technology: Exploring the convergence of technology and pedagogy through mobility, interactivity, AI, and learning tools. *Cogent Eng, 10*(2), 2283282.
[http://dx.doi.org/10.1080/23311916.2023.2283282]

Almulla, M.A. (2020). The Effectiveness of the Project-Based Learning (PBL) Approach as a Way to Engage Students in Learning. *SAGE Open, 10*(3), 2158244020938702.
[http://dx.doi.org/10.1177/2158244020938702]

Anand, A., Singh, P. (2025). Interdisciplinary Studies: Issues, Perspectives, and Challenges From National Education Program 2020. In N. Son (Ed.), *Integrating Personalized Learning Methods Into STEAM Education.* (pp. 233-260). IGI Global Scientific Publishing.

Anapana, G., Donkada, S., Surarapu, N., v, V.R. (2025). Multidisciplinary Integration in Zoology Education: Bridging Biological, Computational, and Environmental Sciences. *Uttar Pradesh J. Zool, 46*(8), 317-331.
[http://dx.doi.org/10.56557/upjoz/2025/v46i84916]

Arora, A.K., Kumar, A., Tyagi, R., Sharma, S., Kumar, A., Kumar, S. (2025). Future of Education With AI-Assisted Technologies. In: *Artificial Intelligence in Peace, Justice, and Strong Institutions* IGI Global Scientific Publishing.
[http://dx.doi.org/10.4018/979-8-3693-9395-6.ch008]

Asif Mir, A., Asif, A. (2024). Analysis of human factors in emerging infectious disease using artificial intelligence and data science. *International Journal of Research and Analytical Reviews, 11*(1), 250-263.

Atenas, J., Havemann, L., Timmermann, C. (2023). Reframing data ethics in research methods education: a pathway to critical data literacy. *Int. J. Educ. Technol. High. Educ, 20*(1), 11.
[http://dx.doi.org/10.1186/s41239-023-00380-y]

Azam, M.A., Khan, K.B., Salahuddin, S., Rehman, E., Khan, S.A., Khan, M.A., Kadry, S., Gandomi, A.H. (2022). A review on multimodal medical image fusion: Compendious analysis of medical modalities, multimodal databases, fusion techniques and quality metrics. *Comput. Biol. Med, 144*, 105253.
[http://dx.doi.org/10.1016/j.compbiomed.2022.105253] [PMID: 35245696]

Babu J, M., Virgin B, A., RS Edwin, M., Priya P, G., Ravichandran, K. (2024). Identifying Learning Difficulties at an Early Stage in Education with the Help of Artificial Intelligence Models and Predictive Analytics. *International Research Journal of Multidisciplinary Scope, 5*(4), 1455-1461.
[http://dx.doi.org/10.47857/irjms.2024.v05i04.01821]

Bergkvist, J. (2023). *Data analytics for unemployment insurance claims: Framework, approaches, and implementation strategies* (Bachelor's thesis). Mälardalen University, Västerås, Sweden.

Bitar, N., Davidovich, N. (2024). Transforming pedagogy: The digital revolution in higher education. *Educ. Sci. (Basel), 14*(8), 811.
[http://dx.doi.org/10.3390/educsci14080811]

Blatti, J.L., Garcia, J., Cave, D., Monge, F., Cuccinello, A., Portillo, J., Juarez, B., Chan, E., Schwebel, F. (2019). Systems thinking in science education and outreach toward a sustainable future. *J. Chem. Educ, 96*(12), 2852-2862.
[http://dx.doi.org/10.1021/acs.jchemed.9b00318]

Chee, K. N., Sanmugam, M. (2023). Embracing Cutting-edge Technology in Modern Educational Settings. Hershey, PA: IGI Global.

Chekalina, V. (2024). *The impact of AI on the illustration and design industries* (Doctoral dissertation). Universidade da Beira Interior, Covilhã, Portugal.

Constantinescu, G.G., Silion, D., Iftene, A. (2024). Using Artificial Intelligence and Mixed Realities to Create Educational Applications of the Future. *Procedia Comput. Sci, 246*, 3313-3322.
[http://dx.doi.org/10.1016/j.procs.2024.09.308]

Dai, C-P. (2023). *Enhancing learning achievements and self-efficacy for preservice teachers using model-based support in simulation-based learning with artificial intelligence-powered virtual agents.* (Phd dissertation). Florida State University, Tallahassee, FL.

Dhivya, K., Kumar, S.N., Victoria, D.R.S., Sherly, S.I., Durgadevi, G. (2025). Advanced Neural Networks for Multimodal Data Fusion in Interdisciplinary Research. In: *Advanced Interdisciplinary Applications of Deep Learning for Data Science* IGI Global Scientific Publishing.

Dogan, G., Sahin, E., Wilkinson, C.F., Moody, A.K., Song, Y. (2024). BlueAI: Designing artificial intelligence for environment science and climate change learning experiences for K–12 students. *J. Comput. Sci. Coll, 40*(1), 1-12.

Dreisbach, C., Koleck, T.A., Bourne, P.E., Bakken, S. (2019). A systematic review of natural language processing and text mining of symptoms from electronic patient-authored text data. *Int. J. Med. Inform, 125*, 37-46.
[http://dx.doi.org/10.1016/j.ijmedinf.2019.02.008] [PMID: 30914179]

Dumitru, C. (2024). Future of Learning: Adaptive Learning Systems Based on Language Generative Models in Higher Education. In: *Impact of Artificial Intelligence on Society* Chapman and Hall/CRC.
[http://dx.doi.org/10.1201/9781032644509-3]

Dwivedi, Y.K., Kshetri, N., Hughes, L., Slade, E.L., Jeyaraj, A., Kar, A.K., Baabdullah, A.M., Koohang, A., Raghavan, V., Ahuja, M., Albanna, H., Albashrawi, M.A., Al-Busaidi, A.S., Balakrishnan, J., Barlette, Y., Basu, S., Bose, I., Brooks, L., Buhalis, D., Carter, L., Chowdhury, S., Crick, T., Cunningham, S.W., Davies, G.H., Davison, R.M., Dé, R., Dennehy, D., Duan, Y., Dubey, R., Dwivedi, R., Edwards, J.S., Flavián, C., Gauld, R., Grover, V., Hu, M-C., Janssen, M., Jones, P., Junglas, I., Khorana, S., Kraus, S., Larsen, K.R., Latreille, P., Laumer, S., Malik, F.T., Mardani, A., Mariani, M., Mithas, S., Mogaji, E., Nord, J.H., O'Connor, S., Okumus, F., Pagani, M., Pandey, N., Papagiannidis, S., Pappas, I.O., Pathak, N., Pries-Heje, J., Raman, R., Rana, N.P., Rehm, S-V., Ribeiro-Navarrete, S., Richter, A., Rowe, F., Sarker, S., Stahl, B.C., Tiwari, M.K., van der Aalst, W., Venkatesh, V., Viglia, G., Wade, M., Walton, P., Wirtz, J., Wright, R. (2023). Opinion Paper: "So what if ChatGPT wrote it?" Multidisciplinary perspectives on opportunities, challenges and implications of generative conversational AI for research, practice and policy. *Int. J. Inf. Manage, 71*, 102642.
[http://dx.doi.org/10.1016/j.ijinfomgt.2023.102642]

Economides, A., & Perifanou, M. (2024). *Higher education students using GenAI tools to design websites. In Proceedings of the Innovating Higher Education Conference, I-HE2024* (pp. 115–129). Maastricht, Netherlands: EADTU.

Elson, J. S., Derrick, D., Ligon, G. (2018). Examining trust and reliance in collaborations between humans and automated agents. In: *In Proceedings of the 51st Hawaii International Conference on System Sciences,*

Honolulu, HI: ScholarSpace.430–439.

Eugenijus, L. (2023). Integrating blended learning and STEM education: Innovative approaches to promote interdisciplinary learning. *Research and Advances in Education, 2*(9), 20-36. [http://dx.doi.org/10.56397/RAE.2023.09.03]

Eziamaka, N., Odonkor, T., Akinsulire, A. (2024). AI-Driven accessibility: Transformative software solutions for empowering individuals with disabilities. *International Journal of Applied Research in Social Sciences, 6*(8), 1612-1641. [http://dx.doi.org/10.51594/ijarss.v6i8.1373]

Fitria, T. N. (2021). Artificial intelligence (AI) in education: Using AI tools for teaching and learning process. *Prosiding seminar nasional & call for paper STIE AAS,* 134-147.

Følstad, A., Araujo, T., Law, E.L.C., Brandtzaeg, P.B., Papadopoulos, S., Reis, L., Baez, M., Laban, G., McAllister, P., Ischen, C., Wald, R., Catania, F., Meyer von Wolff, R., Hobert, S., Luger, E. (2021). Future directions for chatbot research: an interdisciplinary research agenda. *Computing, 103*(12), 2915-2942. [http://dx.doi.org/10.1007/s00607-021-01016-7]

Fui-Hoon Nah, F., Ruilin, Z., Jingyuan, C., Keng, S., Chen, L. (2023). Generative AI and ChatGPT: Applications, challenges, and AI-human collaboration. *J. Inf. Technol. Case Appl. Res, 25*(3), 277-304. [http://dx.doi.org/10.1080/15228053.2023.2233814]

Garcia, K.F., Ong, A.K.S., Gumasing, M.J.J., Delos Reyes, C.R.V. (2025). Engineering students' perceptions and actual use of AI-based math tools for solving mathematical problems. *Acta Psychol. (Amst.), 256*, 105004. [http://dx.doi.org/10.1016/j.actpsy.2025.105004] [PMID: 40215785]

Gilson, A., Safranek, C.W., Huang, T., Socrates, V., Chi, L., Taylor, R.A., Chartash, D. (2023). How does ChatGPT perform on the United States Medical Licensing Examination (USMLE)? The implications of large language models for medical education and knowledge assessment. *JMIR Med. Educ, 9*(1), e45312. [http://dx.doi.org/10.2196/45312] [PMID: 36753318]

Gomez, C., Unberath, M., Huang, C.M. (2023). Mitigating knowledge imbalance in AI-advised decision-making through collaborative user involvement. *Int. J. Hum. Comput. Stud, 172*, 102977. [http://dx.doi.org/10.1016/j.ijhcs.2022.102977]

González-Sendino, R., Serrano, E., Bajo, J. (2024). Mitigating bias in artificial intelligence: Fair data generation *via* causal models for transparent and explainable decision-making. *Future Gener. Comput. Syst, 155*, 384-401. [http://dx.doi.org/10.1016/j.future.2024.02.023]

Groshans, G., Mikhailova, E., Post, C., Schlautman, M., Carbajales-Dale, P., Payne, K. (2019). Digital story map learning for STEM disciplines. *Educ. Sci. (Basel), 9*(2), 75. [http://dx.doi.org/10.3390/educsci9020075]

Guo, A., Kamar, E., Vaughan, J., Wallach, H., Morris, M. (2020). Toward fairness in AI for people with disabilities: a research roadmap. *ACM SIGACCESS Access. Comput,* 125-1. [http://dx.doi.org/10.1145/3386296.3386298]

Ha, M., & Nehm, R. H. (2023). Measuring scientific understanding across international samples: The promise of machine translation and NLP-based machine learning technologies. In *Advancing natural language processing in educational assessment*. New York, NY: Routledge. [http://dx.doi.org/10.4324/9781003278658-16]

Han, B., Buchanan, G., Mckay, D. (2023). Learning in the Panopticon: Examining the potential impacts of AI monitoring on students [Paper presentation]. *Proceedings of the 34th Australian Conference on Human-Computer Interaction,* Canberra, ACT, Australia.9-21. [http://dx.doi.org/10.1145/3572921.3572937]

Hanna, M.G., Pantanowitz, L., Jackson, B., Palmer, O., Visweswaran, S., Pantanowitz, J., Deebajah, M., Rashidi, H.H. (2025). Ethical and Bias Considerations in Artificial Intelligence/Machine Learning. *Mod.*

Pathol, 38(3), 100686.
[http://dx.doi.org/10.1016/j.modpat.2024.100686] [PMID: 39694331]

Haq, I.U., Anwar, A., Rehman, I.U., Asif, W., Sobnath, D., Sherazi, H.H.R., Nasralla, M.M. (2021). Dynamic group formation with intelligent tutor collaborative learning: a novel approach for next generation collaboration. *IEEE Access, 9*, 143406-143422.
[http://dx.doi.org/10.1109/ACCESS.2021.3120557]

Hariyanto, , Kristianingsih, F. X. D, Maharani, R (2025). Artificial intelligence in adaptive education: a systematic review of techniques for personalized learning. *Discov. Educ., 4*, 458.
[http://dx.doi.org/10.1007/s44217-025-00908-6]

Hastings, E.M., Ojha, V., Austriaco, B.V., Karahalios, K., Bailey, B.P. (2023). Composing Team Compositions: An Examination of Instructors' Current Algorithmic Team Formation Practices. *Proceedings of the ACM on Human-Computer Interaction, 7(CSCW2),* 1-24.
[http://dx.doi.org/10.1145/3610096]

He, W., Chen, M. (2024). Advancing urban life: a systematic review of emerging technologies and artificial intelligence in urban design and planning. *Buildings, 14*(3), 835.
[http://dx.doi.org/10.3390/buildings14030835]

Hoque, M.N. (2024). *Designing Human-Centered AI Tools With Interactive Visualization.* (Phd dissertation). University of Maryland, College Park, MD.

Idogho Christian, , Enemali Peter, , Christian, I., Peter, E., Obot, U., Iguoba, V. (2024). Mathematical modeling and simulations using software like MATLAB, COMSOL and Python. *Magna Scientia Advanced Research and Reviews, 12*(2), 062-095.
[http://dx.doi.org/10.30574/msarr.2024.12.2.0181]

Jeyabal, K., Vasuki, M., Kumar, A.D. (2016). From Algorithms to Conceptual Understanding a Review of Instructional Strategies in Math. *Algorithms, 2*(1), 191-205.

Jinuachan, F. G. V. J., P, A., M, A., Jose, D., Roy, T., Prasad, M. (2024). Impact of Digital Literacy, Use of AI tools and Peer Collaboration on AI Assisted Learning - Perceptions of the University students. *Digit. Educ. Review, 45*, 43-49.
[http://dx.doi.org/10.1344/der.2024.45.43-49]

Kasepalu, R., Prieto, L. P., Ley, T., & Chejara, P. (2022). Teacher artificial intelligence-supported pedagogical actions in collaborative learning coregulation: A wizard-of-oz study. *Frontiers in Education, 7*, 736194.

Khedr, A., Hilal, S. (2021). Interactive visualization for statistical modelling through a shiny app in R [Paper presentation]. *2021 International Conference on Data Analytics for Business and Industry (ICDABI),* Sakheer, Bahrain.332-337.
[http://dx.doi.org/10.1109/ICDABI53623.2021.9655841]

Khosravi, H., Shum, S.B., Chen, G., Conati, C., Tsai, Y.S., Kay, J., Knight, S., Martinez-Maldonado, R., Sadiq, S., Gašević, D. (2022). Explainable Artificial Intelligence in education. *Computers and Education: Artificial Intelligence, 3*, 100074.
[http://dx.doi.org/10.1016/j.caeai.2022.100074]

Khurana, D., Koli, A., Khatter, K., Singh, S. (2023). Natural language processing: state of the art, current trends and challenges. *Multimedia Tools Appl, 82*(3), 3713-3744.
[http://dx.doi.org/10.1007/s11042-022-13428-4] [PMID: 35855771]

Kim, J., Lee, H., Cho, Y. H. (2022). Learning design to support student-AI collaboration: perspectives of leading teachers for AI in education. *Educ. Inf. Technol, 27*(5), 6069-6104.
[http://dx.doi.org/10.1007/s10639-021-10831-6]

Kusters, R., Misevic, D., Berry, H., Cully, A., Le Cunff, Y., Dandoy, L., Othmani, A. (2020). Interdisciplinary research in artificial intelligence: challenges and opportunities. *Front. Big Data, 3*, 577974.

Lewis, A. (2022). Multimodal large language models for inclusive collaboration learning tasks. *Proceedings of the 2022 Conference of the North American Chapter of the Association for Computational Linguistics: Human Language Technologies: Student Research Workshop,* Stroudsburg, PA.202-210. [http://dx.doi.org/10.18653/v1/2022.naacl-srw.26]

Li, J., Cao, H., Lin, L., Hou, Y., Zhu, R., & El Ali, A. (2024). User experience design professionals' perceptions of generative artificial intelligence. In *Proceedings of the 2024 CHI Conference on Human Factors in Computing Systems*, pp. 1–18. New York, NY. [http://dx.doi.org/10.1145/3613904.3642114]

Lin, C. C., Huang, A. Y. Q., Lu, O. H. T. (2023). Artificial intelligence in intelligent tutoring systems toward sustainable education: a systematic review. *Smart Learn. Environ, 10*(1), 41. [http://dx.doi.org/10.1186/s40561-023-00260-y]

Liu, M., Yu, D. (2023). Towards intelligent E-learning systems. *Educ. Inf. Technol, 28*(7), 7845-7876. [http://dx.doi.org/10.1007/s10639-022-11479-6]

Lo, C.K., Hew, K.F., Jong, M.S. (2024). The influence of ChatGPT on student engagement: A systematic review and future research agenda. *Comput. Educ, 219*, 105100. [http://dx.doi.org/10.1016/j.compedu.2024.105100]

Lu, R.S., Lin, H.C.K., Yang, Y.C., Chen, Y.P. (2024). Integrating Urban Mining Concepts Through AI-Generated Storytelling and Visuals: Advancing Sustainability Education in Early Childhood. *Sustainability (Basel), 16*(24), 11304. [http://dx.doi.org/10.3390/su162411304]

Luther, T., Kimmerle, J., Cress, U. (2024). Teaming Up with an AI: Exploring Human–AI Collaboration in a Writing Scenario with ChatGPT. *AI, 5*(3), 1357-1376.

Manna, S., Sett, N. (2024). Need of AI in Modern Education: in the Eyes of Explainable AI (xAI). In: *Blockchain and AI in Shaping the Modern Education System* Boca Raton, FL: CRC Press.

Mariën, J. (2021). *Real-time collaboration in Jupyter Notebooks*, (Master's thesis). Ghent University, Ghent, Belgium.

Mariyono, D., Nur Alif Hd, A. (2025). AI's role in transforming learning environments: a review of collaborative approaches and innovations. *Quality Education for All, 2*(1), 267-290. [http://dx.doi.org/10.1108/QEA-08-2024-0071]

Martin, S.M. (2019). *Artificial intelligence, mixed reality, and the redefinition of the classroom.* Bloomsbury Publishing USA.

Mishra, A. (2020). The Role of Data Visualization Tools in Real-Time Reporting: Comparing Tableau, Power BI, and Qlik Sense. *IJSAT-Int. J. Sci. Technol, 11*(3), 1-8.

Mohamed, S.M., Farah, S., Lotfy, A.M., Rizk, K.A., Saeed, A.Y., Mohamed, S.H., Arafa, T. (2025). Knowledge Graphs: The Future of Data Integration and Insightful Discovery. In: *Advanced Research Trends in Sustainable Solutions, Data Analytics, and Security* IGI Global Scientific Publishing. [http://dx.doi.org/10.4018/979-8-3693-7117-6.ch005]

Morales-Gonzalez, C., Harper, M., Rayavaram, P., Yeddanapudi, M., Narain, S., Fu, X. (2024). Evaluating the Efficacy of Productivity Tools in Engineering Education. In: *2024 IEEE Global Engineering Education Conference (EDUCON), IEEE*, pp. 1-10.

Mousavinasab, E., Nahid, Z., Sharareh, R. N. K., Mahnaz, R., Leila, K., Ghazi Saeedi, M. (2021). Intelligent tutoring systems: a systematic review of characteristics, applications, and evaluation methods. *Interact. Learn. Environ, 29*(1), 142-163. [http://dx.doi.org/10.1080/10494820.2018.1558257]

Muawanah, U., Marini, A., Sarifah, I. (2024). The interconnection between digital literacy, artificial intelligence, and the use of E-learning applications in enhancing the sustainability of Regional Languages: Evidence from Indonesia. *Soc. Sci. Humanit. Open, 10*, 101169. [http://dx.doi.org/10.1016/j.ssaho.2024.101169]

Mühlhofer, E., Koks, E.E., Kropf, C.M., Sansavini, G., Bresch, D.N. (2023). A generalized natural hazard risk modelling framework for infrastructure failure cascades. *Reliab. Eng. Syst. Saf, 234*, 109194. [http://dx.doi.org/10.1016/j.ress.2023.109194]

Murikah, W., Nthenge, J. K., Musyoka, F. M. (2024). Bias and ethics of AI systems applied in auditing - A systematic review. *Sci. Afr, 25*, e02281. [http://dx.doi.org/10.1016/j.sciaf.2024.e02281]

Nikitina, S. (2006). Three strategies for interdisciplinary teaching: contextualizing, conceptualizing, and problem-centring. *J. Curric. Stud, 38*(3), 251-271. [http://dx.doi.org/10.1080/00220270500422632]

Novis-Deutsch, N., Etan, C., Hanan, A., Liat, R., Uri, G., Ofir, G., Mann, A. (2024). Interdisciplinary learning in the humanities: Knowledge building and identity work. *J. Learn. Sci, 33*(2), 284-322. [http://dx.doi.org/10.1080/10508406.2024.2346915]

Ouyang, F., Zhang, L. (2024). AI-driven learning analytics applications and tools in computer-supported collaborative learning: A systematic review. *Educ. Res. Rev, 44*, 100616. [http://dx.doi.org/10.1016/j.edurev.2024.100616]

Owusu-Boateng, O. (2025). *Unlocking the Potential: Artificial Intelligence in Education*. Mumbai, India, Deep Science Publishing.

Pagliara, S.M., Bonavolontà, G., Pia, M., Falchi, S., Zurru, A.L., Fenu, G., Mura, A. (2024). The Integration of Artificial Intelligence in Inclusive Education: A Scoping Review. *Information (Basel), 15*(12), 774. [http://dx.doi.org/10.3390/info15120774]

Paithnkar, S. S., & Khaparde, V. S. (2023). Gamification AR and VR as of advanced tasks. In B. R. Babu, S. Singh, K. N. Kandpal, I. M. Shettar, & N. Gogoi (Eds.), *Emerging technologies in libraries: Trends and developments*. Hyderabad, India: BS Publication.

Papadimitriou, S. (2025). *Artificial Intelligence—Based Games as Novel Holistic Educational Environments to Teach 21st Century Skills.* Springer Nature. [http://dx.doi.org/10.1007/978-3-031-77464-5]

Patel, K., Beeram, D., Ramamurthy, P., Garg, P., Kumar, S. (2024). AI-enhanced design: revolutionizing methodologies and workflows. *Development, 2*(1), 135-157.

Pavlik, J. V. (2023). Collaborating with ChatGPT: Considering the implications of generative artificial intelligence for journalism and media education. *Journalism Mass Commun. Educ, 78*(1), 84-93.

Peddinti, S., Sabbani, Y. (2024). Mathematical Modeling of Infectious Disease Spread Using Differential Equations and Epidemiological Insights. *Journal of Nonlinear Analysis and Optimization, 15*, 114-122.

Pedro, F., Subosa, M., Rivas, A., Valverde, P. (2019). Artificial intelligence in education: Challenges and opportunities for sustainable development. *UNESCO Working Paper.*

Perifanou, M., Economides, A.A. (2025). Collaborative Uses of GenAI Tools in Project-Based Learning. *Educ. Sci. (Basel), 15*(3), 354. [http://dx.doi.org/10.3390/educsci15030354]

Pina, E., Ramos, J., Jorge, H., Váz, P., Silva, J., Wanzeller, C., Abbasi, M., Martins, P. (2024). Data Privacy and Ethical Considerations in Database Management. *Journal of Cybersecurity and Privacy, 4*(3), 494-517. [http://dx.doi.org/10.3390/jcp4030024]

Prajapati, M.A.B. (2024). Artificially intelligent in Education:redefining learning in the 21st century. *Journal of Educational Technology, 1*, 12818287.

Pynam, V., Spanadna, R.R., Srikanth, K. (2018). An Extensive Study of Data Analysis Tools (Rapid Miner, Weka, R Tool, Knime, Orange). *Int. J. Comput. Sci. Eng, 5*(9), 4-11. [http://dx.doi.org/10.14445/23488387/IJCSE-V5I9P102]

Pysarchyk, O. (2024). Creating effective learning environment: learner-centred pedagogy, assessment and feedback practices and web-based tools in ESP education. *European Science, 27*(2), 163-192.

Ramaswami, G., Susnjak, T., Mathrani, A., Umer, R. (2023). Use of predictive analytics within learning analytics dashboards: A review of case studies. *Technology. Technology, Knowledge and Learning, 28*(3), 959-980.
[http://dx.doi.org/10.1007/s10758-022-09613-x]

Rane, N., Choudhary, S., Rane, J. (2023). Education 4.0 and 5.0: Integrating artificial intelligence (AI) for personalized and adaptive learning. In *Proceedings Title or Book Title if applicable* (page numbers if available). Publisher if applicable.

Robledo-Rella, V., Toh, B-Y. (2024). Artificial intelligence in physics courses to support active learning. *Proceedings of the 2024 10th International Conference on e-Society, e-Learning and e-Technologies (ICSLT),* New York, NY, USA.68-75.
[http://dx.doi.org/10.1145/3678610.3678631]

Rudolph, J., Tan, S., Tan, S. (2023). War of the chatbots: Bard, Bing Chat, ChatGPT, Ernie and beyond. The new AI gold rush and its impact on higher education. *Journal of Applied Learning and Teaching, 6*(1), 364-389.

Seo, K., Tang, J., Roll, I., Fels, S., Yoon, D. (2021). The impact of artificial intelligence on learner–instructor interaction in online learning. *Int. J. Educ. Technol. High. Educ, 18*(1), 54.
[http://dx.doi.org/10.1186/s41239-021-00292-9]

Shinde, B. G., Shivthare, S. (2024). Impact Of Data Visualization In Data Analysis To Improve The Efficiency Of Machine Learning Models. *J. Adv. Zool, 45*(S-4), 107-112.

Singh, A., Lakhera, G., Ojha, M., Mishra, A.K., Kaushik, S. (2024). Natural language processing for HR Chatbots and virtual assistants: A comprehensive study on enhancing EHRM operations. *Practical Approaches to Agile Project Management* IGI Global.
[http://dx.doi.org/10.4018/979-8-3693-3318-1.ch020]

Stanhope, L., Ziegler, L., Haque, T., Le, L., Vinces, M., Davis, G.K., Zieffler, A., Brodfuehrer, P., Preest, M., M Belitsky, J., Umbanhowar, C., Jr, Overvoorde, P.J. (2017). Development of a Biological Science Quantitative Reasoning Exam (BioSQuaRE). *CBE Life Sci. Educ, 16*(4), ar66.
[http://dx.doi.org/10.1187/cbe.16-10-0301] [PMID: 29196427]

Stanton, D. (2025). *Project Management with AI.* John Wiley & Sons.

Strielkowski, W., Grebennikova, V., Lisovskiy, A., Rakhimova, G., Vasileva, T. (2024). AI-driven adaptive learning for sustainable educational transformation. *Sustain. Dev. (Bradford), 33*(2), 1921-1947.
[http://dx.doi.org/10.1002/sd.3221]

Sukhdeve, D.S.R., Sukhdeve, S.S. (2023). Google colaboratory. *Google Cloud Platform for Data Science: A Crash Course on Big Data, Machine Learning, and Data Analytics Services* Springer.
[http://dx.doi.org/10.1007/978-1-4842-9688-2_2]

Szeliski, R. (2022). *Computer vision: algorithms and applications.* Springer Nature.
[http://dx.doi.org/10.1007/978-3-030-34372-9]

Tan, S.C., Lee, A.V.Y., Lee, M. (2022). A systematic review of artificial intelligence techniques for collaborative learning over the past two decades. *Computers and Education: Artificial Intelligence, 3,* 100097.
[http://dx.doi.org/10.1016/j.caeai.2022.100097]

Tan, X., Cheng, G., Ling, M. H. (2025). Artificial intelligence in teaching and teacher professional development: A systematic review. *Comput. Educ. Artif. Intell, 8,* 100355.
[http://dx.doi.org/10.1016/j.caeai.2024.100355]

Tan, X., Lv, X., Jiang, J., Zhang, L. (2024). Understanding real-time collaborative programming: a study of visual studio live share. *ACM Trans. Softw. Eng. Methodol, 33*(4), 1-28.
[http://dx.doi.org/10.1145/3643672]

Tatipang, D.P., Pontillas, M.S., Alrajafi, G., Rovikasari, M., Ndruru, F. (2024). The Interplay Among AI, Language Research, and Academic Publishing: A Systematic Review of Trends and Future Directions. *Journal of English Culture, Language. Literature and Education, 12*(2), 215-237.

Thong, C., Down, A., Kocsis, A. (2023). Education and interdisciplinarity: New keys for future generations and the challenge-based learning. In: *In: New challenges and opportunities in physics education* Springer. [http://dx.doi.org/10.1007/978-3-031-37387-9_23]

Van den Beemt, A., MacLeod, M., Van der Veen, J., Van de Ven, A., van Baalen, S., Klaassen, R., Boon, M. (2020). Interdisciplinary engineering education: A review of vision, teaching, and support. *J. Eng. Educ, 109*(3), 508-555.
[http://dx.doi.org/10.1002/jee.20347]

Veytia Bucheli, M. G., Gómez-Galán, J., Cáceres Mesa, M. L., López Catalán, L. (2024). Digital technologies as enablers of universal design for learning: higher education students' perceptions in the context of SDG4. *Discover Sustainability, 5*(1), 473.
[http://dx.doi.org/10.1007/s43621-024-00699-0]

Wang, X., Liu, Q., Pang, H., Tan, S.C., Lei, J., Wallace, M.P., Li, L. (2023). What matters in AI-supported learning: A study of human-AI interactions in language learning using cluster analysis and epistemic network analysis. *Comput. Educ, 194*, 104703.
[http://dx.doi.org/10.1016/j.compedu.2022.104703]

Wright, J. (2019). Teaching LaTeX / Sarah Lang and Astrid Schmölzer. Tutorials 10: The DuckBoat—News from TeX. *TUGboat, 40*(3), 252–253.

Yadav, S. (2024). Artificial Intelligence (AI) Integration in Higher Education: Navigating Opportunities and Ethical Frontiers in Education with Advanced Technologies. *Impact of Artificial Intelligence on Society* Chapman and Hall/CRC.
[http://dx.doi.org/10.1201/9781032644509-4]

Yilmaz, Ö. (2023). Exploring Technology Integrated Personalized Learning Approaches In Science Education: An Examination Of Emerging Trends And Efficacies. In: *In: International research and reviews in educational sciences* Istanbul, Turkey: Serüven Publishing.

Zhai, C., Wibowo, S., Li, L. D. (2024). The effects of over-reliance on AI dialogue systems on students' cognitive abilities: a systematic review. *Smart Learn. Environ, 11*(1), 28.
[http://dx.doi.org/10.1186/s40561-024-00316-7]

Zheng, L., Fan, Y., Gao, L., Huang, Z., Chen, B., Long, M. (2024). Using AI-empowered assessments and personalized recommendations to promote online collaborative learning performance. *J. Res. Technol. Educ,* 1-27.

Zhu, G., Fan, X., Hou, C., Zhong, T., Seow, P., Shen-Hsing, A. C., Poh, T. L. (2023). Embrace opportunities and face challenges: Using ChatGPT in undergraduate students' collaborative interdisciplinary learning. *arXiv preprint.*
[http://dx.doi.org/10.48550/arXiv.2305.18616]

SUBJECT INDEX

A

B

C

O

P

Q

R

S

T

U

V

W

Y

www.ingramcontent.com/pod-product-compliance
Lightning Source LLC
LaVergne TN
LVHW070946210926
848482LV00002B/118